# Sperm Competition in Humans

Todd K. Shackelford Nicholas Pound
Editors

# Sperm Competition in Humans

## Classic and Contemporary Readings

Todd K. Shackelford
Department of Psychology
Florida Atlantic University
Davie, FL 33314
USA

Nicholas Pound
Centre for Cognition and Neuroimaging
School of Social Sciences and Law
Brunel University
Uxbridge UB8 3PH
United Kingdom

Library of Congress Control Number: 2005931436

ISBN-10: 0-387-28036-7 e-ISBN 0-387-28039-1
ISBN-13: 978-0387-28036-3

Printed on acid-free paper.



Printed in the United States of America. (SPI/SBA)

9 8 7 6 5 4 3 2 1

springeronline.com

# FOREWORD: HUMAN SPERM COMPETITION AND WOMAN'S DUAL SEXUALITY

Randy Thornhill*

Sperm competition in animals was discovered by Geoff Parker and reported in his 1970 paper on the topic in insects. In 1984, Bob Smith published an edited book on sperm competition in animals. Bob's chapter in the book initiated the study of sperm competition in humans; it reviewed the relevant human anatomy, physiology and behavior, and proposed numerous hypotheses about the potential role of sperm competition in humans. Robin Baker and Mark Bellis took a serious look at the role of sperm competition in humans in the late 1980s and early 1990s and published their book on the topic in 1995. Sperm competition has been an important type of sexual selection in all major taxa examined by researchers, according to three book-length reviews of the status of sperm-competition research: Smith (1984*a*), Birkhead & Møller (1998), and Simmons (2001).

The important role of sperm competition in evolution across major taxa includes the Mammalia (for a recent review of mammalian research on sperm competition, see Gomendio *et al.*, 1998). The early expectation was that sperm competition will be rare, possibly even evolutionarily unimportant, in mammals, because, in general, female mammals do not have sperm-storage organs and sperm are short-lived (Parker, 1984), patterns in sharp contrast with many other taxa, for example, insects and birds. Seemingly also relevant to low opportunity for sperm competition in mammals is that mammals have estrus and hence the time window for conception is restricted. In order for sperm competition to occur, there must be competition between ejaculates of different males for the egg(s) of a single female. In the absence of sperm storage, only ejaculates placed near ovulation, i.e., during estrus, are capable of engaging in sperm competition. Yet, sperm competition occurs in a wide range of mammals, apparently without sperm storage and with narrow fertility windows and short-lived sperm (Gomendio *et al.*, 1998). Evidence for sperm competition in these mammal species comes from two sources. First, females sometimes produce a single brood of offspring with multiple paternity. Second, research finds male adaptations of physiology, anatomy, and behavior that seem to be designed functionally to solve the problem of sperm competition.

*Department of Biology, The University of New Mexico, Albuquerque, NM 87131-0001 (thorn@unm.edu).

Although multiple paternity is known in human twins (Smith, 1984b), demonstrating that sperm competition can occur, the real research issue is whether sperm competition has been a selective force in shaping human adaptation. General questions of interest to human sperm-competition researchers, such as Smith, Baker, Bellis and others since the early 1990s, are as follows. Do men have adaptations functionally designed to compete against a sexual rival's sperm or to prevent a rival's insemination of the mate? Do women have adaptations that function to promote sperm competition and thereby to produce sons with highly competitive ejaculates, or to produce offspring of both sexes with superior genotypes, if sperm-competition ability covaries positively with general fitness?

In the remainder of the foreward, I will discuss recent research findings about the design of men and women's sexuality that may be essential to consider for significant research progress in answering the two general questions posed above. This book is a collection of much of the scholarly work on human sperm competition, and if the important work itself is not in this book, it is discussed in some detail in it. This book also contains important criticisms of the human sperm competition literature. In particular, Chapter 1 by Pound *et al.* is a thoughtful critique and overview of the state of the science of human sperm competition. Virtually all findings reported to date, although many are consistent with an important role for sperm competition in human evolutionary history, do not provide the most convincing evidence because reasonable alternative hypotheses may explain the findings (see Chapter 1). This book tells us the current status of the field of research, and that future research is needed to better identify the role, if any, of sperm competition in human evolution. Two major research findings over the last several years may provide the knowledge base for a more sophisticated study of human sperm competition.

Specifically, it will be important in future research to incorporate the now-considerable evidence that woman has two sexualities, what I refer to as "estrus" and "extended sexuality," and that men know (perceive and respond to) her estrus, but very imperfectly, because of the coevolutionary race between women to conceal estrus and men to detect it (Thornhill, in press). Woman's estrous sexuality is seen only surrounding ovulation and hence when there is significant conception probability. Human sperm competition can occur only during estrus. Woman's extended sexuality is all her sexuality outside the fertile phase of the menstrual cycle, including that during pregnancy and adolescent subfertility. Human sperm competition research has not recognized fully the importance of distinguishing the dual nature of woman's sexuality. This is understandable because the conventional wisdom in anthropology, psychology and reproductive biology is that woman lost estrus during her evolutionary history. It is believed generally that she, unlike all other female mammals, has no unconscious knowledge of her peak fertility in the reproductive cycle; she does not perceive the peak nor respond to it sexually. This knowledge is the basis of the high sexual motivation surrounding ovulation in the typical estrous female mammal. Relatedly, traditional consensus opinion claims that men have no knowledge of peak female fertility, whereas males of all other mammals typically do as a result of perception and response to olfactory or other cues from the estrous female. Recent research points a quite different picture of the sexuality of both sexes that may have far-reaching consequences for research on human sperm competition and human sexuality in general.

The recent research does not imply that woman's estrus is not concealed. Instead, it is present, but arguably concealed by direct selection for concealment from the main pair-bond (in-pair) partner in the service of extra-pair copulation (EPC) with a male(s) with superior genetic quality, as first hypothesized by Benshoof and Thornhill (1979) and Symons (1979). This concealment or crypsis, is imperfect, because males are selected to detect high fertility days in the menstrual cycle. Accordingly, females and males in our species have long been and are still engaged in a coevolutionary race to conceal cycle fertility (women) and to know it (men). Each antagonist in this race is adapted, but incompletely so (as is typical of coevolutionary races in general), to the opposite sex. Awareness of this coevolutionary race may inform importantly future research in human sperm competition.

## THE FERTILE WINDOW

For humans, it is not so much how long sperm live, the frequency of polyandry, or even the frequency of so-called "double matings" (the overlap of ejaculates from different males within a woman over a 5–day period, the often-assumed longevity of sperm inside a woman). By definition, sperm competition can occur only if conception can occur. Ejaculates of different men placed inside a woman in her luteal menstrual cycle phase, even if the sperm lives inside her for up to 3 weeks, will not compete, because she has virtually zero probability of conception. The luteal phase begins with ovulation and extends until day 1 of her next menstrual cycle (first day of menses). Fertility drops to zero when ovulation occurs and a woman is infertile through the next menses, which lasts a mode of 5 days. In the days thereafter, this conception probability is above zero and continues to climb, reaching a peak a day or two prior to her next ovulation. Ovulation typically occurs on day 14 or 15 in young women with regular cycles (about 85% of young women). Ovulation is later in women with irregular cycles. The fertility window of the six days prior to ovulation is the case regardless of cycle regularity (Wilcox *et al.*, 2000, 2001).

For sperm competition to occur, competing ejaculates have to co-occur inside a woman during her fertile window. The 5–day window for double-mating is too long. Apparently, sperm live only a few days inside a woman, with a rare maximum of 6 days (Gomendio *et al.*, 1998). Moreover, double-matings outside the fertile window don't count—they cannot generate sperm competition. There is no convincing evidence of any sperm storage by woman, or by most other female mammals for that matter (Gomendio *et al.*, 1998). Actually, the important fertile window for sperm competition is even more restricted. It would correspond to days of high conception risk and hence the few days prior to ovulation. The 3–day window of high conception probability in the fertile window corresponds to men's sperm longevity inside a woman. I hypothesize that this longevity is the result of selection on males to produce sperm that can survive through the peak-fertility phase of the mate in the absence of sperm storage. Gomendio *et al.* (1998) provide evidence that the time from beginning of estrus to ovulation is correlated positively with sperm longevity across many non-human mammals. The effective window for sperm competition, then, is 3–6 days of a woman's 28–30-day menstrual cycle. Outside the effective window, polyandry is irrelevant with respect to sperm competition.

## WOMAN'S DUAL SEXUALITY

There is strong evidence that woman has a dual sexuality. During the fertile phase of the menstrual cycle, she shows mate preferences that are distinctly different than that at infertile cycle phases. Recent studies have found that at high cycle fertility, but not at infertile cycle times, women prefer the body scent, faces and behaviors of symmetric men over the same traits of asymmetric men; prefer relatively high degrees of facial testosteronization in men, indicating a fertile-phase-specific preference for another marker (in addition to symmetry) of potential male genetic quality; prefer scents (andostenone) related to high testosteronization; prefer relatively high degrees of male skin coloration (melanin- and hemoglobin-based) that may correspond to elevated testosterone; prefer relatively high degrees of mental functioning in men (creative intelligence); and show relatively high levels of disgust about incestuous and other maladaptive matings. The effects listed are reviewed by Thornhill and Gangestad (2003), Fessler and Navarrete (2003), Thornhill *et al.* (2003) and Haselton and Miller (in press). A number of these effects are seen primarily or solely in women's preferences for short-term, rather than long-term, mates, as expected if the preferences function in pursuing sires for offspring, as opposed to long-term partners. Moreover, multiple studies report that high-fertility phase women seem to modify their behavior to reduce the risk of rape and hence insemination by a male who may be an unsuitable sire (reviewed in Bröder & Hohmann, 2003). Also, the effects listed are not seen in women using hormonal contraception (e.g., the pill). This means that the changed mid-cycle sexuality of women proximately depends on normal ovulatory hormonal factors acting around ovulation.

Another line of this recent empirical work has studied normally ovulating women's sexual thoughts, attractions and fantasies across the menstrual cycle. In general, non-partner men, rather than the in-pair partner, are the focus of women's sexual interests at peak fertility, according to two separate studies (Gangestad *et al.*, 2002; Haselton & Gangestad, in press). One other study, however, did not find this; instead, it found that high-fertility women focus more sexual interest toward the main partner than toward non-partner men. The exception may be driven by the women studied being in a new romantic relationship in which their EPC was not perceived as beneficial because of heightened male investment early in the relationship (see Pillsworth *et al.*, 2004 for discussion).

Other studies have looked at moderators of women's extra-pair sexual attraction. One study found that paired women's extra-pair sexual interest depends upon their in-pair partner's degree of symmetry and thus his potential genetic quality. Fertile-phase women paired with relatively symmetric males showed more interest in their in-pair partners than in non-partner men, but fertile-phase women paired with asymmetric men showed the reverse (Gangestad *et al.*, 2003). An additional study has shown that main male, in-pair partner's physical attractiveness, and hence putative genetic quality, moderates a woman's interest within and outside the pair-bond. Women with attractive men as in-pair partners focus their estrous eroticism toward him, whereas women with unattractive in-pair partners focus theirs toward extra-pair men (Haselton & Gangestad, in press).

Collectively, the research summarized above indicates that women perceive and respond to peak fertility in the cycle in relation to changes in cycle hormones. Moreover, it suggests that the peak-fertility sexuality of women is functionally organized to obtain a sire of superior genetic quality. Thus, both in terms of self-knowledge of estrus and of adaptive design of estrus, women at mid-cycle may be no different than other estrous female mammals.

Bullivant *et al.* (2004) label women's sexuality at the fertile time of her menstrual cycle "the sexual phase." This label is problematic because woman has extended sexuality (see below). The sexuality of woman at peak fertility in the cycle is most appropriately designated as estrus because of its apparent homology and function (Thornhill, in press). This sexuality appears to be homologous with estrus in all other female vertebrates. Homology means similarity in traits across species due to descent from a common ancestor with the trait. Homology does not imply identical traits because traits evolve. Estrus appears to have its phylogenetic origin in the species of fish that was ancestral to all the vertebrate groups (fishes, amphibians, reptiles [including birds] and mammals). This phylogenetic inference is based on the apparent presence of estrus in all of these groups. This presence is seen in the similarity of the hormonal underpinnings of female sexual motivation at peak fertility in the reproductive cycle. Also, the neurobiology of female sexual motivation at peak fertility, although far less known comparatively than estrous hormonal patterns, supports homology across the vertebrates (see Nelson, 2000 for a review of homology of female sexuality).

Not only has estrus never been lost by selection against it since its origin in the first vertebrate, it probably has the same general function throughout the vertebrates. Studies of estrous female guppies, frogs, salamanders, junglefowl, barn swallows, and a variety of non-human female mammals in estrus, including woman, reveal their design for preferring mates of actually or potentially superior genetic quality. The function of estrus as sire choice is hypothesized to be the reason for its evolutionary maintenance after its phylogenetic origin on the Tree of Life. This maintenance involved lineage-specific selection that molded estrus for the lineage-specific problems of sire choice in each taxon.

Contrary to the typical interpretation in mammalogy, estrus is not an adaptation to get just any sperm. Sperm are obtained by estrous phase female behavior, but this is incidental to estrus' evolved function of pursuit of sires of superior genetic quality. The view that the heightened sexual motivation of estrous females is the indiscriminate pursuit of sperm, widely held in mammalian reproductive biology (e.g., Nelson, 2000), almost certainly is wrong, given that evolutionary theory claims that females have evolved to be choosy about their offsprings' sires and that studies of a variety of female mammals (from marsupials to non-human primates and woman) in estrus find them to prefer males with traits that connote potential or actual genetic quality.

Similarly, estrus is not a signal of peak fertility (directly selected adaptation for the function of communicating peak fertility). Sexual selection on males will guarantee females get inseminated at peak female fertility. Finding, doggedly pursuing and effectively inseminating fertile-phase females is the only evolutionarily stable strategy of males (e.g., Pagel, 1994).

Previous discussions of woman's sexuality in the literature have been confused by the popular view that the loss of estrus is the same as the absence of female sexual swellings. It sees female sexual swellings in non-human, Old World primates as functioning as signals of ovulation, and thus woman has no estrus because sexual swelling was lost in her evolutionary history. Estrus and sexual swellings, however, are not equivalent. The females of most species of non-human mammals lack swellings, but all the species have estrus. Also, as just mentioned, swellings do not likely function as signals of cycle fertility. Instead, they are a form of female sexual ornamentation that, like all such ornamentation in both sexes, likely honestly signals individual quality. There is increasing evidence that woman's ornamentation—the estrogen-facilitated gynoid fat displays of breasts, hips and thighs; certain facial features; and skin—function as signals of individual quality (Thornhill & Grammer, 1999; Grammer *et al.*, 2002). Hence, woman's ornaments are not permanent, deceptive signals of cycle-related fertility, as some have proposed. Instead, they are permanent (across the reproductive life of the woman), honest signals of residual reproductive value to mates (see Marlowe, 1998, on breasts as signals of future reproductive potential).

Another popular view that is apparently erroneous and has confused prior thinking about woman's sexuality is the notion that loss of estrus is equal to concealed ovulation. Estrus in woman is not lost, but is concealed partially, arguably by design to mask it in the service of female EPC for good genes at the fertile phase of the cycle. Equating the loss of estrus with concealed ovulation is based on the invalid view that estrus functions to reveal ovulation.

The recent empirical findings indicating that woman has extended sexuality adaptations are as exciting as those indicating she has estrous adaptations. Extended sexuality—mating motivation outside the fertile phase of the reproductive cycle—is seen in most Old World primates and pair-bonding birds, and here and there in species in many other taxa. Comparative data (partially reviewed by Rodriguez-Girones & Enquist, 2001) reveal that extended female sexuality generally evolves in species in which males provide females with non-genetic, material benefits and probably functions to obtain those benefits. The benefits gained depend on the species and range from food, social alliances, and protection of self and offspring, including protection of offspring from maltreatment by males in the group. Findings reveal that woman's extended sexuality may be a distinct adaptation that functions to secure material benefits. For example, normally ovulating women at infertile phases of their menstrual cycle, in contrast to estrous women, prefer men with limited facial testosteronization and such men appear to be more willing to provide benefits to mates than men with greater degrees of testosteronization (see review in Penton-Voak & Perrett, 2001). Also, the preferences of fertile-phase females for symmetry (mentioned above) seem to be contrary to material-benefit gain by women. Symmetric men, like highly testosteronized men, invest less in their romantic relationships (see Thornhill & Gangestad, 2003). Thus, estrous preferences of women appear to trade-off male material benefits for male genetic quality. It is not that female choice is relaxed during extended sexuality. Women are choosy across the cycle, but the functional significance of their choices seems to differ greatly between infertile and fertile cycle phases.

Extended female sexuality requires accompanying estrus in the female's sexual repertoire. Extended sexuality allows the female to gain material-benefits from mating with a male(s) without risk of conception by a sire of inferior genetic quality, and estrus can assure offspring of high genetic quality.

## DO WOMEN PROMOTE SPERM COMPETITION?

These findings have important implications for testing the hypothesis that woman has adaptations that function to promote sperm competition. The occurrence of polyandry is evidence for such design only if it occurs during estrus and is pursued by women, rather than being the result of the combination of estrous female preference for one male and sexual coercion by another during estrus. Are women motivated to be inseminated by multiple partners during estrus? Bellis and Baker (1990) interpreted some of their findings to suggest that women double-mate during mid-cycle, specifically that they mate with both the in-pair partner and an extra-pair partner. The above studies, however, suggest that estrous women may strive to avoid insemination by the in-pair partner and selectively pursue an extra-pair partner primarily when the in-pair partner is of low genetic quality. Hence, if women seek superior sperm competitors during estrus, it is reasonable to suggest that they would mate with multiple men with above-average symmetry and testosterone level. Ejaculate quality in men, and presumably, therefore, sperm competitiveness, positively covaries with men's symmetry and facial attractiveness (Manning *et al*., 1998; Soler *et al*., 2003). Double-mating by combining in-pair and extra-pair men during a single estrus might achieve an optimum sperm competition when the in-pair partner and the extra-pair partner are both of high genetic quality. In general, however, mating with multiple, high-genetic quality men is expected to better accomplish a high degree of sperm competition. The strong prediction, then, is that if women are designed to promote sperm competition, they will pursue insemination by multiple men of superior genetic quality during estrus. Studies to date have not addressed this possibility, but only reveal estrous woman's preferences for markers of high genetic quality.

One study indirectly suggests that women during estrus may not promote sperm competition. Schwarz and Hassebrauck (2004) report that women's score on the Sociosexual Orientation Inventory (Gangestad & Simpson, 1990) is correlated negatively with conception probability in the menstrual cycle. That is, fertile-phase women are the most sexually restricted in terms of their attitude about multiple partners, which supports the hypothesis that estrus functions as sire choice. More research, however, focused specifically on women's polyandrous tendencies during estrus is needed to test the prediction that estrous women have a specialized motivation to promote sperm competition.

The research I discussed on estrous women's preferences assumes that estrus is designed to obtain intrinsically good genes (i.e., genetic benefits to offspring that would enhance the reproductive value of offspring for all or most females). That this assumption is reasonable is seen in the general pattern of women overall changing their preference at the fertile cycle phase. If female preferences at estrus were designed only for seeking complementary genes based on the female's own genotype, then the main effects of preference for putative markers of men's genetic quality would not be found consistently by researchers. For example, a female preference for dissimilar major histocompatibility (MHC) alleles may involve complementary good genes. It is possible that women prefer mates with dissimilar alleles, but simultaneously want multiple men with such alleles to inseminate them and thereby achieve for offspring both dissimilar alleles and highly competitive sperm traits. There is no evidence, however, that women at estrus have a preference for dissimilar MHC alleles (Thornhill *et al.*, 2003). It remains conceivable, of course, that women have an estrous preference for complementary alleles at other loci. If so, then in addition, women may place ejaculates from men carrying preferred alleles in competition by estrous polyandry. Research to examine this must take into account that sperm competition can only occur during estrus.

Another type of good-genes preference potentially relevant to sperm competition is for diverse genes. This kind of preference is hypothesized widely to be potentially important in various non-human animals. It allegedly functions to diversify the genotypes of a brood and thereby create at least some offspring that can deal with unpredictable environments, such as parasites coevolving rapidly to penetrate host defensive adaptation. Sperm competition necessarily will occur if females mate with multiple males at the fertile time of the reproductive cycle. Whether sperm competition is incidental to any female preference for genetic diversity or is a result of direct selection on females to promote sperm competition is an empirical question. If females mate with more males than necessary to achieve optimum genetic diversity, then promoting sperm competition could account for the excess mating. In woman, the multi-offspring brood is achieved by sequential births (barring twins) and hence if she is adapted to pursue diverse genes, then the pursuit would be to achieve multiple fathers across reproductive bouts. Sperm competition could play a role only if women pursue multiple partners at estrus in each bout.

## ESTROUS WOMEN'S EPC AND MEN'S MATE GUARDING

I have mentioned two lines of evidence that woman has extra-pair copulation adaptations that function to secure sperm of a male of superior genetic quality during estrus. Women at estrus shift their preference to male traits that may connote superior genetic quality. They think sexually about non-partner men when in estrus, especially if the in-pair partner is of apparent low genetic quality. This evidence makes reasonable the hypothesis that pair-bonded men have faced sperm competition regularly throughout human evolutionary history. But do men have adaptations for this problem? Specifically, do pair-bonded men respond toward their estrous mate in ways that would offset the problem of sperm competition generated by an unfaithful mate seeking superior genes? Two recent studies suggest that the answer is yes. Gangestad *et al.* (2002) and Haselton and Gangestad (in press) have found that in-pair men's mate guarding peaks during the fertile phase of the menstrual cycle. Any insemination of an estrous woman is a paternity threat for her in-pair partner, whether it be due to rape or consensual female choice. These studies indicate that the increase in the in-pair partner's mate guarding during estrus is not just a response to prevent a woman's rape, but includes mate-guarding behaviors that seem to be focused on preventing her consensual EPC by increasing the in-pair partner's value to his mate.

Men appear to guard estrous partners more than non-estrous partners, but what are they guarding? The sperm-competition hypothesis requires that they are guarding their paternity against the risk of its loss from conception by a sperm competitor. This interpretation is certainly reasonable when coupled with the evidence that women seek EPC during estrus. But why do men mate guard at times other than estrus? This guarding cannot be the result of direct selection to prevent sperm competition, because there is no sperm competition during woman's extended sexuality.

There is a large literature on men's guarding of their mates from other men (see Buss, 2003, for a review). For example, there is evidence that men guard young mates more than older mates, and non-gravid mates more than pregnant mates. But young mates have higher reproductive value; even in the absence of sperm competition, they are expected to be guarded or tended more. Non-gravid mates are worth more to men in terms of getting an additional offspring underway. There is evidence that fathers guard young daughters more than older daughters; certainly, this is unrelated to preventing the father's sperm from being in competition. Instead, its function is to maximize the daughter's mate value for a suitable marriage. I hypothesize that fathers also guard daughters more during estrus.

The literature on men's mate guarding includes numerous studies showing that men are more jealous about sexual infidelity by a mate than about her emotional infidelity (see Buss, 2003 for a review). This is interpreted as evidence that men are designed to prevent sperm competition. Men, however, are more focused on sexual things than on about any other category of things. The sperm-competition hypothesis would be supported strongly if men's concern about sexual infidelity of their mate, in contrast to their concern about her emotional infidelity, showed a menstrual-cycle effect, peaking at estrus.

The recent research on men's increased mate guarding during estrus indicates that the mate's peak cycle-fertility is known to men and that they respond to it in ways that would reduce the chances of sperm competition. Other recent research also supports that men have knowledge about high female fertility in the cycle. Women's estrus seems to have the attractivity component seen generally in mammalian estrus. Three separate studies of normally ovulating women have found that men rate the body scent of fertile-phase women as more attractive than that of infertile-phase women, and one of these studies showed also that there is no menstrual cycle variation in the body-scent attractiveness to men of women using hormonal contraception (Singh & Bronstad, 2001; Thornhill *et al.*, 2003; Kuukasjärvi *et al.*, 2004).

Although men have far more knowledge of women's high fertility in the cycle than earlier scholars thought, men are not as astute at detecting fertility as most other male mammals (e.g., the pig or the mouse) in which male sexual interest is only or primarily focused on estrous females. Also, men's incomplete knowledge is demonstrated by their great interest in copulation across the menstrual cycle, not just at peak fertility. In human evolutionary history, the absence of complete knowledge is hypothesized to have maintained the selection on males for copulating and mate guarding regardless of the partner's cycle phase.

High copulation rates by male animals, including men, are interpreted typically as adaptation against sperm competition. This can apply only when females are fertile in their reproductive cycle, because it is only at this time that sperm competition can be a selective agent. A more parsimonious hypothesis is that high copulation rates and all male copulations outside the fertile phase are the result of incomplete male knowledge of high female fertility in the reproductive cycle (Thornhill, 1984).

Men's limited knowledge of the timing of estrus (relative to other male mammals and male vertebrates in general) indicates selection on females for crypsis of high fertility in the cycle and selection on males to circumvent the crypsis and know peak fertility. The recent research findings imply that the genetically superior sire for an offspring may or may not be the main pair-bond partner of a female. This leads to condition-dependent extra-pair copulation (EPC) behavior by women that functions during fertility in the menstrual cycle to maximize the genetic quality of offspring when the in-pair partner is of relatively low genetic quality. Similarly, the research implies that woman will exhibit concealed peak fertility in her menstrual cycle, and that this is an adaptation for EPC at peak fertility through its design for disguising a woman's EPC pursuits from her main partner. Thus, woman, like any other female mammal in estrus, will perceive and respond to her peak fertility in the cycle, but woman will be cryptic about her motivations to secure the best sire through EPC at peak fertility. Furthermore, males always gain from knowing (perceiving and responding to) the peak fertility of females. Hence, the recent research implies a continuous coevolutionary race in humans to hide peak fertility (selection on females) and detect it (selection on males).

Co-evolutionary-race scenarios are rich with predictions that can test the sperm competition hypothesis. How is men's copulatory behavior adjusted in relation to estrus and extended sexuality, given that men have some knowledge of estrus? Gallup and Burch (2004) have provided evidence that man's penis may have functional design for removing a competitor's sperm; in this scenario, the action of the penis depends on the depth of thrusting and other copulatory actions. The co-evolutionary race between females to control conception and men to circumvent that control suggests that men may have evolved specialized copulatory behaviors for sperm removal that would be manifested primarily during estrus. Also, is the male motivation underlying human cunnilingus focused on the fertility window to discern peak fertility within it, as is certainly the case in non-human mammals? Strongly supportive of the sperm-competition hypothesis would be the result that male-initiated cunnilingus or male interest in cunnilingus with the partner peaks at estrus, especially in couples in which the man perceives his partner as having extra-pair interest. There is evidence that men with partners with multiple, simultaneous boyfriends show a greater peak in mate guarding during estrus than men with partners who report dating only one man (Gangestad *et al.*, 2002). Perhaps cunnilingus would reveal the same pattern. Cunnilingus during estrus also might function to detect by taste/olfaction another man's ejaculate, or to promote sperm-retaining female orgasm.

Co-evolutionary theory makes cryptic female choice a salient female adaptation in the intersexual race, if sperm competition occurs and females strive to obtain the best sire. Do men work harder during estrus than extended sexuality to achieve the optimal timing of the mate's copulatory orgasm that will retain his sperm (see Baker and Bellis, 1995, on the timing of men's and women's orgasm in relation to ejaculate retention)? Does man's semen change at estrus, either in quantity or chemically, in ways implying defense against the risk of sperm competition?

Other questions relevant to human sperm competition derive from co-evolutionary theory. Are men designed to conceal the pair-bond mate's estrus from other men to protect paternity? As mentioned, paternity would be reduced by an estrous partner's mating consensually or as a result of sexual coercion. The recent findings, mentioned earlier, that woman appears to have adaptations that defend against rape means that rape was a recurrent threat to female reproductive success in human evolutionary history. This, in turn, suggests that rape may have been a recurrent context for sperm competition in our history. Hence, human sperm competition may arise from both consensual EPC and coerced matings by males other than the in-pair partner. Rapists may target disproportionately estrous-phase women, given men's partial knowledge of cycle-related fertility. The apparently quite high conception rate following rape, compared to consensual mating (Gottschall & Gottschall, 2003), is consistent with the hypothesis that rapists have some knowledge of the fertile window and differentially force insemination during it.

Co-evolutionary theory also leads to the expectation that there will be considerable individual variation in women's ability to hide estrus and men's ability to detect it. Ongoing co-evolutionary races generate high phenotypic variance, including considerable maladaptation, on both sides of the contest. How might this relate to individual differences in mating and romantic relationship tactics and competencies, including those of men that may defend against sperm competition?

In discussing some of the recent findings about women's changes in mate preference across the menstrual cycle, David Buss (2003, p. 249) states that human "[m]ating research is now entering an 'ovulation revolution.'" I suggest a much broader program is ahead, one that will address woman's dual sexuality as well as the co-evolutionary races of males and females in relation to estrus. Certainly, most hypotheses about the role of sperm competition in the human mating system are informed by separating estrus and extended sexuality and by co-evolutionary theory. The future study of human sperm competition will benefit from more widely adopting this broader theoretical perspective on human sexuality.

## REFERENCES


Baker, R. R. & Bellis, M. A. (1995). *Human Sperm Competition: Copulation, Masturbation and Infidelity.* Chapman and Hall, London, U.K.

Bellis, M. A. & Baker, R. R. (1990). Do females promote sperm competition? Data for humans. *Animal Behaviour* **40**, 997–999.

Benshoof, L. & Thornhill, R. (1979). The evolution of monogamy and concealed ovulation in humans. *Journal of Social and Biological Structure,* **2**, 95–106.

Birkhead, T. R. & Møller, A. P., eds. (1998). *Sperm Competition and Sexual Selection.* Academic Press, London, U.K.

Bröder, A. & Hohmann, N. (2003). Variations in risk taking behavior over the menstrual cycle: An improved replication. *Evolution and Human Behavior* **24**, 391–398.

Bullivant, S. B., Sellergren, S. A., Stern, K., Spencer, N. A., Jacob, S., Mennella, J. A. & McClintock, M. K. (2004). Women's sexual experience during the menstrual cycle: Identification of the sexual phase by noninvasive measurement of luteinizing hormone. *Journal of Sex Research* **41**, 82–93.

Buss, D. M. (2003). Sexual strategies: A journey into controversy. *Psychological Inquiry* **14**, 219–226.

Buss, D. M. (2003). *The Evolution of Desire* (rev. ed.). New York: Basic Books.

Fessler, D. M. T. & Navarrete, C. D. (2003). Domain-specific variation in disgust sensitivity across the menstrual cycle. *Evolution and Human Behavior* **24**, 406–417.

Gallup, G. G. Jr & Burch, R. L. (2004). Semen displacement as a sperm competition strategy in humans. *Evolutionary Psychology* **2**, 12–23.

Gangestad, S. W. & Simpson, J. A. (1990). Toward an evolutionary history of female sociosexual variation. *Journal of Personality* **58**, 69–96.

Gangestad, S. W., Thornhill, R. & Garver, C. E. (2002). Changes in women's sexual interests and their partners' mate retention tactics across the menstrual cycle: Evidence for shifting conflicts of interest. *Proceedings of the Royal Society of London B* **269**, 975–982.

Gangestad, S. W., Thornhill, R. & Garver-Apgar, C. E. (2003). Changes in women's sexual interests and their partners' mate retention tactics across the menstrual cycle: The moderating effects of mate's symmetry. Paper presented at 2003 Human Behavior and Evolution Society Annual Meeting, Lincoln, NE.

Gomendio, M., Harcourt, A. H. & Roldán, R. S. (1998). Sperm competition in mammals. In *Sperm Competition and Sexual Selection* (eds., T. R. Birkhead & A. P. Møller, Eds.), pp. 667–756. Academic Press, New York.

Gottschall, J. A. & Gottschall, T. A. (2003). Are per-incident rape-pregnancy rates higher than per-incident consensual pregnancy rates? *Human Nature* **14**, 1–20.

Grammer, K., Fink, B., Juette, A., Ronzal, G. & Thornhill, R. (2002). Female faces and bodies: N-dimensional feature space and attractiveness. In *Advances in Visual Cognition* (eds. G. Rhodes & L. Zebrowitz, Eds.), Vol. 1: Facial Attractiveness: Evolutionary, cognitive and social perspectives, pp. 91–125. Greenwood Publishing, Westport, CT.

Haselton, M. G. & Gangestad, S. W. (In press). Conditional expression of female desires and male mate retention efforts across the ovulatory cycle. *Behavioral Ecology*.

Haselton, M. & Miller, G. F. (In press). Women's fertility across the cycle increases the short-term attractiveness of creative intelligence compared to wealth. *Human Nature*.

Kuukasjärvi, S., Eriksson, C. J. P., Koskela, E., Mappes, T., Nissinen, K. & Rantala, M. J. (2004). Attractiveness of women's body odors over the menstrual cycle: The role of oral contraception and received sex. *Behavioral Ecology* **15**, 579–584.

Manning, J. T., Scutt, D., Wilson, J. & Lewis-Jones, D. I. (1998). The ratio of the 2nd to the 4th digit length: A predictor of sperm numbers and concentrations of testosterone, luteinizing hormone and oestrogen. *Human Reproduction* **13**, 3000–3004.

Marlowe, F. (1998). The nubility hypothesis: The human breast as an honest signal of residual reproductive value. *Human Nature* **9**, 263–271.

Nelson, R. J. (2000). *An Introduction to Behavioral Endocrinology*, 2nd ed. Sinauer Associates, Inc., Sunderland, MA.

Pagel, M. (1994). Evolution of conspicuous estrous advertisement in old-world monkeys. *Animal Behaviour* **47**, 1333–1341.

Parker, G. A. (1984). Sperm competition and the evolution of animal mating strategies. In *Sperm Competition and the Evolution of Animal Mating Strategies* (R.L. Smith, Ed.), pp. 1–60. Academic Press, New York, NY.

Penton-Voak, I. S. & Perrett, D. I. (2001). Male facial attractiveness: Perceived personality and shifting female preferences for male traits across the menstrual cycle. *Advances in the Study of Behavior* **30**, 219–259.

Pillsworth, E. G., Haselton, M. G. & Buss, D. M. (2004). Ovulatory shifts in female sexual desire. *Journal of Sex Research* **41**, 55–65.

Rodriguez-Girones, M. A. & Enquist, M. (2001). The evolution of female sexuality. *Animal Behaviour* **61**, 695–704.

Schwarz, S. & Hassebrauck, M. (2004). Shifting attitudes toward sexual activities across the menstrual cycle: Are women more choosy during the fertile days of their cycle. Paper presented at 2004 Annual Meeting of the Human Behavior and Evolution Society, Berlin, Germany.

Simmons, L. W. (2001). Sperm Competition and its Evolutionary Consequences in the Insects. Princeton University Press, Princeton, NJ.

Singh, D. & Bronstad, P. M. (2001). Female body odour is a potential cue to ovulation. *Proceedings of the Royal Society of London B* **268**, 797–801.

Smith, R. L., editor. (1984*a*). *Sperm Competition and the Evolution of Animal Mating Systems*. Academic Press, London, U.K.

Smith, R. L. (1984*b*). Human sperm competition. In *Sperm Competition and the Evolution of Animal Mating Systems* (R. L. Smith, Ed.), pp. 601–660. Academic Press, London, U.K.

Soler, C., Núñez, M., Gutiérrez, R., Núñez, J., Medina, P., Sancho, M., Álvarez, J. & Núñez, A. (2003). Facial attractiveness in men provides clues to semen quality. *Evolution and Human Behavior* **24**, 199–207.

Symons, D. (1979). *The Evolution of Human Sexuality*. Oxford University Press, Oxford, U.K.

Thornhill, R. (1984). Alternative hypotheses for traits believed to have evolved by sperm competition. In *Sperm Competition and the Evolution of Animal Mating Systems* (R. L. Smith, Ed.), pp. 151–178. Academic Press, New York, NY.

Thornhill, R. (In press). The evolution of woman's estrus, extended sexuality and concealed ovulation and their implications for understanding human sexuality. In *The Evolution of Mind* (S. W. Gangestad and J. A. Simpson, Eds.). Guilford Publications, Inc. New York, NY.

Thornhill, R. & Gangestad, S. W. (2003). Do women have evolved adaptation for extra-pair copulation? In *Evolutionary Aesthetics* (E. Voland & K. Grammer, Eds.), pp. 341–368. Springer-Verlag, Heidelberg, Germany.

Thornhill, R., Gangestad, S. W., Miller, R., Scheyd, G., Knight, J. & Franklin, M. (2003). MHC, symmetry, and body scent attractiveness in men and women. *Behavioral Ecology* **14**, 668–678.

Thornhill, R. & Grammer, K. (1999). The body and face of woman: One ornament that signals quality? *Evolution and Human Behavior* **20**, 105–120.

Wilcox, A. J., Dunson, D. & Baird, D. D. (2000). The timing of the "fertile window" in the menstrual cycle: Day-specific estimates from a prospective study. *British Medical Journal* **321**, 1259–1262.

Wilcox, A. J., Dunson, D. B., Weinberg, C. R., Trussell, J. & Baird, D. D. (2001). Likelihood of conception with a single act of intercourse: Providing benchmark rates for assessment of post-coital contraceptives. *Contraception* **63**, 211–215.

# PREFACE

In species with internal fertilization, sperm competition occurs when the sperm of two or more males simultaneously occupy the reproductive tract of a female and compete to fertilize an egg (Parker, 1970). A large body of empirical research has demonstrated that, as predicted by sperm competition theory, males and females in many species possess anatomical, behavioral, and physiological adaptations that have evolved to deal with the adaptive challenges associated with sperm competition. Moreover, in recent years, evolutionary biologists and psychologists have begun to examine the extent to which sperm competition may have been an important selective pressure during human evolution.

Some research has suggested that male humans, like males of many bird, insect, and rodent species, might be able to adjust the number of sperm they inseminate according to the risk of sperm competition. Other research has examined whether such responses might be accompanied by psychological changes that motivate human males to pursue copulations when the risk of sperm competition is high. Furthermore, there is research suggesting that aspects of human penile anatomy might function to enhance success in sperm competition. Much of this work has been controversial; some of the findings have been disputed and others have been greeted with skepticism. However, the idea that some aspects of human psychology and behavior might best be understood as adaptations to sperm competition remains intriguing and, in certain cases, very persuasive.

This volume brings together a key set of classic and contemporary papers that have examined possible adaptations to sperm competition in humans. In addition to classic papers by Robin Baker and Mark Bellis, this volume includes later work by other researchers – some developing Baker and Bellis' ideas, some refuting their findings. As is to be expected in any comparatively new area of investigation, there are conflicting findings and unresolved issues. This, however, should encourage rather than discourage future research in this field. This collection of papers is essential reading for students of evolutionary psychology, human sexuality, and researchers considering conducting work in this area. Moreover, some of the work included in this volume has attracted a great deal of media attention with highly selective and often inaccurate reporting of findings in newspapers and in documentaries. Consequently, this collection of papers should be of prime interest to anyone who has been intrigued by these media reports and wants to know the full story with all its controversies, inconsistencies, and exciting advances.

## REFERENCES

Parker, G. A. 1970. Sperm competition and its evolutionary consequences in the insects. *Biol. Rev.* **45**, 525–567.

# ACKNOWLEDGMENTS

For their outstanding support of this project, the Editors thank Andrea Macaluso, Helen Buitenkamp, and Sharon Panulla at Springer. For their help compiling this volume we would like to thank Rebekah Barr, Sarah-Jane Compton-Williams, Sarah Hart, and Lisy Knepper.

We would like to thank all the authors of the papers in this volume. First, for their invaluable contributions to sperm competition research. And second, for kindly allowing us to reprint their work in this volume.

Todd K. Shackelford thanks, for their scholarly support and encouragement, John Alcock, Robin Baker, Mark Bellis, Iris Berent, Jesse Bering, Tim Birkhead, Dave Bjorklund, April Bleske-Rechek, Becky Burch, David Buss, Martin Daly, Harald Euler, Gordon Gallup, Steve Gangestad, Aaron Goetz, Martie Haselton, Steve Hecht, Erika Hoff, Sabine Hoier, Lee Kirkpatrick, Craig LaMunyon, Randy Larsen, Brett Laursen, Rick Michalski, Dave Perry, Steve Platek, Monica Rosselli, Dave Schmitt, Bob Smith, Randy Thornhill, Robin Vallacher, Charles White, and Margo Wilson. Special thanks to Nick Pound, for making this project more fun than work. And to my mother, Lyndell, for her support and encouragement. Finally, I thank Viviana Weekes-Shackelford, for her unwavering support and encouragement, professional and personal.

Nicholas Pound thanks Andrew Clark, Martin Daly, Matt Gage, Steve Josephson, Chris Macfarlane, Ian Penton-Voak, and Margo Wilson, for suggestions, advice, criticism, encouragement, and lots of other things. I'd like to thank Todd Shackelford for making sure we completed this project. And special thanks to Elaine Carrera to whom I'm immensely grateful for her understanding and support.

# CONTENTS

# PART I: INTRODUCTION AND OVERVIEW

# 1. SPERM COMPETITION IN HUMANS

Nicholas Pound[1], Todd K. Shackelford[2] & Aaron T. Goetz[2]

## WHAT IS SPERM COMPETITION?

Sperm competition is the competition that can occur between the sperm of different males to fertilize a female's gamete(s) (Parker, 1970). In species with internal fertilization, there is the potential for sperm competition to occur whenever a female mates with multiple males within a sufficiently short period of time so that live sperm from two or more males are simultaneously present in her reproductive tract. The outcome of such competition, notwithstanding mating order effects, depends on a "raffle" or "lottery" principle (i.e., a particular male can increase the probability of siring a female's offspring by inseminating more sperm) (Parker, 1982, 1990*b*; 1998). The costs of ejaculate production, however, are non-trivial (Dewsbury 1982; Nakatsuru & Kramer, 1982; Olsson, Madsen, & Shine, 1997; Shapiro, Marconato, & Yoshikawa, 1994), and repeated ejaculation can lead to sperm depletion (Ambriz *et al.*, 2002; Preston, Stevenson, Pemberton, & Wilson, 2001). Consequently, for males there is a trade-off between ejaculate production costs and the potential benefits of delivering large numbers of sperm in any particular ejaculate.

### Sperm Number and Risk of Sperm Competition

In light of the trade-off males must make, one of the first hypotheses generated by sperm competition theory was that males will deliver more sperm when the risk of sperm competition is high, both across (Parker, 1982; Chapter 2, this volume) and within species (Parker, 1990*b*). Across species, therefore, high levels of sperm competition should select for increased investment in sperm production. Consistent with this, it has been found that in primates (Harcourt, Harvey, Larson, & Short, 1981; Harvey &

[1] Centre for Cognition & Neuroimaging, School of Social Sciences & Law, Brunel University, Uxbridge, UB8 3PH, United Kingdom.

[2] Florida Atlantic University, Department of Psychology, 2912 College Avenue, Davie, Florida 33314, USA.

Harcourt, 1984; Short, 1979), birds (Møller, 1988), ungulates (Ginsberg & Rubenstein, 1990), frogs (Jennions & Passmore, 1993), and butterflies (Gage, 1994), testis size relative to body size (and therefore investment in sperm production) is correlated positively with the incidence of female polyandry (i.e., the frequency with which females mate with multiple males). Recent work, in addition, has demonstrated experimentally that exposure to mating environments with high levels of sperm competition can produce significant increases in testis size after only 10 generations in yellow dung flies (*Scathophaga stercoraria*) (Hosken & Ward, 2001).

The prediction that males should deliver more sperm when the risk of sperm competition is high also applies within species, to individual males who should adjust the number of sperm they inseminate from one copulation to the next according to variations in sperm competition risk. In recent years, experimental evidence has accumulated indicating that in various species individual males are capable of such prudent sperm allocation (Parker, Ball, Stockley, & Gage, 1997; Wedell, Gage, & Parker, 2002; Chapter 3, this volume). However, since sperm competition risk cannot be assessed directly, males must rely on cues (auditory, olfactory, tactile, or visual stimuli) that reliably predict whether a female's reproductive tract (in the case of internal fertilizers) or the spawning area (in the case of external fertilizers) contains or will soon contain sperm from a rival male.

There is experimental evidence that males of various species respond adaptively to cues of increased sperm competition risk, such as male mating status, in species where it predicts the likelihood of mating with an already-mated female (Cook & Wedell, 1996), and female mating status where it is detectable (Gage & Barnard, 1996). In addition, males of various species appear to be sensitive to the operational sex ratio, or the mere presence of one or more rival males, during a particular mating event. Mealworm beetles (*Tenebrio molitor*) (Gage & Baker, 1991) and mediterranean fruitflies (*Ceratitis capitata*) (Gage, 1991) inseminate more sperm when mating in the presence of rival males. Furthermore, field crickets (*Gryllodes supplicans*) and house crickets (*Acheta domesticus*) increase the number of sperm they inseminate in proportion to the number of rivals present (Gage & Barnard, 1996).

Other species that are known to adjust the number of sperm they deliver either in response to the presence of rival males, or when group spawning, include bucktooth parrotfish (*Sparisoma radians)* (Marconato & Shapiro, 1996), blue crabs (*Callinectes sapidus*) (Jivoff, 1997), bluehead wrasse (*Thalassoma bifasciatum*) (Shapiro *et al.*, 1994), and rainbow darters (*Etheostoma caeruleum*). For a comprehensive review of these findings, see Wedell *et al.* (2002; Chapter 3, this volume). Of perhaps most relevance to work on responses of human males to cues of sperm competition risk, are findings that males of other mammalian species are capable of prudent sperm allocation in response to variations in sperm competition risk. Bellis, Baker, & Gage (1990) reported that male rats (*Rattus norvegicus*) adjust the number of sperm they inseminate depending on the amount of time they have spent with a particular female prior to copulation. In this experiment, rats were housed in mixed-sex pairs but prevented from mating by wire mesh dividing each cage. When allowed to mate, males inseminated less sperm when copulating with a female that they had accompanied during the 5 days preceding her estrus than when mating with a female accompanied by a different male during those 5 days. Bellis *et al.* (1990) interpreted this finding as evidence of prudent sperm allocation, because time spent with a female prior to copulation can be thought of as "guarding" time, and "unguarded" females are more likely to contain sperm from one or more rival males.

More recently, it has been shown that, male rats (*Rattus norvegicus*) inseminate more sperm when mating in the presence of a rival male (Pound & Gage, 2004). Similarly, male meadow voles (*Microtus pennsylvanicus*) inseminate more sperm when they mate in the presence of another male's odors (delBarco-Trillo & Ferkin, 2004).

## HAS SPERM COMPETITION BEEN AN ADAPTIVE PROBLEM FOR HUMANS?

The question of whether sperm competition has been an important selective force during human evolution is somewhat controversial. Smith (1984; Chapter 4, this volume) argued that the comparatively large size of the human penis, and the fact that human testes are larger in relation to body size than are those of monogamous primates (Short, 1981), suggests that sperm competition has been a recurrent feature of human evolutionary history.

Smith (1984) argued that facultative polyandry (i.e., female sexual infidelity) would have been the most common context for the simultaneous presence of live sperm from two or more men within the reproductive tract of an ancestral woman. Other contexts in which sperm competition might have occurred include consensual communal sex, concurrent courtship with multiple males, rape, and prostitution, but Smith (1984) argued that these contexts probably did not occur with sufficient frequency over human evolutionary history to provide selection pressures for adaptations to sperm competition equivalent to female infidelity. Consequently, whether or not sperm competition has been a major mode by which men have competed for reproductive success depends largely on the frequency with which women mated polyandrously in ancestral environments.

The ubiquity and power of male sexual jealousy certainly provides evidence of an evolutionary history of female infidelity, and therefore perhaps also of sperm competition. Male sexual jealousy would only evolve if female sexual infidelity was a recurrent feature of human evolutionary history (see, e.g., Buss, Larsen, Westen, & Semmelroth, 1992; Daly, Wilson, & Weghorst, 1982; Symons, 1979), and female infidelity increases the likelihood that sperm from two or more men will simultaneously occupy the reproductive tract of a single woman. Indeed, based on past and present infidelity rates of men and women, it may be concluded that humans practice social monogamy, the mating system in which males and females form long-term pair bonds but also pursue extra-pair copulations. Extra-pair copulation, and consequently sperm competition, is common in socially monogamous bird species (Birkhead & Møller, 1992); and perhaps also sufficiently common in humans to generate significant levels of sperm competition (Smith, 1984; Chapter 4, this volume; Bellis & Baker 1990; Chapter 9, this volume).

A recent genetic study has shown that the genes coding for proteins involved in the production and function of sperm have been evolving at a much faster rate than most other human genes and at a similar rate to homologous genes in chimpanzees (Wyckoff,

Wang, & Wu, 2000). The rapid evolution of these genes may have occurred as a consequence of the selection pressures generated by female promiscuity, since they appear to be evolving more slowly in gorillas where females are less likely to mate with multiple males.

There is necessarily some circularity involved when trying to understand aspects of human physiology, psychology and behavior as adaptations to sperm competition while also trying to use them as evidence for the very existence of sperm competition in humans. However, this and some subsequent chapters review evidence of physiological, psychological, and behavioral mechanisms that are perhaps most parsimoniously explained if sperm competition was a recurrent feature of human evolutionary history.

**Do Women Generate Sperm Competition?**

Evolutionary accounts of human sexual psychology have tended to emphasize the benefits to men of short-term mating and sexual promiscuity (e.g., Buss & Schmitt, 1993; Symons, 1979). For men to pursue short-term sexual strategies, however, there must be women who mate non-monogamously (Greiling & Buss, 2000). Moreover, if ancestral women never engaged in short-term mating, men could not have evolved a strong desire for sexual variety (Smith, 1984; Schmitt, Shackelford, & Buss, 2001; Schmitt, Shackelford, Duntley, Tooke, & Buss, 2001; Schmitt *et al.*, 2003).

Ancestral women may have benefited from facultative polyandry in several ways (for a review, see Greiling & Buss, 2000). Some of the most important potential benefits include the acquisition of resources, either in direct exchange for sex with multiple men (Symons, 1979) or by creating paternity confusion as a means to elicit investment (Hrdy, 1981). Alternatively, ancestral women may have benefited indirectly by accepting resources and parental effort from a primary mate while copulating opportunistically with men of superior genetic quality (Smith, 1984; Symons, 1979). Furthermore, extra-pair sex might have been useful as insurance against the possibility that a primary mate was infertile, and in unpredictable environments it may be advantageous for women to ensure that offspring are sired by different men and are thus genetically diverse (Smith, 1984). Jennions and Petrie (2000) provide a comprehensive review of the genetic benefits to females of multiple mating.

While multiple mating by women is a prerequisite for sperm competition to occur, not all patterns of polyandry are sufficient to allow for such post-copulatory competition between men. For sperm competition to occur, women must copulate with two or more men within a sufficiently short period of time such that there is overlap in the competitive lifespans of the rival ejaculates. The length of this competitive "window" might be as short as 2–3 days (Gomendio & Roldán, 1993), or as long as 7–9 days (Smith, 1984). Using an intermediate estimate of 5 days, Baker and Bellis (1995) argued that questionnaire data they collected on female sexual behavior indicated that 17.5% of British women "double-mated" in such a way as to generate sperm competition at some point during the first 50 lifetime copulations. It should be noted, however, that in contemporary environments use of barrier contraceptives means that "double-matings" do not always lead to the simultaneous presence of live sperm from two males within the female's reproductive tract, but do represent a pattern of behavior that would typically have led to sperm competition in ancestral environments.

As pointed out by Gomendio *et al.* (1998) there are some problems with Baker and Bellis' (1995) estimate of the frequency with which women engage in "double-matings". In particular, the questionnaire data upon which the estimate was based were derived from a self-selected sample of women from the readership of a popular women's magazine who were probably more sexually active than the female population at large. Gomendio *et al.* (1998) also argued that the competitive lifespan of an ejaculate within the female reproductive tract is in fact probably around 3 rather than 5 days so Baker and Bellis' (1995) definition of "double-mating" may be insufficiently restrictive. In any case, whichever estimate is accepted, there are only limited data available on the frequency with which women in contemporary populations might "double mate" since large-scale studies of sexual behavior have not collected data on this specifically, but many have recorded how often they engage in concurrent sexual relationships, more generally.

Not all concurrent sexual relationships involve copulations with different men within a sufficiently short space of time to be considered double-matings, but it is likely that many do since even using Gomendio *et al.*'s (1998) conservative estimate of the competitive lifespan of an ejaculate (3 days) sperm competition would be very likely to occur if a woman conducts concurrent sexual relationships with two men that involve weekly intercourse with each partner. Consequently, the rate at which women participate in concurrent sexual relationships provides an index of the likelihood of sperm competition in a population. Gomendio *et al.* (1998), for example, argued that survey data indicate that only 2% of women in Britain have engaged in concurrent sexual relationships in the past year and, consequently, that sperm competition is likely to be a relatively infrequent occurrence. However, a major study of sexual behavior in Britain—the National Survey of Sexual Attitudes and Lifestyles conducted between 1999 and 2001 (Johnson *et al.*, 2001)—revealed that 9% of women overall, and 15.2% of those aged 16–24 years, reported having had concurrent sexual relationships with men during the preceding year. While in a large study of sexual behavior in the US, Laumann, Gagnon, Michael, and Michaels (1994), found that while only about 20% of young women reported having multiple partners during the previous year, of those who reported having many (five or more), 83% also report that at least two of these relationships were concurrent.

Sperm competition may arise as a by-product of female behaviors that function to increase the probability that a female will conceive offspring sired by a male other than her primary mate. However, Bellis and Baker (1990; Chapter 9, this volume) argued that women "schedule" their copulations in a way that *actively promotes* sperm competition. Active promotion of successive insemination by two or more men may allow a woman to be fertilized by the most competitive sperm. Bellis and Baker (1990) documented that women are more likely to double-mate when the probability of conception is highest, and when the probability of conception is lower, in contrast, they tend to space their in-pair and extra-pair copulations more widely, making sperm competition less likely. Bellis and Baker (1990) argued that these findings cannot be attributed to men's preferences for copulation with women at peak fertility since such preferences would also increase the frequency of in-pair copulations during the fertile phase of the menstrual cycle.

It is possible, however, that Bellis and Baker (1990) may have been too quick to dismiss the possibility that men prefer to copulate with a woman during the most fertile phase of her menstrual cycle since in-pair copulation frequency may not accurately reflect male preferences and desires. If women mate with extra-pair partners in order to secure genetic benefits for their offspring (see Greiling & Buss, 2000; Gangestad & Simpson, 2000), then while they may seek copulations with extra-pair males when the probability of conception is highest, they may also simultaneously seek to avoid in-pair copulations at this time. Consequently, if in-pair copulations are no more frequent at times of high fertility this may not indicate the absence of increased interest on the part of the male partner. It may just be that any increased interest is offset by female avoidance of in-pair copulations during the high fertility phase of the cycle. Therefore, while Bellis and Baker's (1990) finding that women are more likely to double-mate when the probability of conception is highest is consistent with the hypothesis that women sometimes actively promote sperm competition, it does not rule out the possibility that both in-pair and extra-pair partners prefer to copulate with a woman during her peak fertility.

## Women's Fantasies of Polyandrous Sex

Sexual fantasy may provide a "window" through which to view the evolved psychological mechanisms that motivate sexual behavior (Ellis & Symons, 1990; Symons, 1979). A large empirical literature has addressed sex differences in sexual fantasy, and much of this work has been conducted from an evolutionarily informed perspective (see, e.g., Ellis & Symons 1990; Wilson, 1987, 1997; Wilson & Lang, 1981; and see Leitenberg & Henning, 1995, for a broad review of empirical work on sexual fantasy). This work documents several marked sex differences in the content of sexual fantasies, consistent with hypotheses generated from Trivers' (1972) theory of parental investment and sexual selection. For example, given the asymmetric costs associated with sexual reproduction, sexual access to mates limits reproductive success for males more than for females. Consequently, it has been hypothesized that men more than women will have sexual fantasies that involve multiple, anonymous sexual partners who do not require an investment of time, energy, or resources prior to granting sexual access (e.g., Ellis & Symons, 1990), and empirical investigations have confirmed this hypothesis. Indeed, one of the largest sex differences occurs for fantasies about having sex with two or more members of the opposite sex concurrently, with men more than women reporting this fantasy (see review in Leitenberg & Henning, 1995).

Tests of the hypothesis that men more than women fantasize about concurrent sex with two or more partners have, however, inadvertently provided data on women's polyandrous sexual fantasies. Although this work clearly indicates that men are more likely than women to report fantasies of concurrent sex with multiple partners, polyandrous sex is certainly something that women do fantasize about. In a large survey study, for example, Hunt (1974) found that 18% of women report fantasies of polyandrous sex, imagining themselves as a woman having sex with two or more men concurrently. Wilson (1987) surveyed nearly 5000 readers of Britain's top-selling daily newspaper about their favorite sexual fantasy and performed content analyses on the responses of a random sub-sample of 600 participants. Polyandrous sex was the key element of the favorite sexual fantasy reported by 15% of female participants.

Studies using smaller samples of participants also provide evidence that polyandry is a common theme of women's sexual fantasies, albeit less common than polygyny is in men's fantasies. For example, Rokach (1990) reported that, while sex with more than one partner accounted for 14.3% of the sexual fantasies reported by a sample of 44 men, it also accounted for 10% of the fantasies reported by a sample of 54 women. Person, Terestman, Myers, Goldberg, and Salvadori (1989) and Pelletier and Herold (1988) documented that 27% and 29%, respectively, of the women sampled report fantasies of polyandrous sex. And fully 41% of women sampled by Arndt, Foehl, and Good (1985) report fantasies involving sex with two men at the same time. Davidson (1985) and Sue (1979) report that smaller but still sizable percentages (17% and 15%, respectively) of women recall fantasies involving sex with two or more men concurrently, and Price and Miller (1984) report that polyandrous sex was among the 10 most frequently reported fantasies in a small sample of college women.

If sexual fantasy reflects sexual desires and preferences that might occasionally be acted upon, then the research outlined above indicates that polyandrous sex is not an unlikely occurrence, particularly given the well established finding that women more than men are the "gatekeepers" of sexual access—controlling when, where, and the conditions under which sex occurs (see, e.g., Buss, 1994; Symons, 1979). If, as Symons (1979) has argued (and see Buss, 1994; Ellis & Symons, 1990), sexual fantasy provides a window through which to view evolved human psychology, then human female sexual psychology may sometimes lead to the pursuit of polyandrous sex, with the consequence of promoting sperm competition.

## Evidence of Adaptations to Sperm Competition in Men

Examining adaptive problems and their resultant evolved solutions in non-humans can sometimes provide insights into adaptive problems and their evolved solutions in humans (and vice versa). Shackelford and LeBlanc (2001) argued that because humans share similar adaptive problems with insects (e.g., mate retention) and birds (e.g., partner infidelity), humans, insects, and birds may share similar solutions to these adaptive problems. Consequently, a comparative evolutionary psychological approach to the study of adaptations to sperm competition in other species may lead to the discovery of evolved solutions in humans.

Male humans and male birds, for example, have recurrently faced the adaptive problems of anticipating, preventing, and dealing with the consequences of a partner's infidelity (Shackelford, 2003). By preventing a partner from engaging in extra-pair copulations a male can ensure that his ejaculate remains uncontested within her reproductive tract. However, because a male's preventative measures may not always be successful, he would likely benefit reproductively if he were able to counteract his partner's infidelity by, for example, quickly inseminating her following a detected infidelity and thus perhaps reducing the risk that she will conceive a child sired by another male. Alternatively, in circumstances where a male is able to anticipate, but is unable to prevent, an extra-pair copulation he may be able to improve his chances of siring her offspring by introducing his sperm into her reproductive tract before it contains sperm from a rival. In addition to behavioral adaptations that function to anticipate prevent, and possibly counteract partner infidelity, an evolutionary history of sperm competition would likely have led to the evolution of anatomical and physiological adaptations similar to those seen in other species.

When considering whether men have adaptations to sperm competition, two closely related but distinct questions can be asked. First, do men possess adaptations that evolved to deal with *high* levels of sperm competition? Second, do they possess adaptations that evolved to deal with *variable* levels of sperm competition? Since across species, investment in sperm production will depend on the level of sperm competition (Parker, 1982, 1990*a*, 1990*b*) adaptations to *high* levels of sperm competition will include anatomical, physiological, and behavioral traits that facilitate the delivery of large numbers of highly competitive sperm during copulation. Within species, sperm competition theory predicts that individual males should be able to respond to *variable* levels of sperm competition by allocating their sperm in a prudent fashion and accordingly inseminate more sperm when the risk is high. Consequently, adaptations to *variable* levels of sperm competition are likely to take the form of traits that enable males to adjust the number of sperm they inseminate from one copulation to the next according to variations in the risk of sperm competition. It is possible that adaptations to *variable* levels of sperm competition will be seen in species where overall levels are not especially high—but are sufficiently pronounced to cause the evolution of mechanisms that allow prudent sperm allocation.

Sperm competition can take one of two forms: *contest competition*, in which rival ejaculates actively interfere with each other's ability to fertilize an ovum or ova, and *scramble competition*, which is more akin to a simple race or lottery. In mammals, where the female reproductive tract is extremely large relative to sperm size, sperm competition is likely to take the form of a scramble in which the "raffle principle" applies (Parker, 1982; Chapter 2, this volume). Modeling studies and experimental findings support this view (Gomendio *et al.*, 1998). Male adaptations to scramble competition are likely to take the form of anatomical, physiological, and behavioral features that increase the male's chances of fertilizing an ovum or ova in a competitive environment in which the ability to deliver large numbers of sperm is a crucial determinant of success.

The ability to inseminate a large number of sperm depends on the ability to produce them and, consequently, testis size is predicted to correlate positively with the incidence of polyandrous mating across taxa. Empirical evidence has confirmed this predicted relationship in birds (Møller, 1988), ungulates (Ginsberg & Rubenstein, 1990), frogs (Jennions & Passmore, 1993), and butterflies (Gage, 1994) while in primates, testis size relative to body weight also is correlated positively with the incidence of polyandrous mating (Harcourt, Harvey, Larson, & Short, 1981; Harvey & Harcourt 1984; Short, 1979). Smith (1984) argued that the fact that men have testes that are large relative to monandrous species, such as the gorilla and orangutan suggests that polyandry was an important selection pressure during human evolution. As Gomendio *et al.* (1998) noted, however, human relative testis size is closer to these monandrous primates than to the highly polyandrous chimpanzee. Nevertheless, Gomendio *et al.*'s (1998) conclusion that humans are "monandrous" is not justified. Dichotomizing species into monandrous and polyandrous groups is not useful when there is continuous variation across species in the frequency with which females mate with multiple males. Although human males have probably not experienced levels of sperm competition similar to those found in primate species with very high levels of female promiscuity, it is nevertheless unlikely that sperm competition has been completely absent over human evolutionary history.

Human males have a penis that is longer than in any other species of ape (Short, 1979), but in relation to body weight it is no longer than the chimpanzee penis (Gomendio *et al.*, 1998). Nevertheless, several arguments have been offered to explain how the length and shape of the human penis might reflect adaptation to an evolutionary history of sperm competition. A long penis may be advantageous in the context of scramble competition, which combines elements of a race and a lottery, because being able to place an ejaculate deep inside the vagina and close to the cervix may increase the chance of fertilization (Baker & Bellis, 1995; Short, 1979; Smith, 1984). Additionally, it has been suggested that the length, width, and shape of the human penis indicate that it may have evolved to function as a sperm displacement device.

Gallup and Burch (2004; Chapter 14, this volume) discuss the possibility that semen displacement may function as a sperm competition strategy. Using artificial genitals and simulated semen, Gallup, Burch, Zappieri, Parvez, and Stockwell (2004) empirically tested Baker and Bellis's (1995) hypothesis that the human penis may have been designed to displace semen deposited by other men in the reproductive tract of a woman. Artificial phalluses with a simulated glans and coronal ridge that approximated a real human penis displaced significantly more simulated semen than did a phallus that did not have a glans and coronal ridge. Gallup *et al.* (2004) suggested that when the penis is inserted into the vagina, the frenulum of the coronal ridge makes possible semen displacement by allowing semen to flow back under the penis alongside the frenulum and collect on the anterior of the shaft behind the coronal ridge. Displacement of simulated semen only occurred, however, when a phallus was inserted at least 75% of its length into the artificial vagina, suggesting that successfully displacing rival semen may require specific copulatory behaviors. Following allegations of female infidelity or separation from their partners (contexts in which the likelihood of rival semen being present in the reproductive tract is relatively greater), both sexes report that men thrusted deeper and more quickly at the couple's next copulation (Gallup *et al.*, 2004). Such vigorous copulatory behaviors seem likely to increase semen displacement.

In an independent test of the hypothesis that successfully displacing rival semen may require specific copulatory behaviors, Goetz, Shackelford, Weekes-Shackelford, Euler, Hoier, Schmitt, & LaMunyon (2005; and see Goetz, Shackelford, Weekes-Shackelford, Euler, Hoier, & Schmitt, 2003) investigated whether, and how, men under a high risk of sperm competition might attempt to counteract a female partner's sexual infidelity. Using a self-report survey, men in committed, sexual relationships reported their use of specific copulatory behaviors arguably designed to displace the semen of rival men. These copulatory behaviors included number of thrusts, deepest thrust, average depth of thrusts, duration of sexual intercourse, and number of sexual positions initiated by the male. An increase in these behaviors would afford a man a better chance to displace rival semen. As hypothesized, men mated to women who place them at a high recurrent risk of sperm competition were more likely to perform possible semen-displacing behaviors, suggesting that men might perform specific copulatory behaviors apparently designed to counteract female sexual infidelity by displacing rival semen that may be present in the woman's reproductive tract.

One concern with the hypothesis that the human penis has evolved as a semen displacement device is that, during copulation, a man's penis might frequently remove his own sperm. The risk of this might be reduced by the male refractory period, which may have evolved partly because the inability to maintain an erection following ejaculation could serve to allow for sperm migration and thus minimize self-semen displacement (Gallup & Burch, 2004, Chapter 14, this volume).

In discussing the hypothesis that semen displacement may function as a sperm competition strategy, Gallup and Burch (2004) interpret premature ejaculation as a failure to achieve semen displacement. Gallup and Burch (2004; and Goetz *et al.*, 2005) also discuss the impact semen displacement may have on the reproductive strategies of women. If women use extra-pair copulations to secure genes superior to those of a primary mate, and men use semen-displacing behaviors as anti-cuckoldry tactics, it is predicted that, following an extra-pair copulation, a woman should attempt to postpone copulation with her primary mate, whereas her primary mate should attempt to initiate copulation as soon as possible.

Recent work has provided evidence that men are attuned to women's cuckoldry tactics and respond with preventative measures designed to thwart their partner's sexual infidelity. Gangestad, Thornhill, and Garver (2002), for example, documented that, as women enter the high fertility phase of their menstrual cycle, they are more likely to be sexually attracted to, and to fantasize about, men *other than* their current partners, and that their partners are more proprietary of and attentive to them near ovulation. These results suggest that men are sensitive to their partner's increased interest in extra-pair copulation near ovulation, and respond by increasing their mate guarding behavior to reduce the risk of sperm competition. In addition, Gangestad *et al.* (2002) found that the less women initiated sex with their partners, the more men attempted to monopolize their partner's time. Failing to initiate sex may signal to a woman's partner that she anticipates extra-pair copulation and, therefore, his attempt to monopolize her time may be a tactic to reduce the risk of sperm competition.

Since men adjust their mate guarding efforts according to their partner's fertility (Gangestad *et al.*, 2002), Goetz *et al.* (2003; 2005) hypothesized that men mated to women who placed them at a high recurrent risk of sperm competition would more frequently use mate retention tactics. The results suggested that men mated to women who have traits linked to a higher probability of sexual infidelity more frequently used mate retention tactics apparently designed to thwart potential infidelities. However, because preventative tactics are not always successful, these men also performed specific copulatory behaviors apparently designed to counteract female sexual infidelity by displacing rival semen that may be present in the woman's reproductive tract (Gallup *et al.*, 2004; Goetz *et al.*, 2003; 2005).

As Baker and Bellis (1995) noted, an evolutionary history of sperm competition may be responsible for myriad male behaviors related directly and indirectly to mating. Research informed by sperm competition theory is just beginning to uncover those behaviors. Men's sexual motivation throughout the duration of a mateship, aspects of men's short-term mate selection, male homosexuality, and men's preferences in pornographic images, for example, may have their origins in sperm competition.

A recurrent evolutionary history of sperm competition may explain why men are continually interested in copulating with their partners throughout the duration of a mateship (Klusmann, 2002), a prediction first made by Baker and Bellis (1993*a*; Chapter 10, this volume). According to Baker and Bellis's (1993*a*) "topping-up" model, a woman's primary partner should endeavor to maintain an optimum level of sperm in his partner's reproductive tract as a defense against sperm competition. It has been reported that the "quality" of marital sex declines with time (Chien, 2003), nevertheless, surveying German participants, Klusmann (2002) documented that while sexual desire for one's partner declines in women it remains constant in men for the duration of a mateship, and interpreted the results in accordance with the topping-up model. Of course, sexual desire cannot lead to "topping-up" unless copulation takes place so this interpretation is somewhat problematic given that Klusmann (2002) found that while male sexual desire may remain constant throughout the duration of a mateship, levels of sexual activity decline. However, since women more than men control sexual access their waning interest in sex may lead to a decrease in sexual activity despite constant male desire.

The selection pressures associated with sperm competition may have led to the evolution of mating preferences in males that function to select as short-term sexual partners women who present the lowest risk of current or future sperm competition (Shackelford, Goetz, LaMunyon, Quintus, & Weekes-Shackelford, 2004). The risk of sperm competition for a man increases with a prospective short-term partner's involvement in one or more relationships. Women who are not in a long-term relationship and do not have casual sexual partners, for example, present a low risk of sperm competition and therefore may be perceived as desirable short-term sexual partners. Women who are not in a long-term relationship but who engage in short-term matings may present a moderate risk of sperm competition, since they probably do not experience difficulty obtaining willing sexual partners. Women in a long-term relationship, however, may present the highest risk of sperm competition due to frequent insemination by their primary partner and thus may be least attractive as short-term sexual partners.

As predicted, Shackelford *et al.* (2004) found that men's sexual arousal and reported likelihood of pursuing a short-term sexual relationship was lowest when imagining that the potential short-term partner is married, next lowest when imagining that she is not married but involved in casual sexual relationships, and highest when imagining that she is not married and not involved in any casual sexual relationships. These results suggest that, when selecting short-term sexual partners, men do so in part to avoid sperm competition. However, a number of alternative explanations for these findings are plausible. For example, a man may be less likely to get involved in violent confrontation with a rival male if he pursues females who are not already involved in relationships. Moreover, a female with a partner may already be pregnant and thus unable to conceive. Finally, by preferring unmated women, men could perhaps avoid the costs associated with contracting a sexually transmitted infection (STI). The data, however, seem to be inconsistent with this last explanation since the potential short-term partner most likely to be infected with an STI would probably be the one having casual sex and not the married potential sexual partner who was least preferred. This suggests that avoiding STIs, seems to be less important than avoiding sperm competition when selecting short-term partners.

The possibility that male homosexuality may arise as a by-product of adaptations that evolved in response to the selective pressures associated with sperm competition has been proposed by a number of authors (MacIntyre & Estep, 1993; Pound 2002). MacIntyre and Estep (1993) suggested that male adaptations for sexual promiscuity may be advantageous in environments in which the risk of sperm competition is high, but also will tend to increase the rate at which men engage in homosexual activity. In contrast, Pound (2002) argued that sexual arousal in response to other men may reflect a specific adaptation to sperm competition, because in certain circumstances other men might be cues of increased sperm competition risk.

Pound (2002) argued that sexual arousal may be an adaptive response to increased sperm competition risk since it may motivate the pursuit of copulation and this could lead to the displacement of rival males before they can inseminate a female, or the delivery of more sperm as a consequence of more frequent intromission. Moreover, sexual arousal may a psychological correlate of, or have a causal role in, facultative increases in the number of sperm ejaculated. Pound (2002) hypothesized that men, therefore, should be more aroused by pornography that incorporates cues of sperm competition than by comparable material in which such cues are absent. Content analyses of pornographic images on World Wide Web sites and of commercial "adult" video releases revealed that depictions of sexual activity involving a female and multiple males are more prevalent than those involving a male and multiple females indicating that the former category may be preferred by men. Additionally, an online survey of self-reported preferences and an online preference study that unobtrusively examined image selection behavior yielded corroborative results. Pound (2002) argued that the most parsimonious explanation for these findings is that male sexual arousal in response to visual cues of sperm competition risk reflects the function of psychological mechanisms that would have motivated adaptive patterns of copulatory behavior in ancestral males exposed to cues of female promiscuity.

The idea that men might experience increased sexual motivation in response to cues of sperm competition risk also is supported by anecdotal accounts of men who engage in "swinging" or "partner-swapping." Such men often report that they find the sight of their partner interacting sexually with other men to be sexually arousing (Talese, 1981). Moreover, they report that they experience increased sexual desire for their partner following her sexual encounters with other men, and some indicate that this increase in desire is particularly acute when they have actually witnessed their partner having sexual intercourse with another man (Gould, 1999).

## IS THERE EVIDENCE OF PRUDENT SPERM ALLOCATION BY MEN?

Compared to other primates, human ejaculates do not contain especially large numbers of sperm (Baker & Bellis, 1995; Dixson, 1998) so men do not appear be adapted to particularly high levels of sperm competition. Nevertheless, it may be the case that they possess physiological adaptations that allow them to allocate sperm prudently in the face of variable levels of sperm competition risk.

The number of sperm contained in a man's ejaculate varies considerably from one ejaculate to the next (Read & Schnieden, 1978; Schwartz, Laplanche, Jouannet, & David, 1979; Mallidis, Howard, & Baker, 1991) and although clinicians have tended to treat this intra-individual variability as "noise" or as a barrier to determining a man's "true" semen parameters, there is the intriguing possibility that some of this variability might be caused by temporal variation in exposure to cues of sperm competition risk. Whether or not such variability in ejaculate composition is patterned adaptively in contemporary environments, it is possible that it may reflect the functioning of mechanisms that evolved to allocate sperm prudently in response to variations in the risk of sperm competition in ancestral environments.

Men display prudent sperm allocation in contemporary environments in at least one fundamental sense: sperm are not emitted continuously, but instead are ejected during discrete ejaculatory events that occur in response to sexual stimulation of sufficient intensity and duration. That some of these events are masturbatory would seem to be imprudent but sperm wastage due to masturbation is mitigated to a certain extent by the fact that masturbatory ejaculates typically contain fewer sperm than do copulatory ejaculates (Zavos, 1985; Zavos & Goodpasture, 1989). The only published evidence, however, indicating that men can adjust ejaculate composition in response to adaptively relevant aspects of their sociosexual environment was reported in a series of papers by Baker and Bellis.

In 1989, Baker and Bellis first reported that the number of sperm inseminated by men varied according to hypotheses generated by sperm competition theory (Baker & Bellis, 1989*b*; Chapter 8, this volume). For this study, 10 heterosexual couples provided semen specimens collected via masturbation and others collected during copulation. In each case, participants used non-spermicidal condoms to collect the specimens and also provided the researchers with information about the time since their last ejaculation, the time since their last copulation, and the percentage of time spent together with their partner since the last copulation. Although participants provided multiple specimens, the analysis was restricted to the first specimens provided in each of the two experimental contexts (masturbatory and copulatory). For the 10 copulatory specimens, there was a significant negative rank-order correlation between the percentage of time the couple had spent together since their last copulation and the estimated number of sperm in the ejaculate. No such relationship was identified for masturbatory ejaculates. If percentage of time spent apart from a partner is a reliable cue of the risk of female double-mating, then these findings are consistent with the hypothesis that there is a positive association between the number of sperm inseminated and the risk of sperm competition (Parker 1970, 1982).

What Baker and Bellis (1989*b*) reported, however, was a between-subjects relationship between sperm competition risk and ejaculate composition—that is, an observation that, for a sample of 10 couples, men who had spent the most time apart from their partners since their last copulation produced ejaculates containing the most sperm. Baker and Bellis (1989*b*) did not provide direct evidence of prudent sperm allocation by individual men from one specimen to the next in response to variation in sperm competition risk. It could be the case that men who tend to produce larger ejaculates also tend to spend a greater proportion of their time between copulations apart from their partners. Moreover, this relationship could be mediated by between-male differences in testicular size and associated levels of testosterone production if variability in these variables predicts semen parameters and certain aspects of sexual behavior.

In a follow-up to this initial report, Baker and Bellis (1993*a*; Chapter 10, this volume) attempted to address the aforementioned problem by including in their analyses more than one ejaculate from each couple that participated in this second study. Twenty-four couples provided a total of 84 copulatory ejaculates. To assess whether the number of sperm inseminated by a man depended on the percentage of time spent together since the last copulation with his partner, only those copulatory specimens that were preceded by an ejaculation also produced during an in-pair copulation (IPC) were included in the analyses (IPC–IPC ejaculates). Forty specimens produced by five men were included in the final analysis, and for these a non-parametric test based on ranks indicated a negative association between the number of sperm inseminated and the proportion of time the couple had spent together since their last copulation.

This was the first evidence of ejaculate adjustment by men from one copulation to the next in response to a cue indicating temporal variation in the risk of sperm competition. It should be noted, however, that 33 of the 40 specimens included in the ranking test were produced by just two men. Moreover, although data were presented for the first IPC–IPC ejaculates produced by all 15 couples who provided copulatory specimens, an analysis similar to that presented in the 1989 paper was not reported. We conducted this analysis using the 1993 data, which revealed that, for the first IPC–IPC ejaculate produced by each couple, the negative rank-order correlation between the number of sperm inseminated by a man and the percentage of time spent together with his partner since their last copulation was only marginally significant ($r = -0.50$; $p = 0.058$).

Aside from the small sample size used in Baker and Bellis' (1993*a*) demonstration of prudent sperm allocation by individual men, a number of additional methodological concerns have led some researchers to be skeptical of the findings. The reasons for doubting the reliability, validity, and generalizability of Baker and Bellis' (1993*a*) findings have rarely been articulated in articles published in peer-reviewed journals, however. Of major concern is the possibility that the people who participated in this highly intrusive research about some of their most private behaviors may not be representative of people in general. Recruited from the staff and postgraduate students in a biology department, the participants might have had some knowledge of the experimental hypothesis. It is not clear, however, how such knowledge could affect semen parameters. Knowledge about the experimental hypothesis could have affected the sexual behavior of the participants, and there is some evidence that semen parameters are subject to behavioral influences (Zavos, 1985, 1988; Zavos & Goodpasture 1989; Zavos, Kofinas, Sofikitis, Zarmakoupis, & Miyagawa, 1994; Pound, Javed, Ruberto, Shaikh, & Del Valle, 2002). However, evidence that men are able to adjust their semen parameters in response to the demand characteristics of an experiment would perhaps be more remarkable than evidence of prudent sperm allocation in the face of cues of sperm competition risk.

Baker and Bellis (1993*a*) argued that increases in the number of sperm inseminated by a man in response to a decrease in the proportion of time spent together with his partner since the couple's last copulation reflects prudent sperm allocation in response to a cue of increased sperm competition risk. Several alternative interpretations are possible, however. For example, changes in ejaculate composition may be secondary to changes in female sexual behavior induced by partner absence. Women who have spent a smaller proportion of time together with their partner since the couple's last copulation may behave differently during intercourse and thus provide different stimuli prior to, and at the time of, ejaculation. This may be significant because evidence that ejaculates obtained via uninterrupted coitus have higher semen volume, total sperm number, and sperm motility than those obtained via *coitus interruptus* (Zavos *et al.*, 1994) indicates that sexual stimuli present at the moment of ejaculation may be important determinants of ejaculate composition.

Exactly what the men in Baker and Bellis' (1993*a*) study were doing during periods of partner absence may determine the extent to which the findings are explicable solely by reference to sperm competition theory. During periods of partner absence, men may spend more time associating, and perhaps flirting, with women other than their primary partner. Given that sexual arousal immediately prior to ejaculation can facilitate sperm delivery and thus improve semen quality in humans (Zavos, 1988; Pound, 2002) and in other species (Collins, Bratton, & Hendersen, 1951; Almquist, Hale, & Amann, 1958; Hafs, Kinsey, & Desjardins, 1962; Almquist, 1973), repeated exposure to arousing stimuli during a period of ejaculatory abstinence could have similar effects. During a period in which a man is away from his partner, these arousing stimuli could be non-copulatory but sexually charged encounters with other women.

Finally, changes in semen parameters following a period of partner absence might not function primarily as a response to the risk that a partner contains sperm from a rival male but may be an anticipatory response to the fact that past absence may predict future absence (Gomendio *et al.*, 1998). Thus, increases in the number of sperm delivered might serve simply to maximize the chances of conception during a future period of partner absence during which ovulation might occur.

## HOW MIGHT MEN ACHIEVE PRUDENT SPERM ALLOCATION?

### Psychological and Behavioral Mechanisms

Males in many non-human species are capable of adjusting the number of sperm they inseminate in response to cues of sperm competition risk. Moreover, there is some evidence that this is something that men might also be able to do (Baker & Bellis, 1993*a*). Little attention has been paid, however, to the psychological mechanisms that might be involved in regulating such responses in either human or non-human males. Adaptive changes in semen parameters can serve no function unless they are accompanied by a desire to copulate with a partner when cues of sperm competition risk are present. Accordingly, Shackelford, LeBlanc, Weekes-Shackelford, Bleske-Rechek, Euler, and Hoier (2002; Chapter 13, this volume) investigated the psychological responses of men to cues of sperm competition risk, arguing that there must be psychological mechanisms in men that evolved to motivate behavior that would have increased the probability of success in sperm competition in ancestral environments.

Baker and Bellis (1993a, 1995) operationalized risk of sperm competition as the proportion of time a couple has spent together since their last copulation and examined changes in semen parameters associated with variations in this index which, they argued, is inversely related to the risk of sperm competition. However, Shackelford *et al.* (2002) argued that it is probably time spent apart that has most salience to men themselves and it is this information that is processed by male psychological mechanisms that subsequently motivate a man to inseminate his partner as soon as possible, to combat the increased risk of sperm competition.

Total time since last copulation is not necessarily linked to sperm competition risk. Instead, it is the proportion of time a couple has spent apart since their last copulation—time during which a man cannot account for his partner's activities—that is linked to the risk that his partner's reproductive tract might contain sperm from a rival male (Baker & Bellis, 1995). Nevertheless, total time since last copulation might have important effects on a man's sexual behavior, with him perhaps feeling increasingly "sexually frustrated" as this time increases (whether or not that time has been spent apart or together). Consequently, Shackelford *et al.* (2002) controlled for this potential confound in their examination of the effects, on male sexual psychology, of time spent apart from a partner.

Shackelford *et al.* (2002) suggested that the nature of a couple's relationship may have important effects on the how a man might respond to cues of sperm competition risk. Satisfaction with, and investment in, a relationship are likely to be linked, with the result that a man who is more satisfied may have more to lose in the event of cuckoldry. For this reason, when examining responses of men to increases in the proportion of time spent apart from their partner since their last copulation, Shackelford *et al.* (2002) controlled for the extent to which the participants were satisfied with their relationships.

Consistent with their predictions, Shackelford *et al.* (2002) found a positive association between the proportion of time spent apart since last copulation and (a) male ratings of their partner's attractiveness; (b) ratings of their own interest in having sex with their partner and (c) ratings of their partner's interest in sex. Shackelford *et al.* (2002) argued that no existing theory other than sperm competition theory can account for the predictive utility of the proportion of time spent apart since the couple's last copulation, independent of the total time since last copulation and independent of relationship satisfaction. Additionally, they argued that their findings support the hypothesis that men, like males of other socially monogamous, but not sexually exclusive species, have psychological mechanisms designed to solve adaptive problems associated with a partner's sexual infidelity.

This study by Shackelford *et al.* (2002) examined psychological responses in men that arguably are designed to motivate the pursuit of in-pair copulations as insurance against the possibility that a partner has engaged in an extra-pair copulation. Men, however, are likely to employ a large repertoire of behaviors in order to reduce the likelihood that their partner will engage in extra-pair sex in the first place. Buss and Shackelford (1997; see also Buss, 1988), for example, identified 19 different mate retention tactics used by people to prevent their partner from becoming romantically involved with someone else. These tactics varied from vigilance about a partner's whereabouts to violence against an encroaching rival. Of particular interest is whether some of the more abhorrent male mate retention tactics, such as psychological abuse, physical abuse, and sexual coercion, may be linked to the risk of sperm competition. Identifying such links may be a first step to helping women who suffer at the hands of abusive and controlling men (see Daly & Wilson, 1988, 1996; Dobash & Dobash, 1979; Dutton, 1995; Jacobson & Gottman, 1998; Peters, Shackelford, & Buss, 2002).

Following Thornhill and Palmer (2000), Goetz and Shackelford (2005) hypothesize that sexual coercion of an unwilling partner may be an anti-cuckoldry tactic designed to solve the adaptive problems generated by sperm competition. Reviewing the existing literature on: (a) instances of sexual coercion following accusations of female infidelity (b) the relationship between male sexual jealousy and sexual coercion of a partner and (c) narratives provided by victims of wife rape, Goetz and Shackelford (2005) argue that sexual coercion within a mateship is predictably most likely to occur in response to suspicious or detection of female sexual infidelity.

## Physiological Mechanisms

Increased sexual motivation in response to cues of sperm competition risk may lead to in-pair copulations that might reduce the risk that a woman will conceive a child sired by another man. However, sperm competition theory predicts that males should also be capable of adjusting the number of sperm they inseminate from one copulation to the next according to variations in the risk of sperm competition. The findings of Baker and Bellis (1993a, 1995) suggest that men may be capable of this, but it is not clear how such prudent sperm allocation might be accomplished. The physiological mechanisms involved in the regulation of ejaculate composition are poorly understood, but clues as to their possible nature might be derived from observations of the factors known to affect semen parameters.

In studies in which men provide multiple semen specimens over several days or weeks, there is substantial intra-individual variability in parameters such as ejaculate volume and sperm concentration (Read & Schnieden, 1978; Schwartz *et al.*, 1979; Mallidis *et al.*, 1991), in part because both parameters are affected by the duration of ejaculatory abstinence (Padova, Tita, Briguglia, & Giuffrida, 1988; Sauer, Zeffer, Buster, & Sokol, 1988; Blackwell & Zaneveld, 1992; Matilsky *et al.*, 1993; Rolf, Behre, & Nieschlag, 1996). There also is evidence that the context in which an ejaculate is produced can be important. For example, ejaculates produced during copulation and collected in non-spermicidal condoms are generally superior to those produced via masturbation (Zavos, 1985). Copulatory ejaculates, relative to masturbatory ejaculates, have greater volumes, greater total sperm numbers, and a higher grade of sperm motility (Zavos & Goodpasture, 1989). The percentage of motile and morphologically normal sperm also is higher for copulatory ejaculates, and these ejaculates consequently perform better on various sperm function tests (Sofikitis & Miyagawa, 1993).

The mechanisms that cause copulatory ejaculates to contain more sperm than masturbatory ejaculates are not fully understood, but the difference may be attributable, in part, to the greater intensity and duration of sexual arousal that typically precedes copulatory ejaculation. One study indicated that sexual stimulation, in the form of sexually explicit videotapes, can improve semen parameters for masturbatory ejaculates (Yamamoto, Sofikitis, Mio, & Miyagawa, 2000), but this contradicts a previous finding (van Roijen *et al.*, 1996). An increase in the duration of pre-coital stimulation increases the number of motile sperm with normal morphology in copulatory ejaculates (Zavos, 1988). There also is a positive association between the duration of pre-ejaculatory sexual arousal and sperm concentration for masturbatory ejaculates (Pound *et al.*, 2002). The psychological stress associated with providing a specimen for *in vitro* fertilization can impair total, and motile, sperm concentrations (Clarke, Klock, Geoghegan, & Travassos, 1999), and it is possible that these effects are secondary to impaired levels of sexual arousal.

Relationships between semen quality and the duration of sexual arousal also have been documented in domesticated farm animals when specimens are collected for artificial insemination. Prolonging the duration of pre-ejaculatory arousal through a combination of restraint and "false mounts," during which the male is allowed to mount a female repeatedly but without intromission, can increase the number of sperm ejaculated by bulls (Collins *et al.*, 1951; Almquist *et al.*, 1958; Hafs *et al.*, 1962; Almquist, 1973), the number of sperm in the sperm-rich fraction of the ejaculate in boars (Hemsworth & Galloway, 1979), and the fluid content of the accessory sex glands and ductus deferens in stallions (Weber, Geary, & Woods, 1990).

Given the relationship documented between the duration of pre-ejaculatory sexual arousal and the number of sperm ejaculated in various species, it is possible that males achieve adaptive changes in ejaculate composition through behavioral changes that prolong arousal prior to ejaculation. The idea that males delay intromission and ejaculation in response to cues of sperm competition risk is counter-intuitive, however, because it is known that they are likely to experience increased sexual motivation at such times. Perhaps more importantly, mammalian sperm competition is a race as well as a lottery. It therefore may be costly to prolong ejaculatory latency and thus delay insemination. Whether the increase in sperm numbers with prolonged arousal has an adaptive function is not clear, but this increase may depend on the same physiological mechanisms involved in adaptive increases in sperm numbers in other circumstances. An understanding of how sexual arousal can improve semen quality therefore can shed light on some of the possible sites where adaptive regulation might take place.

Contractions of the smooth muscle in the wall of the vas deferens are critical for sperm delivery (Batra, 1974; Guha, Kaur, & Ahmed, 1975; Hib, Ponzio, & Vilar, 1982). Sexual arousal, in turn, increases the rate of sperm transport in the vas deferens (Prins & Zaneveld, 1979, 1980). Prolonged or intense sexual arousal, therefore, may increase the number of sperm ejaculated by increasing the magnitude and duration of contractions of the vas deferens (Senger, 1997). Evidence that the vas deferens plays a critical role in sperm delivery has led some scholars to suggest that adaptive regulation of ejaculate composition in response to variations in sperm competition risk might be mediated by changes in vas deferens contractility (Baker & Bellis, 1995; Pound, 1999). This hypothesis is supported by the fact that contractions of this muscular duct may be modulated by hormones whose secretion can be regulated by aspects of the social environment that could be predictive of sperm competition risk (Pound, 1999).

## EVIDENCE OF ADAPTATIONS TO SPERM COMPETITION IN WOMEN

Evidence from other species suggests that females are not simply passive receptacles into which sperm are deposited and they may have an active role in biasing paternity when sperm competition occurs (e.g., Pizzari & Birkhead, 2000; Hosken & Ward, 2000). Consequently, if sperm competition was a recurrent feature of human evolutionary history, we would expect to identify adaptations to this form of postcopulatory competition not only in men, but also in women. Where copulatory timing is important, females may influence the outcome of sperm competition via precopulatory choice mechanisms that determine when, and by which male, they are inseminated. However, female influence may still be possible during or after copulation through what is known as "postcopulatory choice" – a term that encompasses the various ways in which females might bias paternity after the initiation of copulation (see, e.g., Eberhard, 1996).

Direct evidence of post-copulatory sperm selection by females is scarce across species, partly since it is difficult to distinguish between sperm competition and active female choice (Birkhead, 1998). However, it has been shown that female feral fowl actively eject sperm from lower ranking-males (Pizzari & Birkhead, 2000). Additionally, in the absence of direct evidence, sperm choice has been inferred to exist in a number of species. For example, Eberhard (1996) has suggested that the existence and use of multiple sperm storage sites within the female reproductive tract in various species allows females to bias paternity. However, while evidence of female control of post-copulatory sperm usage has been obtained using histological techniques in the yellow dung fly, *Scathophaga stercoraria* (Hosken & Ward, 2000) it is methodologically difficult to study mechanisms of sperm selection within mammalian female reproductive tracts.

In humans, Bellis and Baker (1990; Chapter 9, this volume) documented that women are more likely to engage in successive copulations with in-pair and extra-pair partners in a short time interval when the probability of conception is highest, suggesting that women may have psychological adaptations that motivate active promotion of sperm competition, thus allowing women to be fertilized by the most competitive sperm.

It is possible that human female psychology also includes mechanisms designed to motivate the avoidance of sperm competition, under certain conditions. Gangestad *et al.* (2002), for example, documented that as women enter the high conception phase of their menstrual cycle, they are sexually attracted to, and fantasize about, men *other than* their regular partner. A consequence of this might be, not sperm competition, but preferential and exclusive insemination by an extra-pair male during the most fertile phase of the cycle. Avoidance of in-pair copulations at this time may be best understood as a precopulatory female adaptation designed by sperm competition. But because men, in turn, have been selected to be sensitive to their partner's increased interest in extra-pair copulation near ovulation (Gangestad *et al.*, 2002), complete avoidance of in-pair copulation may not be possible and so women may possess postcopulatory adaptations designed to selectively favor sperm from one man over another.

One such postcopulatory adaptation in women may be orgasm. Because the structure that develops into a clitoris or penis is a bipotential organ in a developing embryo, Symons (1979) and Gould (1987) argued that female orgasm is a by-product of male orgasm. Others have hypothesized, however, that female orgasm has an adaptive function (e.g., Alexander, 1979; Baker & Bellis, 1993*b*; Fox, Wolff, & Baker, 1970; Hrdy, 1981; Morris, 1967; Smith, 1984). The leading functional hypothesis is that female coital orgasm was designed in the context of sperm competition as a mechanism of selective sperm retention (Baker & Bellis, 1993*b*; Chapter 11, this volume; Smith, 1984; Chapter 4, this volume). Female orgasm causes the cervix to dip into the seminal pool deposited by the male at the upper end of the vagina, and this may result in the retention of a greater numbers of sperm (see research reviewed in Baker & Bellis, 1993*b*; 1995). Baker and Bellis (1993*b*) and Smith (1984) contend that women may select preferentially the sperm of extra-pair partners who are likely to be of higher genetic quality than in-pair partners.

In a test of this hypothesis, Baker and Bellis (1993*b*) estimated the number of sperm in ejaculates collected by condoms during copulation and by vaginal "flowbacks" (i.e., ejected seminal and vagina fluid) and documented that women influence the number of sperm retained in their reproductive tract through the presence and timing of a coital orgasm. Coital orgasms that occurred between 1 minute before and 45 minutes after their partner ejaculated were linked with significantly greater sperm retention than coital orgasms that occurred earlier than 1 minute before their partner ejaculated. Analyzing women's copulatory behavior, Baker and Bellis (1993*b*) also provided evidence that women with a regular partner and one or more extra-pair partners had significantly fewer high sperm retention orgasms with their regular, primary partner and more high sperm retention orgasms with their extra-pair partners.

Missing from Baker and Bellis's (1993*b*) study, however, is the explicit demonstration of higher sperm retention associated with partners of higher genetic quality. However, work by Thornhill, Gangestad, and Comer (1995; Chapter 15, this volume) suggests that preferential retention of sperm from high quality males might occur. Thornhill *et al.* (1995) documented that women mated to men with low fluctuating asymmetry (indicating relatively high genetic quality) reported significantly more copulatory orgasms than did women mated to men with high fluctuating asymmetry (indicating relatively low genetic quality). Women mated to men with low fluctuating asymmetry did not simply have more orgasms, but specifically reported more copulatory orgasms likely to result in greater sperm retention according to Baker & Bellis' (1993*b*). Another indicator of high genetic quality and related to fluctuating asymmetry is physical attractiveness. Replicating Thornhill *et al.*'s (1995) work, Shackelford, Weekes-Shackelford, LeBlanc, Bleske, Euler, and Hoier (2000) found that women mated to more physically attractive men were more likely to report having a copulatory orgasm at their most recent copulation than were women mated to less attractive men.

While the hypothesis that female orgasm is an adaptation for postcopulatory female choice between rival ejaculates is plausible, there are some problems with Baker & Bellis's interpretation of their flowback data (Pound & Daly, 2000). Baker and Bellis (1993*b*) found that flowbacks were smaller when women orgasm between 1 minute before and 45 minutes after their partner ejaculates than if they orgasm earlier than 1 minute before or not at all. From this they infer that greater sperm retention occurs with orgasms that occur after male ejaculation. This inference, however, is based on the assumption that the number of sperm ejaculated by the male is identical regardless of whether or when the woman has an orgasm. This assumption may be false, since the duration of pre-ejaculatory sexual arousal in men has been shown to be correlated positively with the number of sperm ejaculated (Zavos, 1988; Pound, 1999). Moreover, it has yet to be demonstrated that female orgasm influences conception rates. If female orgasm causes the cervix to dip into the seminal pool, causing greater numbers of sperm to be retained, it would follow that the likelihood of conception will increase accordingly, but this has not been tested empirically. The observation that men are often concerned with whether their partner achieves orgasm and the observation that women often fake orgasms to appease their partner further suggests that female orgasm may have adaptive value (see Thornhill *et al.*, 1995).

## IS THERE EVIDENCE OF CONTEST COMPETITION BETWEEN HUMAN EJACULATES?

Apart from the remarkable feat of traversing a hostile reproductive tract to fertilize an ovum or ova, sperm do some astonishing things. Sperm of the common wood mouse (*Apodemus sylvaticus*) have a hook that allows the sperm to adhere to one another to form a motile "train" of several thousand sperm (Moore, Dvorakova, Jenkins, & Breed, 2002). These trains display greater motility and velocity than single sperm, facilitating fertilization. This cooperative behavior between sperm of a single male reveals that sperm are capable of complex behavior. Might sperm display equally complex behavior in the presence of rival sperm?

Baker and Bellis (1988; Chapter 5, this volume) proposed that, in mammals, post-copulatory competition between rival male ejaculates might involve more that just scramble competition and that rival sperm may interfere actively with each other's ability to fertilize ova. Mammalian ejaculates contain sperm that are polymorphic (i.e., existing in different morphologies or shapes and sizes). Previously interpreted as the result of developmental error (Cohen, 1973), Baker and Bellis (1988) suggested that sperm polymorphism was not due to meiotic errors, but instead reflected a functionally adaptive "division of labor" between sperm. They proposed two categories of sperm: "egg-getters" and "kamikaze" sperm. Egg-getters comprise the small proportion of sperm programmed to fertilize ova. Most of the ejaculate, they argued, is composed of kamikaze sperm that function to prevent other males' sperm from fertilizing the ova by forming a barrier at strategic positions within the reproductive tract. Preliminary evidence for the "Kamikaze Sperm Hypothesis" came from the observation that the copulatory plugs of bats are composed of so-called "malformed" sperm (Fenton, 1984), and from documentation that, in laboratory mice, different proportions of sperm morphs are found reliably at particular positions within the female reproductive tract (Cohen, 1977).

Harcourt (1989; Chapter 6, this volume) challenged Baker and Bellis's (1988) Kamikaze Sperm Hypothesis. Harcourt argued that "malformed" sperm were unlikely to have adaptive functions, citing evidence from Wildt *et al.* (1987) that, in lions, inbreeding results in an increase in the proportion of deformed sperm. Harcourt (1989) argued that, if deformed sperm were produced by an adaptation, inbreeding would not increase the expression of the trait, but instead would decrease it and the presence of malformed sperm in the copulatory plugs of bats is a consequence of the malformed sperm's poor mobility and, therefore, plug formation was not a designed function of deformed sperm. Following Cohen (1973), Harcourt (1989, p. 864) concluded that "abnormal sperm are still best explained by errors in production."

Baker and Bellis (1989; Chapter 7, this volume) responded to Harcourt's (1989) objections and elaborated on their Kamikaze Sperm Hypothesis proposing a more active role for kamikaze sperm, speculating that evolutionary arms races between ejaculates could result in kamikaze sperm that incapacitate rival sperm with acrosomal enzymes or by inducing attack by female leucocytes. They later suggested that in addition to sperm that are specialized to "seek-and-destroy" rival sperm, there may be also another type of kamikaze sperm that act as "blockers" that inhibit rival sperm progression (Baker & Bellis, 1995).

In a test of the hypothesis, Baker and Bellis (1995, p. 274) reported that, when ejaculates from two different men were mixed *in vitro*, there was increased agglutination, reduced sperm velocity, and increased sperm mortality relative to homospermic control mixtures. They interpreted these findings as an indication that, when encountering sperm from another male, some sperm impede the progress of rival sperm (blockers) and some sperm attack and incapacitate rival sperm (seek-and-destroyers). The Kamikaze Sperm Hypothesis and the reported interaction of rival sperm have generated substantial criticism, however (see, e.g., Birkhead, Moore, & Bedford, 1997; Short, 1998).

Moore, Martin, and Birkhead (1999; Chapter 12, this volume) attempted to replicate this finding that sperm from different males appear to incapacitate each other in a variety of ways but found no significant increases in sperm aggregation or mortality, not decreases in velocity, in heterospermic versus homospermic mixtures. It should be noted, however, that whereas Moore *et al.* (1999) looked for incapacitation effects after incubating sperm mixtures for up to 3 hours, Baker & Bellis (1995) reported increased sperm mortality and decreased velocity occurring 3–6 hours after mixing. Moore *et al.* (1999) offered theoretical reasons for this shorter interactive window (i.e., because one to three hours is the time that sperm normally remain in the human vagina), but perhaps this interval was too restrictive. Upon insemination, sperm have one of two initial fates: some are ejected from the vagina and some travel quickly from the vagina to the cervix and uterus. Perhaps the majority of sperm warfare takes place in the cervix and uterus, locations in the reproductive tract where sperm are able to interact for a prolonged period. If this is the case, Baker and Bellis's (1995) longer, 3–6 hour interactive window would be more valid ecologically. In addition, both Baker and Bellis (1995) and Moore *et al.* (1999) investigated sperm interactions *in vitro*, and one cannot be sure that sperm in a petri dish behave precisely as they do in the human vagina. Unfortunately, the technology to investigate sperm interaction *in vivo* (i.e., in the vagina) is not yet widely available.

Aside from Moore *et al.*'s (1999) failure to replicate Baker and Bellis's (1995) findings, additional skepticism was generated by the their failure to clearly specify how sperm can differentiate self-sperm from non-self-sperm. Given that sperm consist of a diminutive single-cell structure, a self-recognition system that must differentiate between not just different genes (because even sperm from a single male contain different combinations of genes), but different *sets* of competing genes (i.e., genes from another male) may be unlikely. Moore *et al.*'s (1999) failure to replicate Baker and Bellis's (1995) findings and the absence of a clear self-recognition system are not fatal to the Kamikaze Sperm Hypothesis, but such concerns are cause for serious skepticism about its plausibility. Recently, however, Kura and Nakashima (2000) have used mathematical models to examine the conditions that could lead to the evolution and maintenance of a "soldier" class of sperm. They concluded that these are not stringent, and "soldier" sperm may be particularly likely to evolve in contexts where there is substantial variance in the number of sperm from different males available to compete for fertilizations, due to factors such as mating order effects, between male-differences in ejaculate size, and copulation timing in relation to ovulation.

## CONCLUDING REMARKS AND VOLUME OVERVIEW

Since Smith (1984) suggested that sperm competition might have been an important selective pressure during human evolution many researchers have been interested in the possibility that aspects of human anatomy, physiology, psychology, and behavior are perhaps best understood as adaptations to this form of postcopulatory competition. This volume is intended to bring together, in one place, some of their most important work and, consequently, may of use to researchers and students in fields, such as psychology, biology, reproductive health, sociology, anthropology, medicine, sociobiology, animal behavior, and sexuality.

The volume begins with an overview, by Randy Thornhill, of theoretical and empirical work on human sperm competition and women's "dual sexuality." In Part I, in addition to our own overview of the field we reprint Geoff Parker's classic paper on how sperm competition can lead to the evolution of tiny, numerous sperm (Parker, 1982; Chapter 2, this volume). This is followed by a review of empirical findings indicating that in many species male ejaculate expenditure is dynamic and can be adjusted in response to cues of sperm competition risk (Wedell, Gage & Parker, 2002, Chapter 3, this volume).

Part II includes classic readings in human sperm competition. First, we include the chapter by Smith (1984, Chapter 4, this volume) that first seriously proposed that sperm competition may have been an important selective pressure during human evolution and examined some possible implications of this. It is followed by five papers (Baker & Bellis, 1988; 1989*a*; 1989b; 1993*a*; Harcourt, 1989) that focus primarily on possible male adaptations to sperm competition, and another two (Bellis & Baker, 1990; Baker & Bellis, 1993*b*) that focus primarily on female adaptations .

Part III includes four more recent papers examining issues relating to human sperm competition. Three of these (Gallup & Burch, 2004; Moore *et al.*, 1999; Shackelford *et al.*, 2002) focus primarily on possible psychological, physiological and anatomical adaptations in males to sperm competition. Finally, we include a paper (Thornhill *et al.*, 1995) reporting evidence that suggests that the human female orgasm may have a role in selective sperm retention.

We have made an effort to include classic and contemporary readings that address adaptations in both men and women that may have evolved in response to an evolutionary history of sperm competition. That more of these readings focus on male adaptations is an accurate reflection of the historical and current state of research and theory in the field. In the same way that much of the experimental work in non-human species has focused on the physiological and behavioral responses of males to cues of increased sperm competition risk, some of the most notable research involving humans also has focused on the adaptive responses of men. Cryptic female choice in the context of sperm competition has recently received more attention in non-humans, and also in humans with work focusing on female infidelity and the functional significance of the female orgasm (see Bellis & Baker, 1990; Baker & Bellis, 1993*a*; 1993*b*; Gallup & Burch, 2004; Shackelford *et al.*, 2002; Thornhill *et al.*, 1995; each included in this volume).

The possibility that some aspects of human anatomy, physiology, psychology and behavior are perhaps best understood as adaptations to an evolutionary history of sperm competition has generated a great deal of interest among behavioral scientists and in the media. However, enthusiasm can sometimes lead researchers to adopt an insufficiently critical stance when evaluating work on human behavior and psychology, so we would like to encourage readers to temper their enthusiasm with some caution. In this chapter, we have pointed out some problems with some of the work that has examined possible adaptations to sperm competition in humans. Moreover, in this volume we have included as readings theoretical critiques (Harcourt, 1989) and empirical rebuttals (Moore *et al.*, 1999) of some specific claims that have been made by Baker and Bellis.

## NOTE

Some of the material in this chapter has been adapted from Shackelford, T. K., Pound, N. & Goetz, A. T. (2005) Psychological and physiological adaptations to sperm competition in humans. *Review of General Psychology, 9,* 228–248.

## REFERENCES


Alexander, R. D. (1979). Sexuality and sociality in humans and other primates. In A. Katchadourian (Ed.) *Human sexuality* (pp. 81–97.) Berkeley: University of California Press.

Almquist, J. O. (1973). Effects of sexual preparation on sperm output, semen characteristics and sexual activity of beef bulls with a comparison to dairy bulls. *Journal of Animal Science,* **36**, 331–336.

Almquist, J. O., Hale, E. B. & Amann, R. P. (1958). Sperm production and fertility of dairy bulls at high collection frequencies with varying degrees of sexual preparation. *Journal of Dairy Science,* **41**, 733.

Ambriz, D., Rosales, A. M., Sotelo, R., Mora, J. A., Rosado, A. & Garcia, A.R. (2002). Changes in the quality of rabbit semen in 14 consecutive ejaculates obtained every 15 minutes. *Archives of Andrology,* **48**, 389–395.

Arndt, W. B., Jr., Foehl, J. C. & Good, F. E. (1985). Specific sexual fantasy themes: A multidimensional study. *Journal of Personality and Social Psychology,* **48**, 472–480.

Baker, R. R. & Bellis, M. A. (1988). "Kamikaze" sperm in mammals? *Animal Behaviour,* **36**, 936–939.

Baker, R. R. & Bellis, M. A. (1989*a*). Elaboration of the kamikaze sperm hypothesis: A reply to Harcourt. *Animal Behaviour,* **37**, 865–867.

Baker, R. R. & Bellis, M. A. (1989*b*). Number of sperm in human ejaculates varies in accordance with sperm competition theory. *Animal Behaviour,* **37**, 867–869.

Baker, R. R. & Bellis, M. A. (1993*a*). Human sperm competition: Ejaculate adjustment by males and the function of masturbation. *Animal Behaviour,* **46**, 861–885.

Baker, R. R. & Bellis, M. A. (1993*b*). Human sperm competition: Ejaculate manipulation by females and a function for the female orgasm. *Animal Behaviour,* **46**, 887–909.

Baker, R. R. & Bellis, M. A. (1995). *Human sperm competition*. London: Chapman & Hall.

Batra, S. K. (1974). Sperm transport through vas deferens: review of hypotheses and suggestions for a quantitative model. *Fertility and Sterility,* **25**, 186–202.

Bellis, M. A. & Baker, R. R. (1990). Do females promote sperm competition: Data for humans. *Animal Behavior,* **40**, 197–199.

Bellis, M. A., Baker, R. R. & Gage, M. J. G. (1990). Variation in rat ejaculates consistent with the Kamikaze Sperm Hypothesis. *Journal of Mammalogy,* **71**, 479–480.

Birkhead, T. R. (1998). Cryptic female choice: Criteria for establishing female sperm choice. *Evolution,* **52**, 1212–1218.

Birkhead, T. R. & A. P Møller. (1992). *Sperm competition in birds: Evolutionary causes and consequences.* San Diego, CA, Academic Press.

Birkhead, T. R., Moore, H. D. M. & Bedford, J. M. (1997). Sex, science, and sensationalism. *Trends in Ecology and Evolution,* **12**, 121–122.

Blackwell, J. M. & Zaneveld, L. J. (1992). Effect of abstinence on sperm acrosin, hypoosmotic swelling, and other semen variables. *Fertility and Sterility*, **58**, 798–802.
Buss, D. M. (1988). From vigilance to violence: Tactics of mate retention among American undergraduates. *Ethology and Sociobiology*, **9**, 191–317.
Buss, D. M. (1994). *The evolution of desire*. New York: Basic Books.
Buss, D. M., Larsen, R., Westen, D. & Semmelroth, J. (1992). Sex differences in jealousy: Evolution, physiology, and psychology. *Psychological Science*, **3**, 251–255.
Buss, D. M. & Shackelford, T. K. (1997). From vigilance to violence: Mate retention tactics in married couples. *Journal of Personality and Social Psychology*, **72**, 346–361.
Buss, D. M. & Schmitt, D. P. (1993). Sexual strategies theory: An evolutionary perspective on human mating. *Psychological Review*, **100**, 204 –232.
Chien, L. (2003). Does quality of marital sex decline with duration? *Archives of Sexual Behavior*, **32**, 55–60.
Clarke, R. N., Klock, S. C., Geoghegan, A. & Travassos, D. E. (1999). Relationship between psychological stress and semen quality among in-vitro fertilization patients. Human Reproduction, **14**, 753–758.
Cohen, J. (1973). Cross-overs, sperm redundancy and their close association. *Heredity*, **31**, 408– 413.
Cohen, J. (1977). *Reproduction*. London: Butterworth.
Collins, W. J., Bratton, R. W. & Henderson, C. R. (1951). The relationship of semen production to sexual excitement of dairy bulls. *Journal of Dairy Science*, **34**, 224–227.
Cook, P. A. & Wedell, N. (1996). Ejaculate dynamics in butterflies: A strategy for maximizing fertilization success? Proceedings of the Royal Society of London B, **263**, 1047–1051.
Daly, M. & Wilson, M. (1988). *Homicide*. Hawthorne, NY: Aldine de Gruyter.
Daly, M. & Wilson, M. (1996). Male sexual proprietariness and violence against wives. *Current Directions in Psychological Science*, **5**, 2–7.
Daly, M., Wilson, M. & Weghorst, S. J. (1982). Male sexual jealousy. *Ethology and Sociobiology*, **3**, 11–27.
Davidson, J. K. (1985). The utilization of sexual fantasies by sexually experienced university students. *Journal of American Health*, **34**, 24–32.
delBarco-Trillo, J. & Ferkin, M.H. 2004 Male mammals respond to a risk of sperm competition conveyed by odours of conspecific males. *Nature*, **431**, 446– 449.
Dewsbury, D. A. (1982). Ejaculate cost and male choice. *American Naturalist*, **119**, 601–610.
Dixson, A. F. (1998). *Primate sexuality*. Oxford University Press.
Dobash, R. E. & Dobash, R. P. (1979). *Violence against wives*. New York: Free Press.
Dutton, D. G. (1995). *The domestic assault of women* (rev. ed.). Vancouver: University of British Columbia Press.
Eberhard, W. G. (1996). *Female control*. Princeton, NJ: Princeton University Press.
Ellis, B. J. & Symons, D. (1990). Sex differences in sexual fantasy: An evolutionary psychological approach. *Journal of Sex Research*, **27**, 527–555.
Fenton, M. B. (1984). The case of vepertilionid and rhinolophid bats. In R. L. Smith (Ed.) *Sperm competition and the evolution of animal mating systems* (pp. 573–587). London: Academic Press.
Fox, C. A., Wolff, H. S. & Baker, J. A. (1970). Measurement of intra-vaginal and intra-uterine pressures during human coitus by radio-telemetry. *Journal of Reproduction and Fertility*, **22**, 243–251.
Gage, A. R. & Barnard, C. J. (1996). Male crickets increase sperm number in relation to competition and female size. *Behavioral Ecology and Sociobiology*, **38**, 349–353.
Gage, M. J. G. (1991). Risk of sperm competition directly affects ejaculate size in the Mediterranean fruit fly. *Animal Behaviour*, **42**, 1036–1037.
Gage, M. J. G. (1994). Associations between body–size, mating pattern, testis size and sperm lengths across butterflies. *Proceedings of the Royal Society of London B*, **258**, 247–254.
Gage, M. J. G. & Baker, R. R. (1991). Ejaculate size varies with sociosexual situation in an insect. *Ecological Entomology*, **16**, 331–337.
Gangestad, S. W. & Simpson, J. A. (2000). The evolution of human mating: Trade–offs and strategic pluralism. *Behavior and Brain Sciences*, **23**, 573–587.
Gangestad, S. W., Thornhill, R. & Garver, C. E., (2002). Changes in women's sexual interests and their partner's mate–retention tactics across the menstrual cycle: Evidence for shifting conflicts of interest. *Proceedings of the Royal Society of London*, **269**, 975–982.
Gallup, G. G. Jr & Burch, R. L. (2004). Semen displacement as a sperm competition strategy in humans. *Evolutionary Psychology*, **2**, 12– 23.

Gallup G. G., Burch, R. L., Zappieri, M. L., Parvez, R. A. & Stockwell, M. L. (2004). The human penis as a semen displacement device. *Evolution and Human Behavior,* **24**, 277–289

Ginsberg, J. R. & Rubenstein, D. I. (1990). Sperm competition and variation in zebra mating behavior. *Behavioral Ecology and Sociobiology,* **26**, 427–434.

Goetz, A. T. & Shackelford, T. K. (2005). *Wife rape as an anti-cuckoldry tactic.* Manuscript in preparation. Department of Psychology, Florida Atlantic University.

Goetz, A. T., Shackelford, T. K., Weekes–Shackelford, V. A., Euler, H. A., Hoier, S., Schmitt, D. P. & LaMunyon, C. W. (2005). Mate retention, semen displacement, and human sperm competition: A preliminary investigation of tactics to prevent and correct female infidelity. *Personality and Individual Differences,* **38**, 749–763.

Goetz, A. T., Shackelford, T. K., Weekes–Shackelford, V. A., Euler, H. A., Hoier, S. & Schmitt, D. P. (2003). Mate retention, semen displacement, and human sperm competition: Tactics to prevent and correct female infidelity. Paper presented in symposium, "Paternal certainty and anti–cuckoldry tactics (Steven Platek & Todd K. Shackelford, Co–Chairs), *15th Annual Meeting of the Human Behavior and Evolution Society.* Lincoln, NE: University of Nebraska.

Gomendio, M., Harcourt, A. H. & Roldán, E. R. S. (1998). Sperm competition in mammals. In T. R. Birkhead and A. P. Møller (Eds.), *Sperm Competition and sexual selection* (pp. 667–756). New York: Academic Press.

Gomendio, M. & Roldán, E. R. S. (1993). Mechanisms of sperm competition: Linking physiology and behavioral ecology. *Trends in Ecology and Evolution,* **8**, 95–100.

Gould, S. J. (1987). Freudian slip. *Natural History,* **96**, 14–21.

Gould, T. (1999). *The lifestyle.* New York: Firefly.

Greiling, H. & Buss, D. M. (2000). Women's sexual strategies: the hidden dimension of extra-pair mating. *Personality and Individual Differences,* **28**, 929–963.

Guha, S. K., Kaur, H. & Ahmed, A. H. (1975). Mechanics of spermatic fluid transport in the vas deferens. *Medical and Biological Engineering,* **13**, 518–522.

Hafs, H. D., Kinsey, R. C. & Desjardins, C. (1962). Sperm output of dairy bulls with varying degrees of sexual preparation. *Journal of Dairy Science,* **45**, 788–793.

Harcourt, A. H. (1989). Deformed sperm are probably not adaptive. *Animal Behaviour,* **37**, 863–865.

Harcourt, A. H., Harvey, P. H., Larson, S. G. & Short, R. V. (1981). Testis weight, body weight, and breeding system in primates. *Nature,* **293**, 55–57.

Harvey, P. H. & Harcourt, A. H. (1984). Sperm competition, testis size, and breeding systems in primates. In R. L. Smith (Ed.), *Sperm competition and the evolution of animal mating systems* (pp. 589–600). San Diego: Academic Press.

Hemsworth, P. H. & Galloway, D. B. (1979). The effect of sexual stimulation on the sperm output of the domestic boar. *Animal Reproduction Science,* **2**, 387–394.

Hib, J., Ponzio, R. & Vilar, O. (1982). Contractility of the rat cauda epididymidis and vas deferens during seminal emission. *Journal of Reproduction and Fertility,* **66**, 47–50.

Hosken, D. J, & Ward, P. I. (2000). Copula in yellow dung flies (*Scathophaga stercoraria*): investigating sperm competition models by histological observation. *Journal of Insect Physiology,* **46**, 1355–1363.

Hosken, D. J. & Ward, P. I. (2001). Experimental evidence for testis size evolution via sperm competition. *Ecology Letters,* **4**, 10–13.

Hrdy, S. B. (1981). *The woman that never evolved.* Cambridge: Harvard University Press.

Hunt, M. (1974). *Sexual behavior in the 70's.* Chicago: Playboy Press.

Jacobson, N. & Gottman, J. (1998). *When men batter women.* New York: Simon & Schuster.

Jennions, M. D. & Passmore, N. I. (1993). Sperm competition in frogs: Testis size and a sterile male experiment on *Chiromantis xerampelina* (Rhacophoridae). *Biological Journal of the Linnean Society,* **50**, 211–220.

Jennions, M. D. & Petrie, M. (2000). Why do females mate multiply? A review of the genetic benefits. *Biological Reviews of the Cambridge Philosophical Society,* **75**, 21–64.

Jivoff, P. (1997). The relative roles of predation and sperm competition on the duration of the post-copulatory association between the sexes in the blue crab, *Callinectes sapidus. Behavioral Ecology and Sociobiology,* **40**, 175–185.

Johnson, A. M., Mercer, C. H., Erens, B., Copas, A. J., McManus, S., Wellings, K., Fenton, K.A., Korovessis, C., Macdowall, W., Nanchahal, K., Purdon, S. & Field, H. (2001). Sexual behaviour in Britain: Partnerships, practices, and HIV risk behaviours. *Lancet,* **358**, 1835–1842.

Klusmann, D. (2002). Sexual motivation and the duration of partnership. *Archives of Sexual Behavior,* **31**, 275–287.

Kura, T. & Nakashima, Y. (2000). Conditions for the evolution of soldier sperm classes. *Evolution,* **54**, 72–80.

Laumann, E. O., Gagnon, J. H., Michael, R. T. & Michaels, S. (1994). *The social organization of sexuality.* Chicago: University of Chicago Press.

Leitenberg, H. & Henning, K. (1995). Sexual fantasy. *Psychological Bulletin,* **117**, 469–496.

MacIntyre, F. & Estep, K. W. (1993). Sperm competition and the persistence of genes for male homosexuality. *Biosystems,* **31**, 223–233.

Mallidis, C., Howard, E. J. & Baker, H. W. G. (1991). Variation of semen quality in normal men. *International Journal of Andrology,* **14**, 99–107.

Marconato, A. & Shapiro, D. Y. (1996). Sperm allocation, sperm production and fertilization rates in the bucktooth parrotfish. *Animal Behaviour,* **52**, 971–980.

Matilsky, M., Battino, S., Benami, M., Geslevich, Y., Eyali, V. & Shalev, E. (1993). The effect of ejaculatory frequency on semen characteristics of normozoospermic and oligozoospermic men from an infertile population. *Human Reproduction,* **8**, 71–73.

Møller, A. P. (1988). Testes size, ejaculate quality and sperm competition in birds. *Biological Journal of the Linnean Society,* **33**, 273–283.

Moore, H., Dvorakova, K., Jenkins, N. & Breed, W. (2002). Exceptional sperm cooperation in the wood mouse. *Nature,* **418**, 174–177.

Moore, H. D., Martin, M. & Birkhead, T. R. (1999). No evidence for killer sperm or other selective interactions between human spermatozoa in ejaculates of different males in vitro. *Proceedings of the Royal Society of London B,* **266**, 2343–2350.

Morris, D. (1967). *The naked ape.* New York: McGraw–Hill.

Nakatsuru, K. & Kramer, D. L. (1982). Is sperm cheap: Male fertility and female choice in the lemon tetra (*Pisces characidae*). *Science,* **216**, 753–755.

Olsson, M., Madsen, T. & Shine, R. (1997). Sperm really so cheap? Costs of reproduction in male adders, *Vipera berus*. *Proceedings of the Royal Society of London B,* **264**, 455–459.

Padova, G., Tita, P., Briguglia, G. & Giuffrida, D. (1988). Influence of abstinence length on ejaculate characteristics. *Acta Europaea Fertilitatis,* **19**, 29–31.

Parker, G. A. (1970). Sperm competition and its evolutionary consequences in the insects. *Biological Reviews,* **45**, 525–567.

Parker, G. A. (1982). Why are there so many tiny sperm? Sperm competition and the maintenance of two sexes. *Journal of Theoretical Biology,* **96**, 281–294.

Parker, G. A. (1990*a*). Sperm competition games: Raffles and roles. *Proceedings of the Royal Society of London, B,* **242**, 120–126.

Parker, G. A. (1990*b*). Sperm competition games: Sneaks and extra–pair copulations. *Proceedings of the Royal Society of London Series B,* **242**, 127–133.

Parker, G. A. (1998). Sperm competition and the evolution of ejaculates: towards a theory base. In *Sperm Competition and Sexual Selection* (pp. 3–54) (Birkhead, T. R & Møller, A. P., Eds.), Academic Press.

Parker, G. A., Ball, M. A., Stockley, P. & Gage, M. J. G. (1997). Sperm competition games: A prospective analysis of risk assessment. *Proceedings of the Royal Society of London B,* **264**, 1793–1802.

Pelletier, L. A. & Herold, E. S. (1988). The relationship of age, sex guilt, and sexual experience with female sexual fantasies. *Journal of Sex Research,* **24**, 250–256.

Person, E. S., Terestman, N., Myers, W. A., Goldberg, E. L. & Salvadori, C. (1989). Gender differences in sexual behaviors and fantasies in a college population. *Journal of Sex and Marital Therapy,* **15**, 187–198.

Peters, J., Shackelford, T. K. & Buss, D. M. (2002). Understanding domestic violence against women: Using evolutionary psychology to extend the feminist functional analysis. *Violence and Victims,* **17**, 255–264.

Pizzari, T. & Birkhead, T. R. (2000). Female feral fowl eject sperm of subdominant males. *Nature,* **405**, 787–789.

Pound, N. (1999). Effects of morphine on electrically evoked contractions of the vas deferens in two congeneric rodent species differing in sperm competition intensity. *Proceedings of the Royal Society of London B,* **266**, 1755–1858.

Pound, N. (2002). Male interest in visual cues of sperm competition risk. *Evolution and Human Behavior,* **23**, 443–466.

Pound, N. & Daly, M. (2000). Functional significance of human female orgasm still hypothetical. *Behavioral and Brain Sciences,* **23**, 620–621.

Pound, N. & Gage. M. J. G. (2004). Prudent sperm allocation in Rattus Norvegicus: A mammalian model of adaptive ejaculate adjustment. *Animal Behaviour,* **68**, 819–823.

Pound, N., Javed, M. H., Ruberto, C., Shaikh, M. A. & Del Valle, A. P. (2002). Duration of sexual arousal predicts semen parameters for masturbatory ejaculates. *Physiology and Behavior,* **76**, 685–689.

Preston, B. T., Stevenson, I. R., Pemberton, J. M. & Wilson, K. (2001). Dominant rams lose out by sperm depletion. *Nature,* **409**, 681–682.

Price, J. H. & Miller, P. A. (1984). Sexual fantasies of Black and of White college students. *Psychological Reports,* **54**, 1007–1014.

Prins, G. S. & Zaneveld, L. J. (1979). Distribution of spermatozoa in the rabbit vas deferens. *Biology of Reproduction,* **21**, 181–185.

Prins, G. S. & Zaneveld, L. J. (1980). Radiographic study of fluid transport in the rabbit vas deferens during sexual rest and after sexual activity. *Journal of Reproduction and Fertility,* **58**, 311–319.

Read, M. D. & Schnieden, H. (1978). Variations in sperm count in oligozoospermic or asthenozoospermic patients. *Andrologia,* **10**, 52–55.

Rokach, A. (1990). Content analysis of sexual fantasies of males and females. *Journal of Psychology,* **124**, 427–436.

Rolf, C., Behre, H. M. & Nieschlag, E. (1996). Reproductive parameters of older compared to younger men of infertile couples. *International Journal of Andrology,* **19**, 135–142.

Sauer, M. V., Zeffer, K. B., Buster, J. E. & Sokol, R. Z. (1988). Effect of abstinence on sperm motility in normal men. *American Journal of Obstetrics and Gynecology,* **158**, 604–607.

Schwartz, D., Laplanche, A., Jouannet, P. & David, G. (1979). Within–subject variability of human semen in regard to sperm count, volume, total number of spermatozoa and length of abstinence. *Journal of Reproduction and Fertility,* **57**, 391–395.

Schmitt, D. P, *et al.* (118 co–authors). (2003). Universal sex differences in the desire for sexual variety: Tests from 52 nations, 6 continents, and 13 islands. *Journal of Personality and Social Psychology,* **85**, 85–104.

Schmitt, D. P., Shackelford, T. K. & Buss, D. M. (2001). Are men really more oriented toward short–term mating than women? A critical review of theory and research. *Psychology, Evolution, and Gender,* **3**, 211–239.

Schmitt, D. P., Shackelford, T. K., Duntley, J., Tooke, W. & Buss, D. M. (2001). The desire for sexual variety as a tool for understanding basic human mating strategies. *Personal Relationships,* **8**, 425–455.

Senger, P. L. (1997). *Pathways to pregnancy and parturition*. Pullman, WA: Current Conceptions.

Shackelford, T. K. (2003). Preventing, correcting, and anticipating female infidelity: Three adaptive problems of sperm competition. *Evolution and Cognition,* **9**, 90–96.

Shackelford, T. K., Goetz, A. T., LaMunyon, C. W., Quintus, B. J. & Weekes–Shackelford, V. A. (2004). Sex differences in sexual psychology produce sex similar preferences for a short-term mate. *Archives of Sexual Behavior,* **33**, 405–412.

Shackelford, T. K. & LeBlanc, G. J. (2001). Sperm competition in insects, birds, and humans: Insights from a comparative evolutionary perspective. *Evolution and Cognition,* **7**, 194–202.

Shackelford, T. K., LeBlanc, G. J., Weekes–Shackelford, V. A., Bleske–Rechek, A. L., Euler, H. A. & Hoier, S. (2002). Psychological adaptation to human sperm competition. *Evolution and Human Behavior,* **23**, 123–138.

Shackelford, T. K., Weekes–Shackelford, V. A., LeBlanc, G. J., Bleske, A. L., Euler, H. A. & Hoier, S. (2000). Female coital orgasm and male attractiveness. *Human Nature,* **11**, 299–306.

Shapiro, D. Y., Marconato, A. & Yoshikawa, T. (1994). Sperm economy in a coral reef fish, *Thalassoma bifasciatum*. *Ecology,* **75**, 1334–1344.

Short, R. V. (1979). Sexual selection and its component parts, somatic and genital selection as illustrated by man and the great apes. *Advances in the Study of Behavior,* **9**, 131–158.

Short, R. V. (1981). Sexual selection in man and the great apes. In C. E. Graham (Ed.) *Reproductive biology of the great apes* (pp. 319–341). New York: Academic Press.

Short, R. V. (1998). Review of Human Sperm Competition: Copulation, Masturbation and Infidelity, by R. R. Baker and M. A. Bellis. *European Sociobiology Society Newsletter,* **47**, 20–23.

Smith, R. L. (1984). Human sperm competition. In R. L. Smith (Ed.), *Sperm competition and the evolution of animal mating systems* (pp. 601–660). New York: Academic Press.

Sofikitis, N. V. & Miyagawa, I. (1993). Endocrinological, biophysical, and biochemical parameters of semen collected via masturbation versus sexual intercourse. *Journal of Andrology,* **14**, 366–373.

Sue, D. (1979). Erotic fantasies of college students during coitus. *Journal of Sex Research,* **15**, 299–305.

Symons, D. (1979). *The evolution of human sexuality*. New York: Oxford University Press.

Talese, G. (1981). *Thy neighbor's wife*. New York: Ballantine.

Thornhill, R., Gangestad, S. W. & Comer, R. (1995). Human female orgasm and mate fluctuating asymmetry. *Animal Behaviour,* **50**, 1601–1615.

Thornhill, R. & Palmer, C. T. (2000). *A natural history of rape*. Cambridge, MA: MIT Press.

Trivers, R. L. (1972). Parental investment and sexual selection. In B. Campbell (Ed.), *Sexual selection and the descent of man* (pp. 139–179). London: Aldine.

van Roijen, J. H., Slob, A. K., Gianotten, W. L., Dohle, G. R., vander Zon, A. T. M., Vreeburg, J. T. M. & Weber, R. F. A. (1996). Sexual arousal and the quality of semen produced by masturbation. *Human Reproduction,* **11**, 147–151.

Weber, J. A., Geary, R. T. & Woods, G. L. (1990). Changes in accessory sex glands of stallions after sexual preparation and ejaculation. *Journal of the American Veterinary Medical Association,* **196**, 1084–1089.

Wedell, N., Gage, M. J. G. & Parker, G. A. (2002). Sperm competition, male prudence and sperm–limited females. *Trends in Ecology and Evolution,* **17**, 313–320.

Wildt, D. E., Bush, M., Goodrowe, K. L., Packer, C., Pusey, A. E., Brown, J. L., Joslin, P. & O'Brien, S. J. (1987). Reproductive and genetic consequences of founding isolated lion populations. *Nature,* **329**, 328–331.

Wilson, G. D. (1987). Male–female differences in sexual activity, enjoyment and fantasies. *Personality and Individual Differences,* **8**, 125–127.

Wilson, G. D. (1997). Gender differences in sexual fantasy: An evolutionary analysis. *Personality and Individual Differences,* **22**, 27–31.

Wilson, G. D. & Lang, R. J. (1981). Sex differences in sexual fantasy patterns. *Personality and Individual Differences,* **2**, 343–346.

Wyckoff, G. J., Wang, W. & Wu, C. (2000). Rapid evolution of male reproductive genes in the descent of man. *Nature,* **403**, 304–308.

Yamamoto, Y., Sofikitis, N., Mio, Y. & Miyagawa, I. (2000). Influence of sexual stimulation on sperm parameters in semen samples collected via masturbation from normozoospermic men or cryptozoospermic men participating in an assisted reproduction programme. *Andrologia,* **32**, 131–138.

Zavos, P. M. (1985). Seminal parameters of ejaculates collected from oligospermic and normospermic patients via masturbation and at intercourse with the use of a Silastic seminal fluid collection device. *Fertility and Sterility,* **44**, 517–520.

Zavos, P. M. (1988). Seminal parameters of ejaculates collected at intercourse with the use of a seminal collection device with different levels of precoital stimulation. *Journal of Andrology,* **9**, P36.

Zavos, P. M. & Goodpasture, J. C. (1989). Clinical improvements of specific seminal deficiencies via intercourse with a seminal collection device versus masturbation. *Fertility and Sterility,* **51**, 190–193.

Zavos, P. M. Kofinas, G. D., Sofikitis, N. V., Zarmakoupis, P. N. & Miyagawa, I. (1994). Differences in seminal parameters in specimens collected via intercourse and incomplete intercourse (coitus interruptus). *Fertility and Sterility,* **61**, 1174–1176.

# 2. WHY ARE THERE SO MANY TINY SPERM? SPERM COMPETITION AND THE MAINTENANCE OF TWO SEXES

Geoffrey A. Parker[1]

## ABSTRACT

It is suggested that sperm competition (competition between the sperm from two or more males over the fertilization of ova) may account for the fact that sperm are so small and so numerous. In the entire absence of sperm competition, selection may favour an increase in sperm size so that the sperm contributes nutriment to the subsequent viability and success of the zygote. However, an extremely low incidence of sperm competition is adequate to prevent sperm size increasing. Vertebrate sperm should remain at minimal size provided that double matings (one female mated by two males) occur more often than about 4 times the ratio of sperm size:ovum size. The classical theory that sperm are small simply because of the difficulties of ensuring that ova do get fertilized may also explain sperm size, and both effects (sperm competition and ensuring fertilization) are likely to contribute to the stability of anisogamy. Large numbers of sperm can be produced because sperm are tiny and the optimal allocation of reproductive reserves to ejaculates is not trivially small even when double matings are rather rare. It is suggested that of its total mating effort, a male vertebrate should spend a fraction on sperm that is roughly equivalent to a quarter of the probability of double mating.

## 1. INTRODUCTION

Despite the fact that anisogamy is the rule in multicellullar animals and plants, biologists have devoted rather little attention to an interpretation of why evolution has produced and maintained males and females. Why not, say, five sexes, each producing its

---

[1] Department of Zoology, University of Liverpool.

Reprinted from *Journal of Theoretical Biology*, 96 (2), Parker, G. A. Why are there so many tiny sperm? Sperm competition and the maintenance of two sexes. pp. 281–294, Copyright (1982), with permission from Elsevier.

own characteristic gamete? Early theories for the evolution of anisogamy (Kalmus, 1932; Kalmus & Smith, 1960; Scudo, 1967) assumed that selection would act to favour efficiency in fertilization at the species or group level, and considered only isogamy versus anisogamy. Recently, Parker, Baker & Smith (1972) proposed that anisogamy might result from disruptive selection acting on a continuous range of variants, each variant producing gametes of a characteristic size. The assumptions of disruptive selective theory are as follows.

(i) A large population of adults releases its gametes into an external medium (e.g. sea water) so that gametes fuse randomly, independent of size.
(ii) There is a fixed energy budget per parent, so that if an adult produces gametes of size $m$, the relative number of gametes produced is proportional to $m^{-1}$
(iii) The viability (or other components of fitness) of a zygote increases with its size. Thus a zygote produced by the fusion of two large gametes (each with high provisioning) survives better than one resulting from the fusion of two intermediate-sized gametes, or one from a large and small gamete.

Provided that in (iii) the size of the zygote exerts an important enough effect on its survival, the evolutionarily stable strategy (ESS; Maynard Smith, 1974) is anisogamy, i.e. a population consisting of males (microgamete producers) and females (megagamete producers) will be stable. This result (originally obtained by computer simulation) has been confirmed analytically by a number of authors (Bell, 1978; Charlesworth, 1978; Maynard Smith, 1978; Hoekstra, 1980). Some empirical support (Knowlton, 1974; Bell, 1978) is available from the fact that in various groups of algae, a trend towards anisogamy (from isogamy) is associated with a trend towards multicellularity (from unicellularity). During the evolution of anisogamy, selection is likely to favour sperm that fuse disassortatively (with ova), and probably ova that fuse disassortatively (with sperm); see Parker (1978). Fisher's principle (1930) explains why the sex ratio stabilizes at unity.

Although the disruptive selection theory forms a basis for the origin and maintenance of the two sexes, assumption (i) above will be adequate as an approximation only for many plants, and certain animals with external fertilization (e.g. a large population of sessile external fertilizers). Animals with internal fertilization seem perhaps most disparate from the concept of the original model. The aim of the present paper is to consider reasons why anisogamy remains stable even when the reproductive pattern changes from external fertilization to internal fertilization. I argue that it is essentially sperm competition that is responsible for maintaining anisogamy. Sperm or ejaculate competition is competition between the sperm of different males over the fertilization of the ova (Parker, 1970*a*). In sessile animals with external fertilization, there will be a high degree of sperm competition if spawning tends to be synchronous. In species with internal fertilization sperm competition may be much reduced, but probably never entirely absent. Without sperm competition, anisogamy may be unstable, because it would pay males to increase the provisioning in each sperm so as to contribute to the survivorship of the zygote.

There appear to be two central questions. Firstly, what keeps sperm small and devoid of any provisioning for the zygote? Secondly, why are so many sperm produced?

## 2. ANISOGAMY IN SMALL GROUPS OF SYNCHRONOUS EXTERNAL FERTILIZERS

I first investigate the robustness of the anisogamy ESS to the effects of group size. Suppose that anisogamy and disassortative fusion have evolved in a population of synchronous external fertilizers. What happens when there are just $n$ males in each spawning group?

Here and elsewhere we seek conditions under which selection will act against small increases in the size of the sperm. Sperm become reduced to the least size because this allows so many of them to be produced; high productivity yields an advantage through sperm competition. We can assume that ovum size must be stabilized at a unique optimum if sperm contribute nothing to the zygote. The anisogamy ESS must conform to a Nash equilibrium in which it will not pay the male to supply provisioning in the sperm, nor will it pay the female to deviate from her unique optimum specified by zero sperm contribution to the zygote. We can test the robustness of the anisogamy ESS by testing whether a mutation will spread that contributes some investment to the zygote.

Let us assume that the provisioning from the ovum contributes an amount $F$ to the survival prospects of the zygote. At the anisogamy ESS, the male contributes nothing via the sperm to zygote survival. Thus investment in each ovum is optimized at $m_{fopt}$ while investment in each sperm is set at an arbitrary minimum level $m_{min}$. Suppose a mutant male could invest $m > m_{min}$ in each sperm and thereby raise the survival prospects of the zygote by an amount $b(m)$. By so doing, the mutant produces less total sperm than a normal male. Normal males produce (relatively) $m_{min}^{-1}$ sperm, whereas the mutant produces $m^{-1}$ sperm. The mutant will therefore obtain less fertilizations than a normal male, but produces zygotes that survive better.

With $n$ males and $f$ females in each group, the expected fitness of a normal male will be:

$$\frac{fF}{n}.$$

A mutant male playing $m > m_{min}$ will obtain fitness:

$$\underbrace{f[F+b(m)]}_{\text{viability of zygote}} \quad \underbrace{\frac{m^{-1}}{(n-1)m_{min}^{-1}+m^{-1}}}_{\text{proportion of fertilizations obtained by mutant}} = \frac{\text{mutant's sperm number}}{\text{total number of sperm}}.$$

Thus for $m_{min}$ to be an ESS against $m$ requires that

$$\frac{F}{n} > [F+b(m)]\frac{m_{min}}{(n-1)m+m_{min}} \qquad (1)$$

and is clearly independent of sex ratio.

The $m_{\min}$ strategy will be locally stable (i.e. resistant to small increases in the amount of provisioning in sperm) if

$$\frac{\mathrm{d}}{\mathrm{d}m}\left\{[F+b(m)]\frac{m_{\min}}{(n-1)m+m_{\min}}\right\}\Bigg|_{m_{\min}}<0\,.$$

The logic behind this assertion is explained in Fig. 1. Differentiation gives the result that result that $m_{\min}$ is stable if

$$b'(m_{\min})<\frac{F(n-1)}{nm_{\min}}. \tag{2}$$

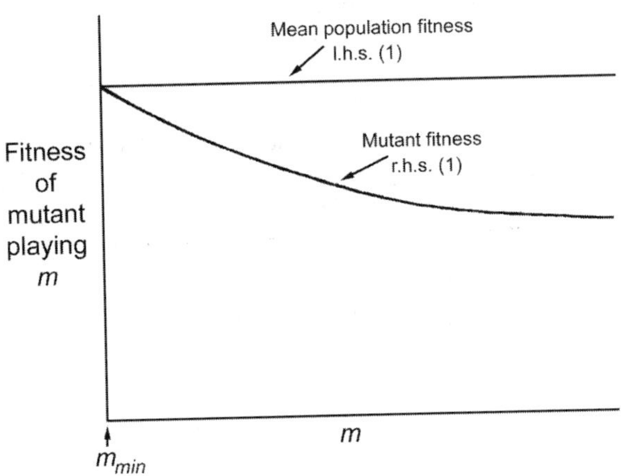


**FIG. 1.** The fitness of a mutant that produces sperm of size $m$, in a population where all other males play $m_{\min}$ (minimal sized sperm with no provisioning for zygote) is given by the right hand side of equation (1). In the case shown here, $m_{\min}$ would be locally stable since all mutants with $m > m_{\min}$ have lower fitness than the rest of the population. The condition for $m_{\min}$ to be locally stable is therefore that the differential coefficient with respect to $m$ of the right hand side of (1), evaluated at $m_{\min}$, is negative.

We can proceed little further until we know more about $b'(m_{\min})$, which is the rate at which provisioning via sperm would contribute to zygote viability. Consider as follows. If the male parent supplies no investment to the zygote, the ESS investment in each ovum for the female parent, $m_{f\text{opt}}$, is given by the tangent method (Smith & Fretwell, 1974) as shown in Fig. 2. Thus if we plot zygote viability $b$ against the provisioning $m_f$ supplied

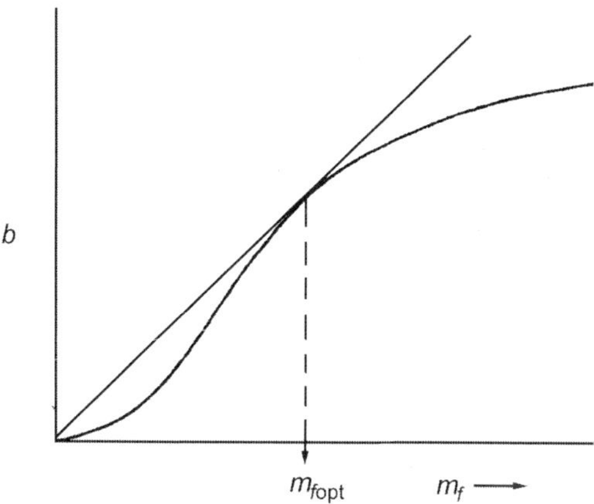


**FIG. 2.** Optimal provisioning $m_{fopt}$ for the female to supply the ovum, assuming that the male will supply nothing to the zygote. The optimum is given by the tangent to $b'(m_f)$ drawn from the origin (see Smith & Fretwell, 1974). This gives the maximum number of surviving offspring by maximizing the gain rate obtainable from limited reserves. Obviously, at $m_{fopt}$, $b'(m_f) = b(m_f)/m_f$ as in equation (3).

by the female to the ovum, assuming zero sperm provisioning, we expect that at $m_{fopt}$ the gradient of the tangent equals the gradient of $b(m_f)$. So we can write

$$b'(m_{fopt}) = \frac{b(m_{fopt})}{m_{fopt}} = \frac{F}{m_{fopt}}. \tag{3}$$

Assuming that provisioning via sperm and via ova would affect zygote survival equivalently, then it is easy to see that $b'(m_{fopt}) = b'(m_{min})$ because females will supply $m_{fopt}$ if males supply $m_{min}$ to the zygote. We can therefore substitute (3) into (2) to give the condition that

$$\frac{m_{fopt}}{m_{min}} > \frac{n}{n-1} \tag{4}$$

for $m_{min}$ to be locally stable.

Rule (4) states that in order for zero provisioning from sperm to be stable, we need roughly the "anisogamy ratio" (ratio of ovum size to sperm size) to be greater than the number $n$ of males in each spawning group divided by $(n - 1)$. Obviously, for large groups, the sperm size can almost equal the ovum size before a mutant with extra provisioning will spread. Even when usually only two males compete for fertilizations, sperm size should not increase from $m_{min}$ unless the anisogamy ratio is less than 2. This result is interesting because it implies that once a state has been attained in which there is a disassortative fusion and where the ovum supplies all the zygotic reserves, it is unlikely to pay males to provision sperm unless there is no sperm competition ($n \to 1$). Then it will always be favourable to increase zygotic reserves by sperm provisioning.

## 3. SPERM SIZE WITH INTERNAL FERTILIZATION

I have established that for synchronous external fertilizers (on which the original anisogamy model of Parker *et al.* was based), sperm competition is essential to maintain anisogamy with zero sperm provisioning. Internal fertilization must reduce dramatically the number of occasions on which sperm competition occurs. How will this affect the stability of the anisogamy ESS?

We retain all the features of the model outlined in section 2, except that sperm competition arises on only proportion $p$ of occasions. Thus with frequency $p$ two males mate with the same female, with frequency $(1-p)$ the female is mated by just one male. When two males mate with the same female, the success of male $i$ in competition with male $j$ is taken as before as

$$\frac{\text{number of } i \text{ sperm}}{\text{total sperm } i + j}.$$

Support for this model as an approximation for vertebrates comes from the work of Martin *et al.* (1974) on chickens and Lanier *et al.* (1979) on rats. For the $m_{\min}$ strategy to be stable against a mutant male that invests $m > m_{\min}$ in each sperm requires that

$$(1-p)F + p\frac{F}{2} > (1-p)[F+b(m)] + p[F+b(m)]\frac{m^{-1}}{m^{-1}+m_{\min}^{-1}} \tag{5}$$

$$\therefore\ F\left(1-\frac{p}{2}\right) > [F+b(m)]\left[(1-p)+p\left(\frac{m_{\min}}{m+m_{\min}}\right)\right]$$

which is directly equivalent to equation (1). By the same technique used in section 1, we can see that $m_{\min}$ will be locally stable if the differential coefficient with respect to $m$ of the RHS of (5) is negative when evaluated at $m_{\min}$. This gives the condition that

$$b'(m_{\min}) < \frac{pF}{m_{\min}(4-2p)} \tag{6}$$

for $m$ to be stable. Remembering again that if the male plays $m_{\min}$, the female must play $m_{f\text{opt}}$ and $b'(m_{\min}) = F/m_{f\text{opt}}$ (Fig. 2; equation (3)), we can substitute into (6) to obtain

$$\frac{m_{f\text{opt}}}{m_{\min}} > \frac{4-2p}{p} \tag{7}$$

for $m_{\min}$ to be stable. As expected, if $p = 1$, the condition is the same as for equation (2) with $n = 2$. If $p$ is small, we obtain the approximation that

$$p > 4\frac{m_{\min}}{m_{f\text{opt}}}$$

to retain the $m_{\text{min}}$ ESS. At high anisogamy ratios, sperm competition can be extremely rare and yet will still be entirely adequate to prevent invasion by mutants with sperm that contain provisioning for the zygote. All that is required is roughly that double matings are more frequent than 4 divided by the anisogamy ratio.

This model appears equally applicable to mobile external fertilizers such as certain fish, in which many spawnings involve a single male and female, but some spawnings involve a "sneak" male as well as the primary male.

In most vertebrates, the ovum is vastly larger than the sperm and the anisogamy ratio commonly exceeds $10^6$. Suppose sperm were so large as to be equivalent to one thousandth the size of an ovum; then anisogamy would be stable provided that double mating occurs for at least 0.4% of litters. Thus provided that mobility and internal fertilization arose at a stage after a disassortative fusion and high anisogamy ratio had evolved, there is no reason to suspect that the reduced potential for sperm competition should lead to a change in the minimal investment characteristic of the sperm. Anisogamy is a remarkably robust ESS.

Essentially, the reason it does not pay to increase sperm provisioning is that a unit increase in investment in each sperm causes significant cost, but insignificant benefit. For example, doubling the sperm size halves the sperm number, which causes significant losses when there is sperm competition. But doubling the sperm size would effect a virtually insignificant increase in the viability of the zygote.

Of course, selection will favour mechanisms in the female to consume what are, for her, excess sperm. Considerable phagocytosis of sperm appears to take place in the female genital tract in vertebrates; the female may therefore profit by male ejaculate expenditure. To the extent that the offspring may benefit from the products of the phagocytosis, the male may also benefit indirectly if the affected offspring are his own. Alternatively, the male may benefit even more directly by adopting various forms of parental care. But it will not pay him to increase his provisioning of the zygote by increasing the amount of reserves bound up in each sperm.

For some groups with internal fertilization, double mating may not lead to approximately equal chances for each male, even when they both transfer equal amounts of sperm. For instance, in insects it appears quite common that the last male to mate displaces much of the previously-stored sperm from the female's sperm stores, and replaces it with his own (e.g. Lefevre & Jonsson, 1962; Parker, 1970*a*,*b*; Waage, 1979). It is obvious that provided sperm displacement is not total, some sperm competition still occurs. Suppose that the last male displaces proportion $z$ of the previous ejaculate on a volumetric basis (some evidence for this comes from Lefevre & Jonsson, 1962; Parker, 1970*b*). Then if sperm are small, there will be relatively more of them left in the $(1 - z)$ volume remaining undisplaced, than if sperm are large. If we apply exactly the same model as for vertebrates, and allow that the mutant male with $m > m_{\text{min}}$ can mate first or last with equal probability, we need that:

$$F\left(1-\frac{p}{2}\right) > (1-p)[F+b(m)] + \frac{p}{2}[F+b(m)]$$
$$\times\left[\frac{zm_{\min}}{(1-z)m+zm_{\min}} + \frac{(1-z)m_{\min}}{zm+(1-z)m_{\min}}\right]$$

if $m_{\min}$ is to be an ESS.

Applying the usual technique, we find that to retain minimal sperm provisioning requires that:

$$\frac{m_{fopt}}{m_{\min}} > \frac{2-p}{pz(1-z)} \tag{8}$$

or if $p$ is small, then approximately

$$p > \frac{m_{\min}}{m_{fopt}} > \frac{2}{z(1-z)}.$$

It is easiest to satisfy (8) when a second male displaces half of the first male's sperm $z(1-z)$ is maximized when $z = ½$). At this level, we need only that double matings are more frequent than 8 divided by the anisogamy ratio in order to be stable. This is admittedly less easy to satisfy than for vertebrates, but not such that anisogamy will be threatened. However, if displacement is very high (or alternatively, very low) then the product $z(1-z)$ becomes very small, and condition (8) progressively less easy to satisfy.

The highest degree of priority achieved by the last male to mate that has so far been recorded for an insect is 0.997 for the bug *Abedus herberti* (Smith, 1979). Even at this exceptional level of sperm displacement, the $m_{\min}$ strategy would be relatively safe, since the ratio of sperm size/ovum size is several orders of magnitude greater than 0.003, the product $z(1-z)$.

In short, variations in the exact pattern of sperm competition are unlikely to affect our general conclusion. Provided that even occasionally the sperm from more than one male compete over fertilizations, anisogamy is likely to be stable and sperm should not contain provisioning for the zygote. They should have minimal size.

## 4. AN ALTERNATIVE HYPOTHESIS

The classical interpretation of small sperm size is that the best chances of ensuring that an ovum gets fertilized occur when there are as many sperm as possible. By making sperm tiny, a maximum number can be produced; this maximizes the chances that one of them will find the egg.

As Cohen (1973) has plausibly argued, it is not easy to accept this solution for vertebrates since ejaculates can often be diluted vastly (for artificial insemination) without loss in fertility. It is also difficult to accept for insects because fertility usually decreases only when the sperm supply becomes very depleted and normally the female would either be dead, or would have remated before this stage is reached (e.g. Parker, 1970*b*). However, no model appears to have been devised to examine the classical proposition that sperm are small simply to provide enough of them to ensure a high probability of fertilization. Note that this does differ from the sperm competition theory for the maintenance of small sperm size. The sperm competition model assumes that the probability of fertilization is independent of sperm numbers over a very wide range, and argues that sperm are small to produce high numbers to outcompete other ejaculates. The classical model ignores sperm competition and suggests that high sperm numbers are necessary for fertilization.

Suppose that the probability $g$ of successful fertilization increases with increasing sperm numbers up to an asymptotic value of 1.0 (see Fig. 3). The maximum number of sperm that can be contained in an ejaculate is $m_{\min}^{-1}$ this gives the highest attainable probability of fertilization with a single mating. We again seek the condition under which the $m_{\min}$ strategy will be an ESS, and again assume that increasing the size of each sperm (by decreasing sperm numbers) can increase the survivorship prospects of the zygote.

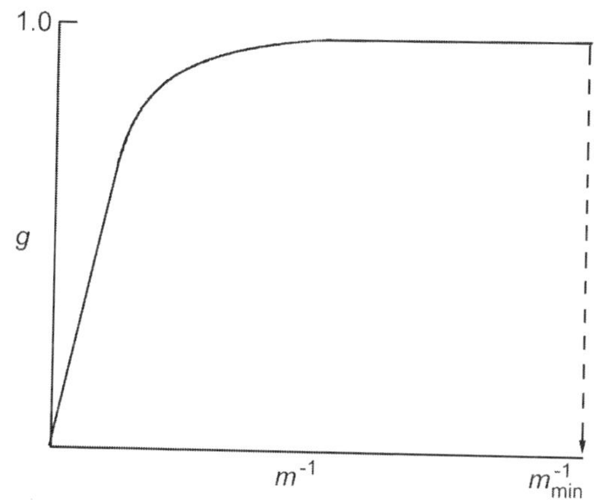


**FIG. 3.** Probability that an ovum is fertilized in relation to increasing sperm numbers in an ejaculate. There is no competition between ejaculates and the maximum number of sperm is proportional to $m_{\min}^{-1}$

Following earlier arguments, $m_{\min}$ will be stable if

$$g\left(m_{\min}^{-1}\right)F > g(m^{-1})[F + b(m)]$$

in which strategy $m$ is again a rare mutant with $m > m_{\min}$. By the usual technique, stability of $m_{\min}$ occurs if

$$\frac{m_{fopt}}{m_{\min}} > \frac{g\left(m_{\min}^{-1}\right)}{g'\left(m_{\min}^{-1}\right) \cdot m_{\min}^{-1}}. \tag{9}$$

If we take the probability of fertilization to be about 1 at sperm number $m_{\min}^{-1}$ (for many species this seems to be a reasonable approximation), then in order for the classical theory to explain the maintenance of small sperm, we need the following rule to hold.

The anisogamy ratio (ovum size/sperm size) must exceed the reciprocal of the product of sperm number ( $m_{\min}^{-1}$ ) and the gradient, $g'(m_{\min}^{-1})$. This gradient is the rate at which sperm number contributes to the probability of the fertilization, when sperm have minimal size $m_{\min}$. The fact that sperm dilution has little effect on the probability of fertilization suggests that $g'(m_{\min}^{-1})$ is very small.

We can see from the case of cattle that condition (9) could possibly account for the maintenance of anisogamy. The number of sperm ejaculated by a bull is 5–15 × $10^9$ (Polge, 1972; Bishop, 1961). The sperm is one twenty thousandth the size of the bovine egg (Bishop & Walton, 1960), giving an anisogamy ratio of 2 ×$10^4$. The maximum probability of conception from a normal insemination appears to be around 0.75.

Substituting into (9), we need

$$g'\left(m_{\min}^{-1}\right) > \frac{0.75}{(5\times10^{9})\times(2\times10^{4})}$$

$$g'\left(m_{\min}^{-1}\right) > 0.75\times10^{-14}$$

for the classical theory to account for the maintenance of anisogamy. Data are available (from artificial insemination studies) about the way in which the probability of conception declines with increasing dilution of the ejaculate. Very roughly, there appears to be only a 1% drop in probability of conception as the number of sperm drops from 17 × $10^6$ to 7 × $10^6$ (Salisbury & Van Denmark, 1961). Thus $g'(m^{-1})$ over this range must be less than $0.01/10^7 = 10^{-9}$. We would therefore except that with a normal ejaculate, containing some thousand times more sperm, the gradient $g'(m_{\min}^{-1})$ must be well below $10^{-9}$. Until further data are available which justify the fitting of an explicit form to $g(m^{-1})$, we are unable to make a firm conclusion as to whether the classical theory is robust enough to explain the extremely small size of the sperm. However, it seems likely that at least part of the reason for having tiny sperm relates to increasing the probability of conception.

## 5. WHY ARE THERE SO MANY SPERM?

Both sperm competition and the problem of increasing the chalice of conception will act to keep sperm size to a minimum. It is therefore easy to see why sperm does not contribute to the reserves necessary for zygote survival; i.e. why the male sex persists even when there is internal fertilization. Sperm stay tiny because it will pay, on a fixed resource budget, to produce as many of them as possible. The models described above tell us why the sperm-producing strategy persists, but tell us rather little about the actual number, or amount of sperm that should be produced. Why do males produce so many sperm?

Cohen (1969, and elsewhere) has proposed an ingenious and startling answer to this question. He suggests that males produce so many sperm because most sperm are defective. Defective sperm might arise from errors in meiosis; such errors could be so prevalent that only a tiny fraction of the sperm in each ejaculate are suitable for fertilization. As evidence, Cohen found a highly significant correlation between mean chiasmata frequency and what he termed "sperm redundancy" (= number of sperm ejaculated/number actually used in fertilization).

However, Cohen's theory does not really answer the question "why so many tiny sperm?". Ova are also products of meiosis, and therefore should suffer the same risk of defectiveness. Cohen argues that this might indeed be the case, and suggests that the high prevalence of oocyte atrysia may be due to removal of these defective female gametes. Thus although he may be correct that gametes become increasingly defective with increasing chiasmata frequency, this is a side issue to the central problem of whether gametes should be tiny and unprovisioned, or large and highly provisioned. The assumption appears to be that because the ova are costly, it pays the female to sort out suitable ones before they are provisioned; sperm on the other hand are not expensive and so it need not pay the male to eliminate defective ones. This therefore prejudges the issue of why sperm are small and numerous.

We can best answer the question "why so many sperm?" by considering how much of his reproductive resources a male should allocate to an ejaculate. Given that each sperm will be tiny because of sperm competition, then if we find a male should invest a significant proportion of reproductive resources on an ejaculate, we can explain why vast numbers of sperm are produced.

In some sessile animals with internal fertilization, there may be no alternative reproductive strategy to profligate gamete production. Thus where males compete only by sperm competition, they will spend all their efforts on sperm and so that the total investment per male on sperm may approximate to that invested per female in ova; gametic masses should be roughly similar. Hence at high anisogamy ratios, sperm numbers will be high relative to ovum numbers.

The equal gametic expenditure rule breaks down if we allow alternative reproductive strategies for males, such as enhanced mobility for mate searching, etc. It also breaks down if sperm competition is not prevalent.

Consider the vertebrate model in which there is internal fertilization, and sperm competition occurs perhaps only rarely when the same female mates with two males with frequency $p$. We must trade off expenditure on sperm against expenditure on an alternative reproductive strategy such as mobility. Let us assume that if a male expends heavily on sperm, he does so at a cost in terms of the number of new females he is likely to encounter, because he has less resources left for mate-searching. Thus a mutant spending proportion $k$ of his total reproductivity resources on sperm will obtain:

$$\frac{1-k}{1-k_*}$$

matings relative to each obtained by a normal male that expends $k_*$ on sperm. In other words, the relative number of females encountered is directly proportional to the distance a male moves relative to other males. (This model may not be accurate if males obtain females by fights, though it may for various reasons serve as a reasonable approximation even then.) Once again, we assume that when two ejaculates compete, success is based on the "raffle" principle, i.e. chances of fertilization are equivalent to (self's sperm number)/(total sperm number).

At the ESS expenditure $k_*$ we require that

$$\underbrace{(1-p)+\frac{p}{2}}_{\text{mean fitness of population strategy } k_*} > \underbrace{(1-p)\left(\frac{1-k}{1-k_*}\right)}_{\text{mutant fitness through single matings}} + \underbrace{\frac{pk}{k_*+k}\left(\frac{1-k}{1-k_*}\right)}_{\text{mutant fitness through double matings}}.$$

For all $k \neq k_*$, the mutant fitness should be less than mean population fitness. Thus if we plot mutant fitness against $k$, the result should be a peak at $k = k_*$. Hence if $k_*$ is an ESS then

$$\frac{\mathrm{d}}{\mathrm{d}k}\left[(1-p)\left(\frac{1-k}{1-k_*}\right)+\frac{pk}{k_*+k}\left(\frac{1-k}{1-k_*}\right)\right]\Bigg|_{k_*} = 0, \text{for } k = k_*.$$

and the second derivative of the left hand side should be negative, indicating that this is indeed a maximum rather than a minimum.

Differentiating, we obtain

$$k_* = \frac{p}{4-p} \tag{10}$$

or, for small $p$,

$$k_* \approx \frac{p}{4}$$

and the second derivative is negative, as required. This result suggests that a male should expend, of his total mating effort, a proportion $k_*$ that is roughly equivalent to a quarter of the probability of double mating.

Consider red deer. Suppose that hinds are mated by two males as infrequently as once in a hundred occasions, which seems a very conservative estimate. Then a stag should spend a quarter of a percent of his total mating effort on sperm. Bearing in mind his immense energetic expenditure on antlers, roaring and fighting, large body size, and mate-searching, the estimate of a quarter percent gametic expenditure seems not to be at all excessive. Thus sperm competition may well account for why there are so many sperm, as well as accounting for the related question of why sperm are so tiny. Even though internal fertilization and reduced sperm competition may usually be associated with a reduction in male gametic expenditure, infrequent double matings can lead to non-trivial expenditures on sperm.

It would be interesting to investigate the degree to which sperm numbers correlated with increased sperm competition in internal fertilizers. Data on the incidence of double matings in nature are not readily available. From Cohen's (1969) work, it is clear that insects produce far fewer sperm than mammals. Sperm storage organs in the female insect are highly developed compared to those in mammals. There is therefore little purpose in introducing vastly more sperm than can adequately fill these stores, unless extra sperm are needed to achieve "sperm flushing" during sperm displacement. In mammals, the situation is quite different; there is a vast genital tract in which sperm survive for (usually) a relatively short time. Here the "raffle principle" is likely to apply; i.e. the more sperm ejaculated, the better the chance of success when another male also mates with the same female.

## ACKNOWLEDGEMENTS

Much of this work was done during a recent Dahlem Conference on Animal Mind–Human Mind; I hope it may in some way compensate for my singularly undistinguished performance there. I am indebted to Miss Jane Farrel for typing.

## NOTE

This manuscript incorporates changes published in an Erratum (1982). *Journal of Theoretical Biology,* **98**, Issue 4, 707.

## REFERENCES

Bell, G. (1978). *J. Theor. Biol.*, **73,** 247.

Bishop, D.W. (1961). In: *Sex and Internal Secretions*, vol. II, 3rd ed. (Young, W. C. ed.), pp 707–795. London: Balliere, Tindall & Cox.

Bishop, M.W.H. & Walton, A. (1960). In: *Marshall's Physiology of Reproduction*, vol. I, part 2, 3rd ed. (Parkes, A. S. ed.), pp. 1–29. London: Longmans.

Charlesworth, B. (1978). *J. Theor. Biol.*, **73**, 347.

Cohen, J. (1969). *Sci. Prog., Lond.*, **57**, 23.

Fisher, R. A. (1930). *The Genetical Theory of Natural Selection.* Oxford: Clarendon Press.

Hoekstra, R. (1980). *J. Theor. Biol.*, **87**, 785.

Kalmus, H. (1932). *Biol. Zentral.*, **52**, 716.

Kalmus, H. & Smith, C. A. B. (1960). *Nature, Lond.*, **186**, 104

Knowlton, N. (1974). *J. Theor. Biol.*, **46**, 283

Lanier, D.L. & Estep, D. Q. & Dewsbury, D. A. (1979). *J. Comp. Physiol. Psych.*, **93**, 781.

Lefevre, G. & Jonsson, U. B. (1962). *Genetics*, **47**, 1719

Martin, P.A., Reimers, T. J., Lodge, J. R. & Dzink, P. J. (1974). *J. Reprod. Fertil.*, **39**, 251.

Maynard Smith, J. (1974). *J. Reprod. Fertil.*, **47**, 209.

Maynard Smith, J. (1978). *The Evolution of Sex.* Cambridge: Cambridge University Press.

Parker, G. A. (1970*a*). *Biol. Rev.*, **45**, 525.

Parker, G. A. (1970*b*). *J. Insect Physiol.*, **16**, 1301.

Parker, G. A. (1978). *J. Theor. Biol.*, **73**, 1.

Parker, G. A., Baker, R. R., & Smith, V. G. F. (1972). *J. Theor. Biol.*, **36**, 529.

Polge, C. (1972). In: *Artificial Control of Reproduction* (Austin, C. R. & Short, R. V. ed.), p. 1. Cambridge University Press.

Salisbury, G.W. & Vandemark, N. L. (1961). *Physiology of Reproduction and Artificial Insemination of Cattle.* 1st ed. San Francisco: Freeman.

Scudo, F. M. (1967). *Evolution*, **21**, 285.
Smith, C. C. & Fretwell, S. C. (1974). *Am. Nat.*, **108**, 499.
Smith, R. L. (1979). *Science*, **205**, 1029.
Waage, J. K. (1979). *Science*, **203**, 227.

# 3. SPERM COMPETITION, MALE PRUDENCE AND SPERM-LIMITED FEMALES

Nina Wedell[1], Matthew J.G. Gage[2] and Geoffrey A. Parker[3]

## ABSTRACT

Sperm are produced in astronomical numbers compared with eggs, and there is good evidence that sperm competition is the force behind the evolution of many tiny sperm. However, sperm production inevitably has costs. Recent research shows that male ejaculate expenditure is dynamic in both time and space, and that males are sensitive to risks of sperm competition and can vary ejaculate size accordingly. We focus on studies showing that males assess mating status and relative fecundity of females, and reveal that modulation of ejaculate investment by males can sometimes result in sperm limitation for females.

---

In the *Selfish Gene* [1] Dawkins wrote 'The word excess has no meaning for a male'. This was the traditional consensus, which considered males to possess an almost limitless supply of sperm. Male and female reproductive success is ultimately achieved only after individuals have successfully passed through a variety of competitive and selective stages. The relative importance of these different stages varies, depending on the evolved mating pattern of the species. Here, we focus on one of the final stages of reproduction: the allocation of optimal numbers of sperm. Bateman's [2] classic experiment with fruitflies illustrates the fundamental difference between males and females: number of mates limits male reproductive success, whereas females are constrained by offspring

---

1 Ecology & Evolution Group, School of Biology, University of Leeds.

2 Centre for Ecology, Evolution and Conservation, School of Biological Sciences, University of East Anglia.

3 Population & Evolutionary Biology Research Group, School of Biological Sciences, University of Liverpool.

Reprinted from *Trends in Ecology & Evolution*, 17 (7), Wedell, N., Gage, M. J. G., & Parker, G. A. Sperm competition, male prudence and sperm-limited females. pp. 313–320, Copyright (2002), with permission from Elsevier.

production. It is certainly true that males possess a greater reproductive potential than do females; however, recent evidence shows that spermatogenesis is far from limitless, and that males have evolved mechanisms for allocating finite numbers of sperm optimally to maximize their lifetime reproductive success.

## WHY ARE THERE SO MANY SPERM?

Anthony van Leeuwenhoek, the pioneering microscopist, was the first to observe spermatozoa accurately and was struck by their numerous nature. His explanation [3] was that 'there cannot be too great a number of adventurers, when there is so great a likelihood to miscarry'. Although sheer chance alone might militate against an ejaculate, females appear to effect various forms of sperm choice. We now recognize two interacting sexually selected forces that are important determinants of male fertilization success. Van Leeuwenhoek's explanation can therefore be refined to: the probability of miscarriage for an individual male depends both upon whether the female encourages or discourages fertilization by his sperm, and the level of competition that his sperm face from rival male gametes.

It has long been realized that females have evolved several mechanisms that encourage or discourage sperm to fertilize. Recent research is consistent with the notion that females might be able to manage the sperm from different males in a sophisticated manner. There is potential for females to encourage or discourage sperm transport using muscular contractions, ciliary currents and other specialized female reproductive tract structures and behaviours [4]. Female feral fowl *Gallus gallus domesticus*, for example, actively eject sperm of lower ranking males [5].

Sperm competition is a widespread phenomenon that occurs when sperm from two or more males compete for a female's ova [6, 7 and 8]( Box 1). There is good evidence that relative sperm numbers are important for sperm competitive success, and comparative studies show that species experiencing higher risks of sperm competition invest in relatively larger testes that produce more sperm (Box 1). Sperm competition is therefore a selective force shaping optimal ejaculate structure.

## ARE SPERM COSTLY?

Males cannot produce limitless numbers of sperm (e.g. [9]), in spite of the potential selection from sperm competition to increase numbers and maintain numerical superiority in the fertilization lottery. Dewsbury [10] highlighted the fact that ejaculate production generates nontrivial costs. In vertebrates, males often need to recover after mating before producing another ejaculate, (frequently with reduced sperm numbers [11]), and there is clear evidence that dietary restriction constrains sperm production in Indian meal moths *Plodia interpunctella* [12]. The direct energetic costs of sperm production are poorly understood. Sperm production reduces the lifespan of *Caenorhabditis elegans* nematodes [13]. In adders *Vipera berus*, body-mass loss during the spermatogenesis stage (when males are physically inactive) is as great as during the subsequent mating phase, when males actively search, court, compete and copulate [14], suggesting that there are significant energetic demands from spermatogenesis.

## SPERM STORAGE AND MECHANISMS OF INSEMINATION

Males have evolved several mechanisms for storing and delivering sperm. In pair-forming birds, which usually have a discrete fertile window because of focused breeding seasonality, there can be constraints on sperm storage. This could explain why variation in ejaculate allocation is achieved predominantly through differences in the rate of insemination (e.g. [15]). In many taxa, males store mature sperm in a specialized zone of the reproductive tract (e.g. the epididymis and vasa deferentia in mammals). In humans, maximum sperm storage is two weeks, after which time matured sperm are released into the urine [16]. Most invertebrates store sperm in seminal vesicles or in specialized enclosed sperm packets (spermatophores). However, although males have evolved various sites for sperm storage, ageing might eventually reduce the fitness of the sperm cell [17].

For males to effect prudence in sperm allocation, they require a mechanism that ensures only limited proportions of sperm are ejaculated at any single mating when matings are frequent (male lions *Panthera leo* can copulate >100 times per day [18]). Clearly, if males ejaculate at high frequencies over a relatively brief period, sperm must be parcelled between matings, requiring mechanisms that prevent all mature sperm from being released. In the blueheaded wrasse *Thalassoma bifasciatum*, the male's sperm duct is divided into numerous small chambers encircled by a thin muscle that enables males to regulate the amount of sperm released at a particular spawning [19]. In mammals, sperm are mobilized by muscular contractions, the rate of which can vary: studies of *Peromyscus* rodents show that morphine, an opioid agonist, inhibits vas deferens contractions in the promiscuous *P. maniculatus* but not in the monogamous *P. californicus* [20]. This suggests that *P. maniculatus* (where males are subject to sperm competition) has evolved a vas deferens that is sensitive to opioid hormones whose secretion depends directly on socio-sexual cues relating to the risk of sperm competition [20].

## EMPIRICAL STUDIES EXPLORING EJACULATE MODULATION

We have considered evidence that: (1) sperm competition and female factors are variable and influence optimal ejaculate structure; (2) spermatogenesis generates nontrivial costs both energetically and in maintaining mature sperm for mating opportunities; and (3) males have evolved mechanisms that enable partitioning of sperm over a series of matings. These combined considerations predict that males should evolve prudence in ejaculate allocation to maximize overall lifetime reproductive success. When males face low sperm competition risks or poor fecundity returns, they should conserve sperm. By contrast, when they face high sperm competition risks or elevated fecundity potential, they should invest in a larger ejaculate to win fertilizations. Here, we review empirical studies exploring these predictions.

## BOX 1. SPERM COMPETITION – RISK MODELS

Sperm competition game models seek evolutionarily stable strategy (ESS) sperm expenditures both within and across species [a]. Ejaculate expenditure is traded off against other aspects of reproductive success (typically finding new mates). 'Risk' models assume that two ejaculates compete in the mating of fraction $q$ of females, and $(1-q)$ females receive just one ejaculate. The models are concerned with individual male allocation in number of sperm ejaculated per copulation (or bout of copulations), that is, how the sperm supply is parcelled out among mating opportunities. Across species, increased competition among the ejaculates of rival males ($>q$) promotes evolution of increased ejaculate expenditure per mating opportunity. Providing sperm compete on an equal basis, the probability of fertilizing a female's eggs increases with a male's proportion of sperm in competition, referred to as the raffle principle. Alternatively, a larger ejaculate volume displaces more rival males' sperm. In both cases, escalated sperm-producing capacities are predicted to evolve in response to sperm competition [a]. Male dungflies, for example, evolved larger testes (and therefore bigger ejaculates) when exposed to increased sperm competition after only ten generations of selection [b]. The predicted correlation between degree of polyandry (an index of sperm competition) and relative testis size (standard measure of investment in sperm, controlling for body size), has been found using comparative methods for most animal groups (e.g. primates, bats, other mammals, birds, frogs, fish and various insects [c–e], indicating that it is a general and robust relationship.

Sperm competition game models also make predictions about sperm allocation within species, predicting that individual males should vary the number of sperm ejaculated per copulation depending on cues relating to sperm competition risk. For instance [a], if males within a species can differentiate between females that have already mated and those that have not, at a given level of risk they should allocate less sperm to virgins; across species, both expenditures increase with risk ( Fig. Ia). Within species, it is important to differentiate between male types (i.e. a male's phenotype), and copulation roles (i.e. what happens in the female), although sometimes these are correlated. Both can affect strategic sperm allocation. A type can be 'big' or 'small', and a 'role' might be 'first to mate' or 'second to mate'. First, consider two male types that represent alternative mating strategies: one type guards females ('guarders'), and the other gains matings by subterfuge ('sneaks'). Guarders always mate with their females, so sneaks always face sperm competition. But guarders might not face competition from sneaks. So if sperm production costs are the same for the two types, sneaks should expend more on ejaculates than do guarders [f]. This prediction is usually met: for example in salmon *Salmo salar*, sneaking precocial parr have relatively bigger testes and denser ejaculates than do anadromous guarders [g], and in Bluegill sunfish *Lepomis macrochirus*, sneakers have higher fertilization success [h]. Similar predictions apply if males are similar in type but play strategies opportunistically, as in socially monogamous birds that mate

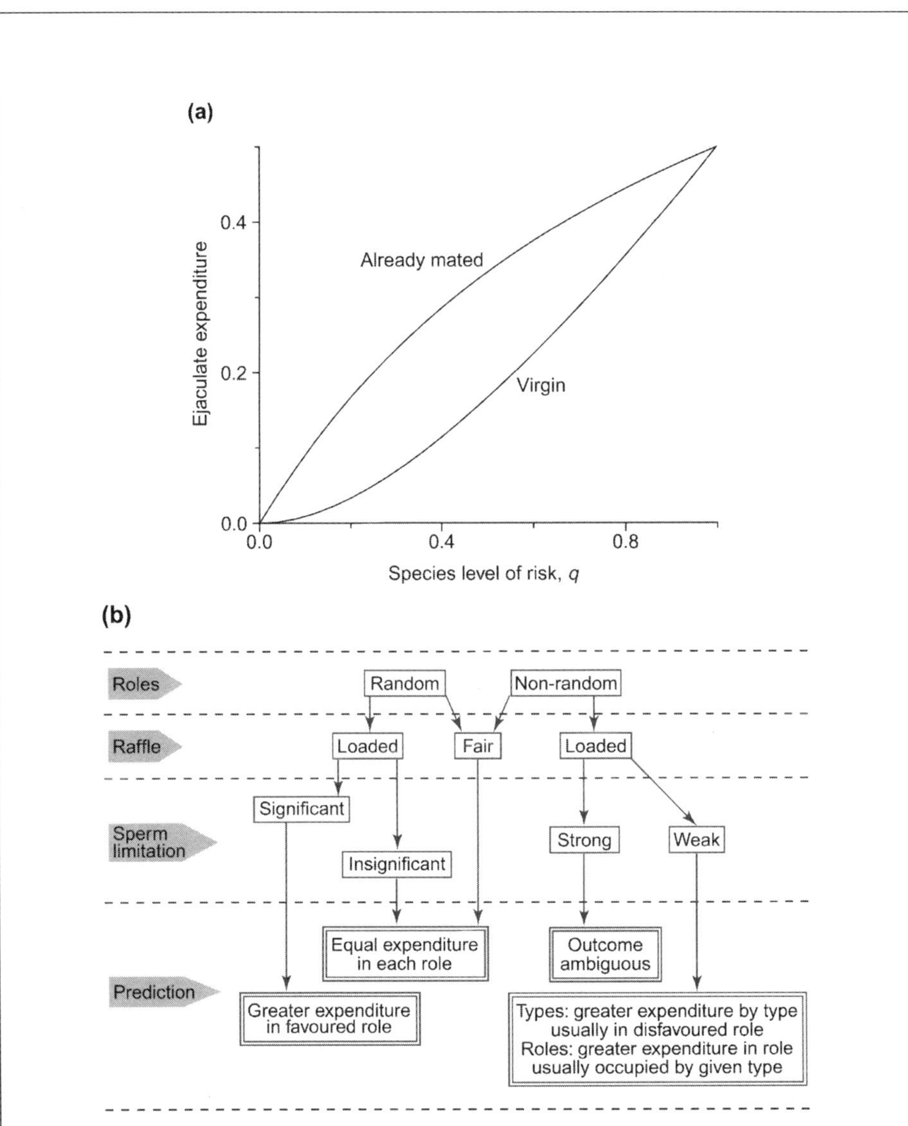


**Fig. I.** Sperm allocation when an ejaculate faces competition with typically only one other ejaculate, if they face competition at all (the risk model). (a) Evolutionarily stable ejaculate expenditure in response to assessment of sperm competition risk when males can differentiate between females that have already mated and those that have not. Sperm compete in a fair raffle. The x-axis is the level of risk (probability q that a female mates twice at a given breeding episode) and is used to give a comparison across species with different levels of risk. At a given level of risk, males within a species are predicted to spend less sperm on virgins [a]. Adapted from [a]. (b) Summary of predictions for the effects of roles (where ejaculates may be favoured or disfavoured in the female tract, e.g. depending whether a male mates first or second), types (where two different male phenotypes tend to occur in different roles), and sperm limitation [l]. Adapted from [l].

opportunistically with the mates of other males [extra-pair copulations (EPCs)]as well as with their own mates. When performing high-risk matings, such as EPCs, males should increase sperm numbers in an attempt to achieve higher paternity. This prediction is supported by, for example male sand martins *Riparia riparia,* producing larger ejaculates with a model female in the presence of rival males [i].

Now imagine that all males are similar, and that a given male is equally likely to mate first or second. If one of these two roles tends to be favoured and the other disfavoured, the raffle is said to be 'loaded', and the ESS is to expend equally in the two roles, even when a male has full information that he is mating first or second [j]. But if the female is sperm limited, the situation changes: a male should spend more in the favoured role [k]. Predictions become more complex when roles are not occupied randomly. Thus, we have two male types, one that is more likely to mate in the disfavoured role and the other in the favoured role. We then have two predictions, depending on whether we consider roles or types [l]. Across types, we expect the type usually mating in the disfavoured role to expend more on ejaculates overall. But across roles, more should be spent in the role that is most usually occupied by a given type. So, if role 1 (mating first) is favoured, and is usually occupied by type A males, and B males usually occupy role 2, then, overall, B should expend more on sperm than does A, but A should spend more when in role 1 than when in role 2, and B should spend more in role 2 than role 1. If females are sperm limited, these predictions are likely to be altered, and the relative expenditures become difficult to predict ( Fig. Ib).

a. Parker, G. A. (1998). Sperm competition and the evolution of ejaculates: towards a theory base. In *Sperm Competition and Sexual Selection* (Birkhead, T. R and Møller, A. P., eds), pp. 3–54, Academic Press.

b. Hosken, D. J. *et al.* (2001). Sexual conflict selects for male and female reproductive characters. *Curr. Biol.* **11**, 489–493.

c. Smith, R. L. (1984). *Sperm Competition and Evolution of Animal Mating Systems*, Academic Press.

d. Birkhead, T. R. and Møller, A. P. (1992). *Sperm Competition in Birds*, Academic Press.

e. Birkhead, T. R. and Møller, A. P. (1998). *Sperm Competition and Sexual Selection*, Academic Press.

f. Parker, G. A. (1990). Sperm competition games – sneaks and extra-pair copulations. *Proc. R. Soc. Lond. Ser. B* **242**, 127–133.

g. Gage, M. J. G. *et al.* (1995). Effects of alternative male mating strategies on characteristics of sperm production in the Atlantic salmon (*Salmo salar*):Theoretical and empirical investigations. *Philos. Trans. R. Soc. Lond. B Biol. Sci.* **350**, 391–399.

h. Fu, P. *et al.* (2001). Tactic-specific success in sperm competition. *Proc. R. Soc. Lond. Ser. B* **268**, 1105–1112.

i. Nicholls, E. H. *et al.* (2001). Ejaculate allocation by male sand martins, *Riparia riparia. Proc. R. Soc. Lond. Ser. B* **268**, 1265–1270.

j. Parker, G. A. (1990). Sperm competition games – raffles and roles. *Proc. R. Soc. Lond. Ser. B* **242**, 120–126.

k. Mesterton-Gibbons, M. (1999). On sperm competition games: incomplete fertilization risk and the equity paradox. *Proc. R. Soc. Lond. Ser. B* **266**, 269–274.

l. Ball, M. A. and Parker, G. A. (2000). Sperm competition games: a comparison of loaded raffle models and their biological implications. *J. Theor. Biol.* **206**, 487–506.

## FUTURE MATING OPPORTUNITIES

A tradeoff between future mating opportunities and sperm competition is predicted to shape optimum ejaculate allocation at a given mating [21]. If male budget more energy on each ejaculate, they will have less energy to expend on finding new females. But meeting new females can be a chance process, so males will be selected to ensure that an amount of sperm is reserved that relates to the probability distribution of females that are likely to be encountered. The number of females (8–80) encountered by male *Brachonius plicatilis* rotifers often exceeds their sperm supply, promoting a sperm budget that is prudently allocated across 13 copulations [22]. In species with a female-biased sex ratio, male mating opportunities, and therefore potential reproductive rates, are increased, promoting sperm economy. For example, the fruitfly *Drosophila pachea* has a female-biased operational sex ratio with males partitioning their limited sperm among successive females [23].

## SPERM COMPETITION AND EJACULATE SIZE

There is clear evidence that males are sensitive to risks of sperm competition and vary their reproductive behaviour accordingly. Males under increased threat of sperm competition put more effort into courtship and mating (e.g. Montagu's harriers *Circus pygargus* [24]), mate guarding (e.g. flour beetles *Tenebrio molitor* [25]), and act more aggressively towards potential rivals (e.g. elephant seals *Mirounga angustirostris* [26]). If males are sensitive to socio-sexual cues that predict sperm competition risk and alter their behaviour, do they also respond at the gametic level and adjust their ejaculate size accordingly?

In taxa in which sperm competition is prevalent, males show greater ejaculate investment as judged by their relative testis size, across both species and populations (Box 1, Box 2). In species in which alternative male mating tactics have evolved that generate different risks of sperm competition, males playing tactics that are associated with higher risks generally invest relatively more in ejaculates than do males that experience lower risks, as predicted by theory (Box 1). Furthermore, males have evolved fixed strategies of sperm allocation that depend on sperm competition risks. For example, male small white butterflies *Pieris rapae* increase the number of sperm transferred on their second mating, because this mating is likely to be with an already mated female, translating to greater risk of sperm competition [27]. Thus, sperm expenditure can be geared to sperm competition risks at a life-history level. However, can an individual male assess sperm competition risks at a given mating and respond by increased ejaculate expenditure when sperm competition risks are elevated and by a decrease when risks are reduced?

## SPERM COMPETITION AND EJACULATE ALLOCATION

### Rival male presence

The risk of sperm competition at a particular mating can be assessed using cues arising from both males and females. The operational sex ratio and presence of rival males provide information about the probability of sperm competition in polyandrous mating systems. For example, female multiple mating increases in *P. interpunctella* meal moths when increased numbers of males are available [28], and population density predicts the level of extra-pair paternity in Bullock's orioles *Icterus galbula bullockii* [29]. Studies of invertebrate and vertebrate taxa showing that risk of sperm competition in the form of rival male presence promotes increased ejaculate size are listed in Table 1. Importantly, however, the presence of rivals does not always imply a proximate risk of sperm competition. For example, in *P. interpunctella*, a female cannot immediately remate physically, because the spermatophore acts as a plug for some hours after insemination; rivals present at mating therefore do not generate a proximate risk of sperm competition and ejaculate size is unaffected [30].

**Table 1.** Studies in which ejaculate size increased in the presence of male rivals

| Species | Refs |
|---|---|
| House cricket *Acheta domesticus* | [51] |
| Decorated cricket *Gryllodes supplicans (sigillatus)* | [51] |
| Spring field cricket *Gryllus veletis* | [52][a] |
| Mealworm beetle *Tenebrio molitor* | [25] |
| Mediterranean fruitfly *Ceratitis capitata* | [53] |
| Blue crab *Callinectes sapidus* | [54] |
| Snow crab *Chionoecetes opilio* | [45] |
| Blueheaded wrasse *Thalassoma bifasciata* | [55] |
| Wrasse *Xyrichtys novacula* | [56] |
| Bucktooth parrotfish *Sparisoma radians* | [57] |
| Rainbow darter *Etheostoma caeruleum* | [36] |
| Sand martin *Riparia riparia* | [58] |
| Domestic fowl *Gallus gallus domesticus* | [b] |
| Rat *Rattus norvegicus* | [59] |

[a] Males increased sperm number as risk increased from zero to one rival male, then decreased ejaculate size as intensity increased to six rivals; ejaculate modulation was not recorded for *G. sigillatus* or *Gryllus texensis*.
[b] T. Pizzari *et al.*, unpublished.

## Female mating status

When detectable, female mated status (Table 2) predicts the risk of sperm competition that a male's ejaculate will subsequently face. In species in which females store sperm (i.e. most insects), males might be able to assess risk via sperm storage cues or other evidence of recent mating. Factors such as female age might correlate with the probability of sperm competition. In most cases, as predicted (Box 1; fig. Ia), males increase sperm number when mating with a mated female. But it is important to note, however, that males do not always increase ejaculate size: they can discriminate against mated females, preferring to mate with virgin females given the opportunity (e.g. great snipe *Gallinago media* [31], or red flour beetles *Tribolium castaneum* [32]). There is also evidence that males might provide smaller ejaculates or terminate copula prematurely in matings with higher risk females (e.g. predatory mites *Macrocheles muscadomesticae* [33]) or females of different mated status (e.g. bumblebees *Bombus terrestris* [34]). The pattern of female sperm utilization might also influence ejaculate investment (Table 2). Many spiders show first-male sperm precedence, and males of the spiny orb weaver *Micrathena gracilis*, for example, discriminate between females, providing few or no sperm when copulating with already mated females, which provide lower fertilization opportunities [35].

**Table 2.** Ejaculate tailoring to female mated status

| Species | Response[a] | Refs |
|---|---|---|
| House cricket *Acheta domesticus* | + | [51] |
| Decorated cricket *Gryllodes supplicans* | + | [51] |
| Bushcricket *Coptaspis sp. 2* | + | [60] |
| Bushcricket *Kawanaphila nartee* | – | [37][b] |
| Indian meal moth *Plodia interpunctella* | + | [30] |
| Small white butterfly *Pieris rapae* | + | [61] |
| Biting midge *Culicoides melleus* | – | [62] |
| Stalk-eyed fly *Cyrtodiopsis whitei* | – | [63][c] |
| Dung fly *Sepsis cynipsea* | + | [64] |
| Bowl and doily spider *Frontinella pyramitela* | – | [65][c] |
| Golden silk spider *Nephila clavipes* | – | [66][c] |
| Spiny orb weaver *Micrathena gracilis* | – | [35][c] |
| Mite *Macrocheles muscaedomesticae* | – | [33][c] |

[a] +, significant increase or –, decrease in ejaculate sperm number.
[b] Sperm number decreases with increasing sperm competition intensity.
[c] First male sperm priority fertilization patterns.

What determines whether, strategically, males should conserve sperm, or invest heavily to win the competition? This depends on mating opportunities – if virgin females are readily available, and mated females relatively infrequent and unprofitable, a male can do best to invest more sperm in virgins, even though relatively few sperm are needed to ensure full fertility. A similar situation arises in species where sperm competition intensity is high. In externally spawning fish, where the intensity of competition is directly assessable, theory predicts that males should reduce ejaculate expenditure as the number of competing males increases, conserving gametes for better opportunities with lower competition (Box 2). In the rainbow darter *Etheostoma caeruleum*, more males abstain from spawning when sperm competition intensity is high [36]. Similarly, in the Australian bushcricket *Kawanaphila nartee* (with internal fertilization and sperm mixing), males respond conservatively to the intensity of sperm competition. Larger females mate more frequently than do smaller females, hence sperm competition intensity covaries with female body size. As predicted (Box 2), males reduce the number of sperm that are provided to larger females [37].

**Female quality**

In species in which females vary in reproductive quality, males achieve greater reproductive success by strategically allocating their sperm to females that provide the biggest fertilization returns. Female fecundity usually depends on condition and age, and several studies have explored male investment relative to these cues. Males prefer larger or younger mates in several species (e.g. insect examples [38]), and there is evidence that males also provide larger ejaculates to heavier females in insects, molluscs, fish, and even humans ( Table 3). By contrast, males might discriminate against smaller, less fecund mates, as in the stinkbug *Thyanta pallidovirens,* in which males even terminate copulation before sperm is transferred when paired with below average-sized and/or previously mated females that provide lower fecundity potential [39]. Males also discriminate between females on the basis of age, providing more sperm to younger females. For example, older female gypsy moths *Lymantria dispar* receive less sperm than do younger ones [40].

It is important to consider that female fecundity and age might co-vary with sperm competition risk (Box 1). For example, older females might provide lower fecundity potentials but generate high risks of sperm competition. Males therefore have to balance reduced ejaculate investment for lower fecundity potential against increased investment for risks of sperm competition. Under such conditions, males of the bushcricket *Requena verticalis* provide older females with more sperm [41]. The cost of finding alternative mating opportunities, the relative costs of mating and sperm production, and the sperm competition risks will interact to favour the ejaculation strategy that provides the greatest overall returns. Similarly, heavier and more fecund females might also attract increased male attention (e.g. *P. interpunctella* meal moths [28]), again demanding a balance between investing sperm in high-fecundity females relative to the greater risks of sperm competition that they might generate.

Female reproductive 'quality' for a male might also decrease with his increasing mating investment to that individual female. The Coolidge effect [42] is a progressive decline in the propensity of a male to mate with a female over successive inseminations, combined with renewed sexual interest with a novel female. This male mating strategy allows sperm reserves to be conserved when new females that present additional reproductive opportunities are encountered.

## BOX 2. SPERM COMPETITION INTENSITY

'Intensity' models are recent sperm competition game models by Parker and co-workers [a, b] for species with typically high levels of sperm competition. They show that, across species, ejaculate expenditure per mating opportunity should increase with the number ($N$) of ejaculates that compete for the eggs of a given female. Within species, they show that males should respond differently to 'intensity' than they do to 'risk'. The risk of sperm competition relates to the probability that a male's ejaculate will compete with rival males' sperm and varies between 0 and 1. The intensity relates to the number of ejaculates that a male competes with. The models were developed originally for external spawners, but apply equally to species with internal fertilization where sperm mix randomly. Although across species, the average ejaculate expenditure increases with the average intensity of sperm competition, as seen in comparative studies on testes size (Box 1), within species the reverse is predicted if males can assess the number of competing ejaculates at a given spawning $i$, $N_i$. During high sperm competition intensity, instead of increasing the number of sperm ejaculated, males should reduce sperm numbers as the number of competitors increases above two (Fig. Ia). The probability of fertilization is assumed to be instantaneous and equal to the number of sperm released by each male divided by the total number of sperm. Males should, under these situations, ejaculate very little sperm when there are no competitors, most in the presence of one competitor, and reduce sperm number when there are two or more competitors, because the benefit of increased sperm expenditure decreases with additional competitors owing to diminishing returns.

Slightly different predictions can be generated under more realistic assumptions about fertilization. If fertilization follows a continuous process, the highest fertilization rate occurs in the first time unit, reducing with each successive unit. Sperm limitation can occur if the total sperm density is low enough: there can still be some eggs left unfertilized after all the sperm have died or become lost from the fertilization site. Then it is possible to generate conditions under which the sperm numbers should decline from $N_i = 1$ (Fig. Ib), suggesting that within a species it can be optimal to inseminate most sperm when there is no sperm competition, although the conditions for this to occur can be quite extreme.

a. Parker, G. A. *et al.* (1996). Sperm competition games: assessment of sperm competition intensity by group spawners. *Proc. R. Soc. Lond. Ser. B* **263**, 1291–1297.

b. Parker, G. A. *et al.* (1997). Sperm competition games: a prospective analysis of risk assessment. *Proc. R. Soc. Lond. Ser. B* **264**, 1793–1802.

c. Ball, M. A. and Parker, G. A. (1997). Sperm competition inter- and intra-species results of a continuous external fertilization model. *J. Theor. Biol.* **186**, 459–466.

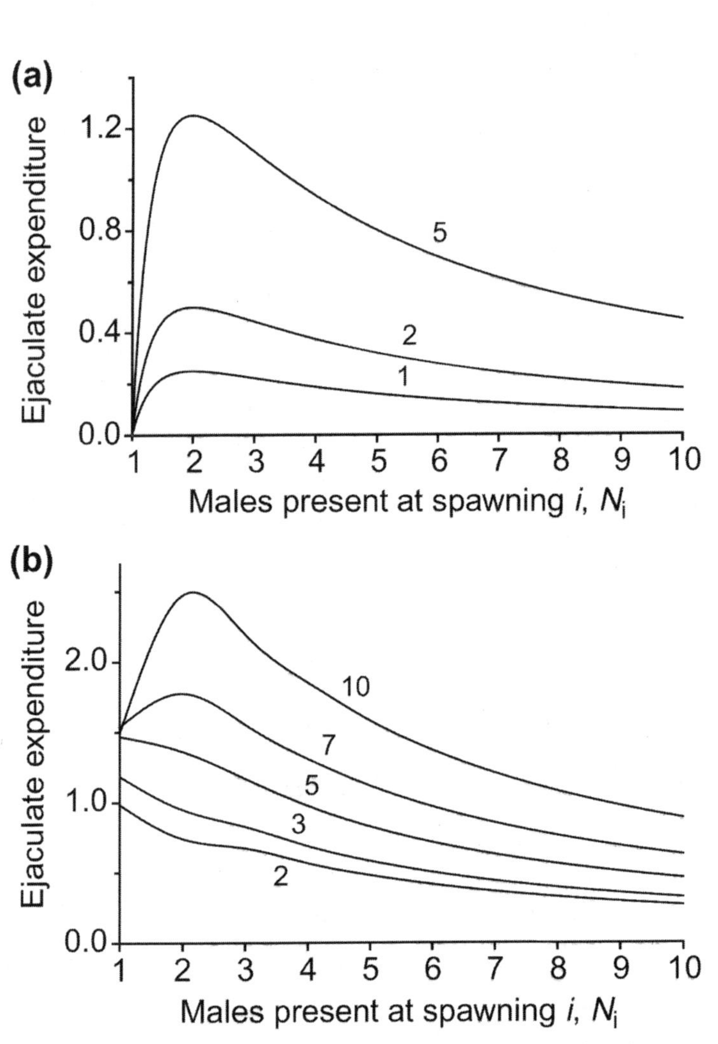


**Fig. I.** Sperm allocation when an ejaculate typically faces sperm competition from several other ejaculates (the intensity model). Each curve shows the evolutionarily stable ejaculate expenditure (relative to the average total reproductive effort per spawning) in relation to the number, $N_i$, of males competing at spawning of type $i$ in the intensity model, assuming that a male can assess $N_i$ at the time of ejaculation. Thus, the x-axis describes the variation within a species, and each curve represents a different species. Sperm compete in a fair raffle. (a) Fertilization is instantaneous; curves are for three species that differ only in the mean number of males present at a spawning: $N$ = 1, 2 and 5 [a]. Adapted from [a]. (b) Fertilization occurs as a continuous process, assuming that the propensity for fusion of gametes is low (see [c] for parameter values and further details). Curves are for five species that differ only in mean number of males present at a spawning: $N$ = 2, 3, 5, 7, and 10. Adapted from [c].

**Table 3.** Ejaculate tailoring to female quality

| Species | Female Factor[a] | Refs |
|---|---|---|
| House cricket *Acheta domesticus* | Size | [51] |
| Bushcricket *Coptaspis sp.2* | Size | [60] |
| Bushcricket *Requena verticalis* | Age | [41] |
| Stink bug *Thyanta pallidovirens* | Size | [39] |
| Gypsy moth *Lymantria dispar* | Age | [40] |
| Indian meal moth *Plodia interpunctella* | Age, Size | [30,67] |
| Small white butterfly *Pieris rapae* | Size | [61] |
| Dung fly *Sepsis cynipsea* | Size, Age | [64] |
| Yellow dung fly *Scatophaga stercoraria* | Size | [68] |
| Weevil *Brentus ancorago* | Size | [38] |
| Mite *Macrocheles muscaedomestićae* | Age | [33] |
| Sea hare *Aplysia parvula* | Size | [69] |
| Spiny lobster *Panulirus argus* | Size | [70] |
| Spiny lobster *Jasus edwadsii* | Size | [70] |
| Blueheaded wrasse *Thalassoma bifasciata* | Size | [55] |
| Wrasse *Xyrichtys novacula* | Size | [56] |
| Bucktooth parrotfish *Sparisoma radians* | Size | [57] |
| Human *Homo sapiens* | Size | [71] |

[a] body size or reproductive age.

## Female sperm limitation

There is evidence that males can become sperm limited [9, 10, 11, 12 and 13] and this can translate to female sperm limitation. For example, in the European cornborer *Ostrina nubilalis*, females mated to recently mated males show reduced fertility [43]. In the turnip moth *Agrotis segetum*, there is a negative correlation between female fertility and the number of previous copulations by the male [44]. Similarly, male snow crabs *Chionoecetes opilio* show prudent sperm allocation and inseminate using only 2.5% of their sperm reserves. This results in some females extruding sub-fertile egg clutches [45]. Female sperm limitation can arise as a direct result of variance in mating opportunities and risks of sperm competition affecting males' ejaculatory strategies. In the blueheaded wrasse *T. bifasciatum*, dominant males are most attractive as mates, but females suffer fertility costs when pairing with these males, because they release fewer sperm [46]. Similarly, in the lekking sandfly *Lutzomyia longipalpis*, females mating with high-ranking males risk sperm limitation [47] and in Soay sheep *Ovis aries*, frequently mating, dominant males lose out in sperm competition toward the end of the rutting season, because of sperm depletion [11]. Females of these species appear to trade good paternal genes from the most attractive males against absolute fertility. Additionally, sperm limitation of preferred males could promote competition among females for early access to successful males, generating increased variance in female reproductive success. In the lekking great snipe *G. media*, females compete for repeated copulation with popular males, but males reject females with whom they have already have mated [31]

## CONCLUSIONS

We have reviewed evidence that sperm competition and female quality have shaped males' optimal sperm delivery strategies, and that males show plastic behaviours at the gametic level. Although sperm competition has resulted in the evolution of extreme sperm numbers, the nontrivial cost of spermatogenesis has also promoted prudent ejaculate allocation by males at each mating. We hope that this review prompts future research to explore this area in more depth, and across broader mating pattern contexts. One clear objective is to understand the underlying mechanisms that enable ejaculate modulation. We show that the phenomenon can exist; future physiological research on the underlying neurological and/or hormonal mechanisms generating sperm allocation will link detailed empirical findings to theoretical considerations.

Having shown that males can exhibit variance in ejaculate allocation, there is the added possibility that male (and female) responses to variance in sperm competition and female quality differ among populations (or even individuals), in the same way that variance in mating pattern varies in space and time within a species. A further fruitful course of research is to quantify the total reproductive fitness consequences of ejaculate modulation. Long-term experiments will reveal how profligate or economical ejaculates influence reproductive success over an individual's lifetime.

We show that male prudence can sometimes generate female sperm limitation. More detailed studies of natural levels of female infertility will establish the reproductive importance of this phenomenon, and therefore the degree of selection on female mating patterns to guard against the potentially severe risk of male-factor infertility [48]. In some species, females appear to trade off preferred paternal genes against reduced fertility [11, 46 and 47]; in others, sperm-constrained males discriminate against certain females [31, 32, 33, 34 and 35], even promoting competition among females [31]. These interactions result in increased variance in female reproductive success with obvious consequences for sexual selection. We urge that the possibility of male and/or mutual mate choice should be considered in future models of sexual selection [e.g. [49], [50]].

We commenced this review of male gamete production and allocation strategies by considering the differences between male and female reproduction according to principles first identified by Bateman [2]. Although these intrasexual differences remain fundamental forces in sexual selection, our review illustrates that possibilities for gamete limitation and mate choice also exist from the less-obvious male perspective. We show that variance in male mating opportunity and expected fertilization returns has generated the evolution of mechanisms allowing prudent sperm allocation in relation to sperm competition risk and potential female fecundity across many animal groups.

## ACKNOWLEDGEMENTS

We are very grateful for helpful comments and unpublished data from Tom Pizzari; four other referees improved the article. N.W. and M.J.G.G. are funded by Royal Society University Research Fellowships.

## REFERENCES

1. Dawkins, R. (1976) *The Selfish Gene*, Oxford University Press.
2. Bateman, A. J. (1948) Intra-sexual selection in *Drosophila*. *Heredity* **2**, 349–368.
3. Ford, B. J. (1991) *The Leeuwenhoek Legacy*, Biopress and Farrand Press.
4. Eberhard, W. G. (1996) Female Control: Sexual Selection by Cryptic Female Choice, Princeton University Press.
5. Pizzari, T. and Birkhead, T. R. (2000) Female feral fowl eject sperm of subdominant males. *Nature* **405**, 787–789.
6. Parker, G. A. (1970) Sperm competition and its evolutionary consequences in the insects. *Biol. Rev.* **45**, 525–567.
7. Birkhead, T. R. and Møller, A. P. (1998) *Sperm Competition and Sexual Selection*, Academic Press.
8. Simmons, L. W. (2001) Sperm Competition and its Evolutionary Consequences in Insects, Princeton University Press.
9. Nakatsuru, K. and Kramer, D. L. (1982) Is sperm cheap? Limited male fertility and female choice in the lemon tetra (Pisces, Characidae). *Science* **216**, 753–755.
10. Dewsbury, D. A. (1982) Ejaculate cost and male choice. *Am. Nat.* **119**, 601–610.
11. Preston, B. T. *et al.* (2001) Dominant rams lose out by sperm depletion. *Nature* **409**, 681–682.
12. Gage, M. J. G. and Cook, P. A. (1994) Sperm size or numbers? Effects of nutritional stress on eupyrene and apyrene sperm production strategies in the moth *Plodia interpunctella* (Lepidoptera: Pyralidae). *Funct. Ecol.* **8**, 594–599.
13. Van Voorhies, W. A. (1992) Production of sperm reduces nematode life-span. *Nature* **360**, 456–458.
14. Olsson, M. *et al.* (1997) Is sperm really so cheap? Costs of reproduction in male adders, *Vipera berus*. *Proc. R. Soc. Lond. Ser. B* **264**, 455–459.
15. Hunter, F. M. *et al.* (2000) Strategic allocation of ejaculates by male Adele penguins. *Proc. R. Soc. Lond. Ser. B* **267**, 1541–1545.
16. Barratt, C. L. R. and Cook, I. D. (1988) Sperm loss in the urine of sexually rested men. *Int. J. Androl.* **11**, 201–207.
17. Siva-Jothy, M. T. (2000) The young sperm gambit. *Ecol. Lett.* **3**, 172–174.
18. Packer, C. and Pusey, A. E. (1983) Adaptations of female lions to infanticide by incoming males. *Am. Nat.* **121**, 716–728.
19. Rasotto, M. B. and Shapiro, D. Y. (1998) Morphology of gonoducts and male genital papilla, in the bluehead wrasse: implications and correlates on the control of gamete release. *J. Fish Biol.* **52**, 716–725.
20. Pound, N. (1999) Effects of morphine on electrically evoked contractions of the vas deferens in two congeneric rodent species differing in sperm competition intensity. *Proc. R. Soc. Lond. Ser. B* **266**, 1755–1758.
21. Parker, G. A. (1982) Why are there so many tiny sperm. Sperm competition and the maintenance of 2 sexes. *J. Theor. Biol.* **96**, 281–294.
22. Gómez, A. and Serra, M. (1996) Mate choice in male *Brachionus plicatilis* rotifers. *Funct. Ecol.* **10**, 681–687.
23. Pitnick, S. (1993) Operational sex ratios and sperm limitation in populations of *Drosophila pachea*. *Behav. Ecol. Sociobiol.* **33**, 383–391.

24. Mougeot, F. *et al.* (2001) Decoy presentations as a means to manipulate the risk of extrapair copulation: an experimental study in a semicolonial raptor, the Montagu's harrier (*Circus pygargus*). *Behav. Ecol.* **12**, 1–7.
25. Gage, M. J. G. and Baker, R. R. (1991) Ejaculate size varies with socio-sexual situation in an insect. *Ecol. Entomol.* **16**, 331–337.
26. LeBœf, B. J. and Peterson, R. S. (1969) Social status and mating activity in elephant seals. *Science* **163**, 91–93.
27. Cook, P. A. and Wedell, N. (1996) Ejaculate dynamics in butterflies: a strategy for maximizing fertilization success? *Proc. R. Soc. Lond. Ser. B* **263**, 1047–1051.
28. Gage, M. J. G. (1995) Continuous variation in reproductive strategy as an adaptive response to population density on the moth *Plodia interpunctella. Proc. R. Soc. Lond. Ser. B* **261**, 331–337.
29. Richardson, D. S. and Burke, T. (2001) Extrapai paternity and variance in reproductive success related to breeding density in Bullock's orioles. *Anim. Behav.* **62**, 519–525.
30. Cook, P. A. and Gage, M. J. G. (1995) Effects of risks of sperm competition on the numbers of eupyrene and apyrene sperm ejaculated by the male moth *Plodia interpunctella* (Lepidoptera: Pyralidae). *Behav. Ecol. Sociobiol.* **36**, 261–268.
31. Sæther, S. A. *et al.* (2001) Male mate choice, sexual conflict and strategic allocation of copulations in a lekking bird. *Proc. R. Soc. Lond. Ser. B* **268**, 2097–2102.
32. Arnaud, L. and Haubruge, E. (1999) Mating behaviour and male mate choice in *Tribolium castaneum* (Coleoptera, Tenebrionidae). *Behaviour* **136**, 67–77.
33. Yasui, Y. (1996) Males of a mite, *Macrocheles muscadomesticae*, estimate a female's value on the basis of her age and reproductive status. *J. Insect Behav.* **9**, 517–524.
34. Sauter, A. and Brown, M. J. F. (2001) To copulate or not? The importance of female status and behavioural variation in predicting copulation in a bumblebee. *Anim. Behav.* **62**, 221–226.
35. Bukowski, T. C. and Christenson, T. E. (1997) Determinants of sperm release and storage in a spiny orbweaving spider. *Anim. Behav.* **53**, 381–395.
36. Fuller, R. C. (1998) Sperm competition effects male behaviour and sperm output in the rainbow darter. *Proc. R. Soc. Lond. Ser. B* **265**, 2365–2371.
37. Simmons, L. W. and Kvarnemo, L. (1997) Ejaculate expenditure by male bush-crickets decreases with sperm competition intensity. *Proc. R. Soc. Lond. Ser. B* **264**, 1203–1208.
38. Johnson, L. K. and Hubbell, S. P. (1984) Male choice: experimental demonstration in a brentid weevil. *Behav. Ecol. Sociobiol.* **15**, 183–188.
39. Wang, Q. and Millar, J. G. (1997) Reproductive behavior of *Thyanta pallidovirens* (Heteroptera: Pentatomidae). *Ann. Entomol. Soc. Am.* **90**, 380–388.
40. Proshold, F. I. (1996) Reproductive capacity of laboratory-reared Gypsy moths (Lepidoptera: Lymantriidae): effect of age of female at time of mating. *J. Econ. Entomol.* **89**, 337–342.
41. Simmons, L. W. *et al.* (1993) Bushcricket spermatophores vary in accord with sperm competition and parental investment theory. *Proc. R. Soc. Lond. Ser. B* **251**, 183–186.
42. Dewsbury, D. A. (1981) Effects of novelty on copulatory behavior – the Coolidge effect and related phenomena. *Psychol. Bull.* **89**, 464–482.
43. Royer, L. and McNeil, J. N. (1993) Male investment in the European corn borer, *Ostrinia nubilalis* (Lepidoptera Pyralidae) – impact on female longevity and reproductive performance. *Funct. Ecol.* **7**, 209–215.
44. Svensson, M. G. E. *et al.* (1998) Mating behaviour and reproductive potential in the turnip moth *Agrotis segetum* (Lepidoptera: Noctuidae). *J. Insect Behav.* **11**, 343–359.
45. Rondeau, A. and Sainte-Marie, B. (2001) Variable mate-guarding time and sperm allocation by male snow crabs (*Chionoecetes opilio*) in response to sexual competition, and their impact on the mating success of females. *Biol. Bull.* **201**, 204–217.
46. Warner, R. R. *et al.* (1995) Sexual conflict – males with the highest mating success convey the lowest fertilization benefits to females. *Proc. R. Soc. Lond. Ser. B* **262**, 135–139.

47. Jones, T. M. (2001) A potential cost of monandry in the lekking sandfly *Lutzomyia longipalpis*. *J. Insect Behav.* **14**, 385–399.

48. Sheldon, B. C. (1994) Male phenotype, fertility, and the pursuit of extra-pair copulations by female birds. *Proc. R. Soc. Lond. Ser. B* **257**, 25–30.

49. Johnstone, R. A. *et al.* (1996) Mutual mate choice and sex differences in choosiness. *Evolution* **50**, 1382–1391.

50. Kokko, H. and Monaghan, P. (2001) Predicting the direction of sexual selection. *Ecol. Lett.* **4**, 159–165.

51. Gage, A. R. and Barnard, C. J. (1996) Male crickets increase sperm number in relation to competition and female size. *Behav. Ecol. Sociobiol.* **38**, 227–237.

52. Schaus, J. M. and Sakaluk, S. K. (2001) Ejaculate expenditures of male crickets in response to varying risk and intensity of sperm competition: not all species play games. *Behav. Ecol.* **12**, 740–745

53. Gage, M. J. G. (1991) Risk of sperm competition directly affects ejaculate size in the Mediterranean fruit fly. *Anim. Behav.* **42**, 1036–1037.

54. Jivoff, P. (1997) The relative roles of predation and sperm competition on the duration of the postcopulatory association between the sexes in the blue crab, *Callinectes sapidus*. *Behav. Ecol. Sociobiol.* **40**, 175–185.

55. Shapiro, D. Y. *et al.* (1994) Sperm economy in a coral reef fish, *Thalassoma bifasciatum*. *Ecology* **75**, 1334–1344.

56. Marconato, A. *et al.* (1995) The mating system of *Xyrichthys novacula*: sperm economy and fertilization success. *J. Fish Biol.* **47**, 292–301.

57. Marconato, A. and Shapiro, D. Y. (1996) Sperm allocation, sperm production and fertilization rates in the bucktooth parrotfish. *Anim. Behav.* **52**, 971–980.

58. Nicholls, E. H. *et al.* (2001) Ejaculate allocation by male sand martins, *Riparia riparia*. *Proc. R. Soc. Lond. Ser. B* **268**, 1265–1270.

59. Bellis, M. A. *et al.* (1990) Variation in rat ejaculates is consistent with the kamikaze sperm hypothesis. *J. Mammal.* **71**, 479–480.

60. Wedell, N. (1998) Sperm protection and mate assessment in the bushcricket *Coptaspis* sp. 2 (Orthoptera: Tettigoniidae). *Anim. Behav.* **56**, 357–363.

61. Wedell, N. and Cook, P. A. (1999) Butterflies tailor their ejaculate in response to sperm competition risk and intensity. *Proc. R. Soc. Lond. Ser. B* **266**, 1033–1039.

62. Linley, J. R. and Hinds, M. J. (1975) Quantity of the male ejaculate influences by female unreceptivity in the fly, *Culicoides melleus*. *J. Insect Physiol.* **21**, 281–285.

63. Lorch, P. D. *et al.* (1993) Copulation duration and sperm precedence in the stalk-eyed fly *Cyrtodiopsis whitei* (Diptera: Diopsidae). *Behav. Ecol. Sociobiol.* **32**, 775–781.

64. Martin, O. Y. and Hosken, D. J. Copula duration and ejaculate transfer are positively correlated in the common dungfly, *Sepsis cynipsea*. *Anim. Behav.* (in press).

65. Suter, R. B. (1990) Courtship and the assessment of virginity by male Bowl and Doily spiders. *Anim. Behav.* **39**, 307–313.

66. Cohn, J. (1990) Is it the size that counts? Palp morphology, sperm storage, and egg hatching frequency in *Nephila clavipes* (Araneae, Araneidae). *J. Arachnol.* **18**, 59–71.

67. Gage, M. J. G. (1998) Influence of sex, size and symmetry on ejaculate expenditure in a moth. *Behav. Ecol.* **9**, 592–597.

68. Parker, G. A. *et al.* (1999) Optimal copula duration in yellow dungflies: effects of female size and egg content. *Anim. Behav.* **57**, 795–805.

69. Yusa, Y. (1994) Factors regulating sperm transfer in an hermaphroditic sea hare, *Aplysia parvula* Mörch, 1863 (Gastropoda: Opisthobranchia). *J. Exp. Mar. Biol. Ecol.* **181**, 213–221.

70. MacDiarmid, A. B. and Butler, M. J. (1999) Sperm economy and sperm limitation in spiny lobsters. *Behav. Ecol. Sociobiol.* **64**, 14–24.

71. Baker, R. R. and Bellis, M. A. (1993) Human sperm competition – ejaculate adjustment by males and the function of masturbation. *Anim. Behav.* **46**, 861–885.

# PART II: CLASSIC READINGS IN HUMAN SPERM COMPETITION

# 4. HUMAN SPERM COMPETITION

Robert L. Smith[1]

## I. INTRODUCTION

Alexander (1979) has invited ". . . biologists to contribute to the analysis of human behavior on all legitimate fronts. . ." He considered it ". . . especially relevant that [they] take up the problem of relating human attributes to evolutionary history." The analysis of human sexual behavior surely qualifies as a legitimate topic in evolutionary biology. This chapter represents a contribution to the argument that sperm competition does occur in humans and has been a selective force in the evolution of certain human characteristics.

There has been considerable controversy over what may be the "natural" sexual inclinations (promiscuous, polygynous, serially polygynous, monogamous, or some mixture of these) of human males (*e.g.*, Trivers 1972; Wilson 1975; Alexander 1977; Short 1977, 1979, 1981; Daly and Wilson 1978; Symons 1979; Lovejoy 1981; Barash 1982; Harvey and Harcourt, this volume), but relatively much less debate over the sexual predilections of human females (Hrdy 1981). Females are widely assumed to be monogamous, with little formal recognition of alternative female strategies (but see Hrdy 1981, and Knowlton and Greenwell, this volume). The compromise view of human male mating strategy proposes mixed tactics (Trivers 1972) where males attempt to pair-bond with one or more females by high investment, and opportunistically (more or less promiscuously) mate with other females. All combinations of male tactics from rape (Shields and Shields 1983, Thornhill and Thornhill 1983) to high investment (Trivers 1972) and the environmental and social circumstances that occasion their expression have received analysis in the literature. As Hrdy (1981) observed: "The sociobiological literature stresses the travails of males – their quest for different females, the burdens of intra-sexual competition, the entire biological infrastructure for the double standard. No doubt this perspective has led to insights concerning male sexuality. But it has also

---

[1] Department of Entomology, University of Arizona.

Reprinted from *Sperm competition and the evolution of animal mating systems*, R. L. Smith (Ed.), New York: Academic Press. Smith, R. L. Human sperm competition. pp. 601–659, Copyright (1984), with permission from Elsevier.

effectively blocked progress toward understanding female sexuality – defined here as the readiness of a female to engage in sexual activity."

The biological irony of the double standard is that males could not have been selected for promiscuity if historically females had always denied them opportunity for expression of the trait. If strict monogamy were the singular human female mating strategy, then only rape would place ejaculates in position to compete and the potential role of sperm competition as a force in human evolution would be substantially diminished.

Here I shall explore the literature for evidence of the evolutionary significance of sperm competition in humans. I present data on the circumstances that would place ejaculates from different human males together in the reproductive tract of a female during a single reproductive cycle. I summarize the evidence that human sperm competition actually occurs. And, finally, I speculate on how selection within the context of potential or actual sperm competition may have operated in human evolutionary history to shape some aspects of human anatomy, physiology, behavior, and culture.

## II. HUMAN SPERM COMPETITION IN CONTEXT

Human sperm are motile for up to 7–9 days in the reproductive tract of the female (Morris 1977, Porter and Finn 1977). Therefore, any circumstance that places ejaculates from two or more males in the vagina or uterus of a female Within a 7–9 day period creates the potential for contested fertilization of her ova. The contexts that may cause human sperm to compete are communal sex, rape, prostitution, courtship, and facultative polyandry.

### A. Communal Sex

Communal sex is used here as a collective term to include all of the consensual sexual arrangements among females and their primary mates that create contexts for sperm competition. Included are "orgies," "wife sharing," "wife swapping," and variations on these themes. This is a relatively insignificant context that seems to have received publicity greatly disproportionate to its historical and modern importance. The cross-cultural index shows 60.9% of indexed societies have no communal sex of any kind, and in the 35.4% that permit some form (Fig. IH), it is typically highly regulated, reciprocal, involves wife sharing only with close relatives, and/or results in some benefit to the consenting male spouse. In only 3.6% of societies was there a high level of permissiveness for these activities (Broude and Green 1976). A 1970s survey (conducted during a period of sexual experimentation in the United States) on "swinging," *i.e.*, various forms of mate exchange, showed about 1% of respondents had engaged in these activities with any frequency, and only 3.5% had tried them once or twice (Athanasiou 1973). The infamous orgies of classical Greece and Rome apparently always involved prostitutes, and never wives (Bullough and Bullough 1978), so appropriately are included under prostitution rather than communal sex.

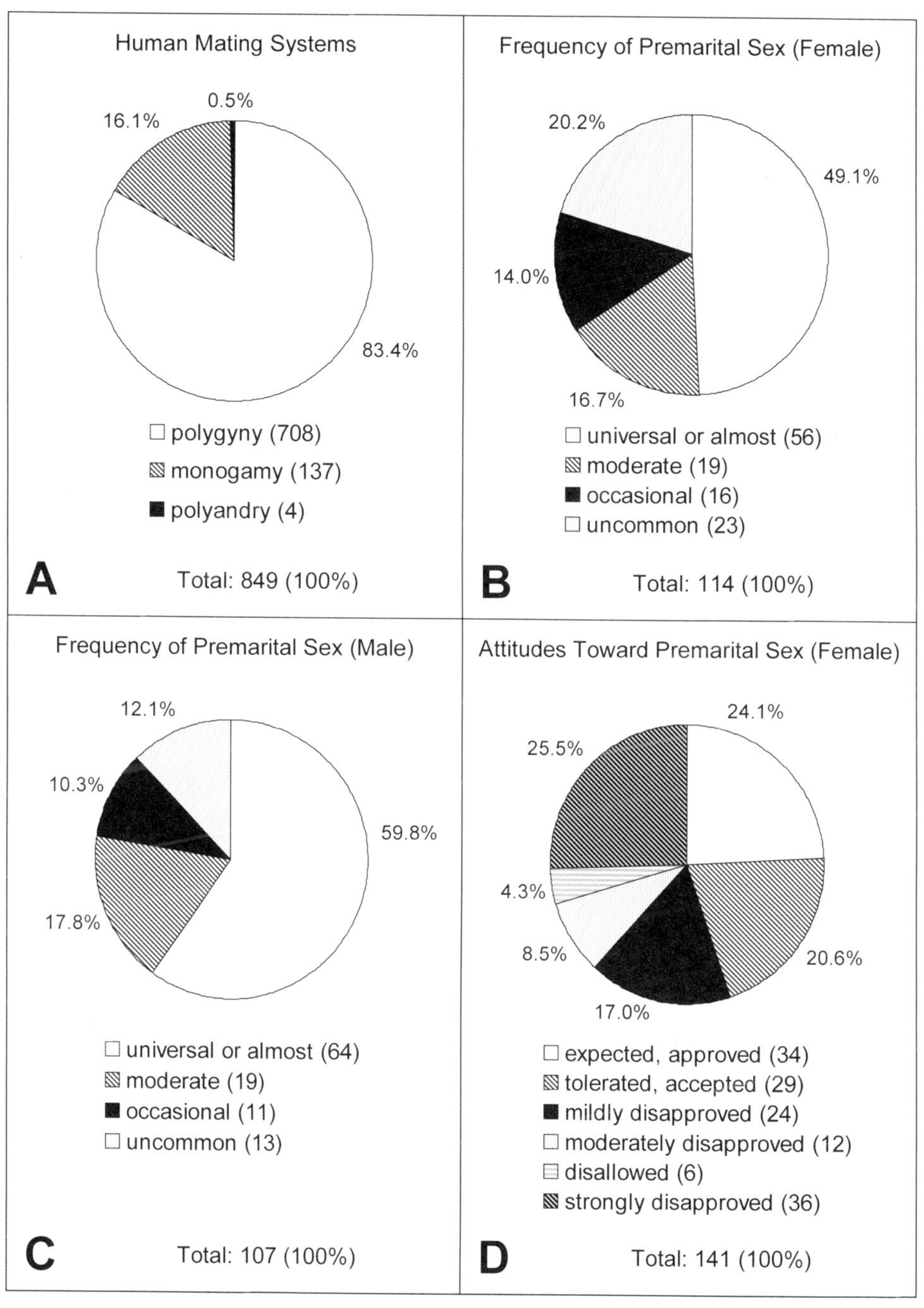


**Figure 1.** Cross-cultural indices of human mating systems, and sexual practices and attitudes. A. Data from Murdoch (1967), B-F and H-K. Data from Broude and Green (1976). G. Data from Burley and Symanski (1981). LK. Data from Gaulin and Schlegel (1980).

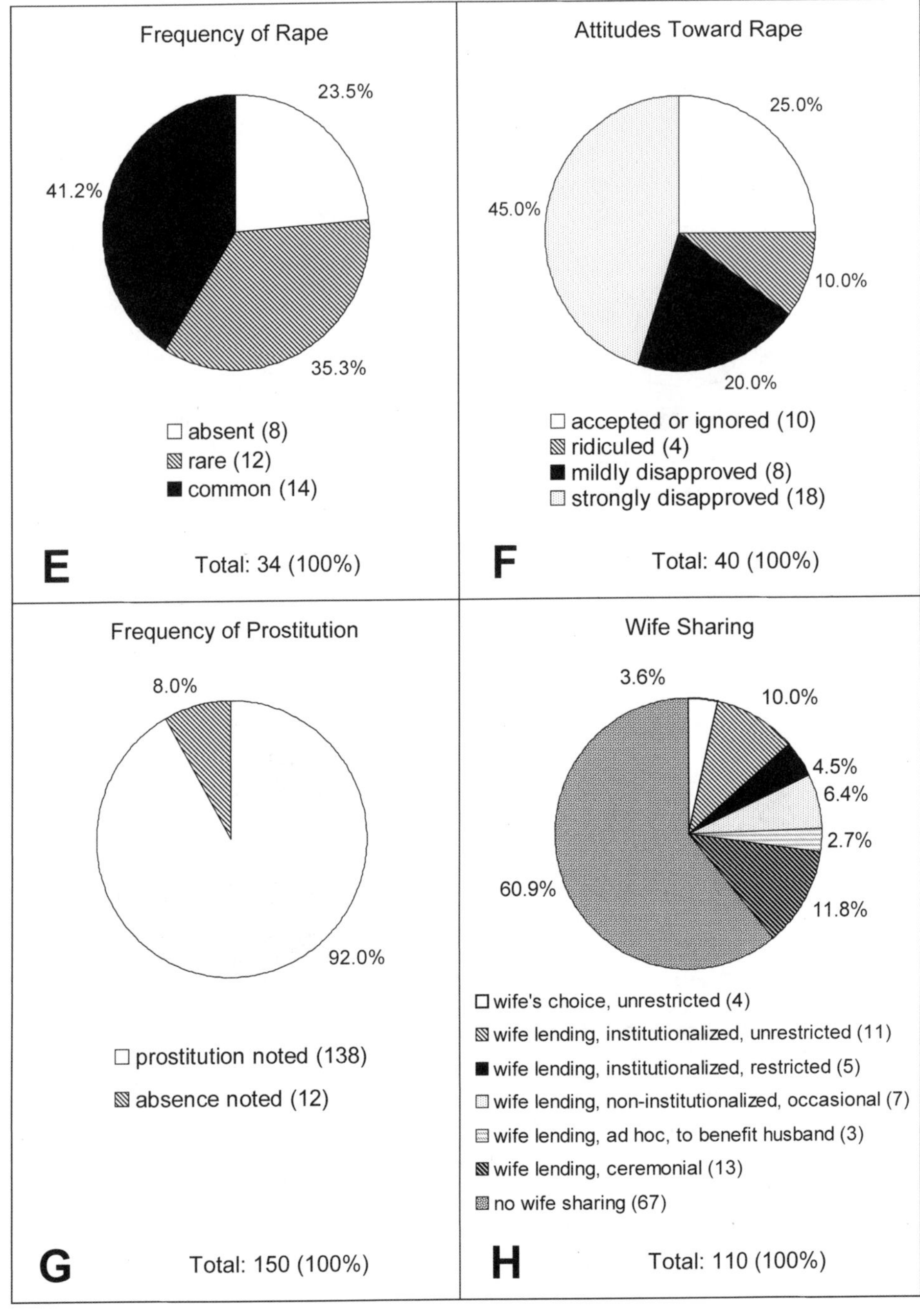


**Figure 1. (cont)** Cross-cultural indices of human mating systems, and sexual practices and attitudes.

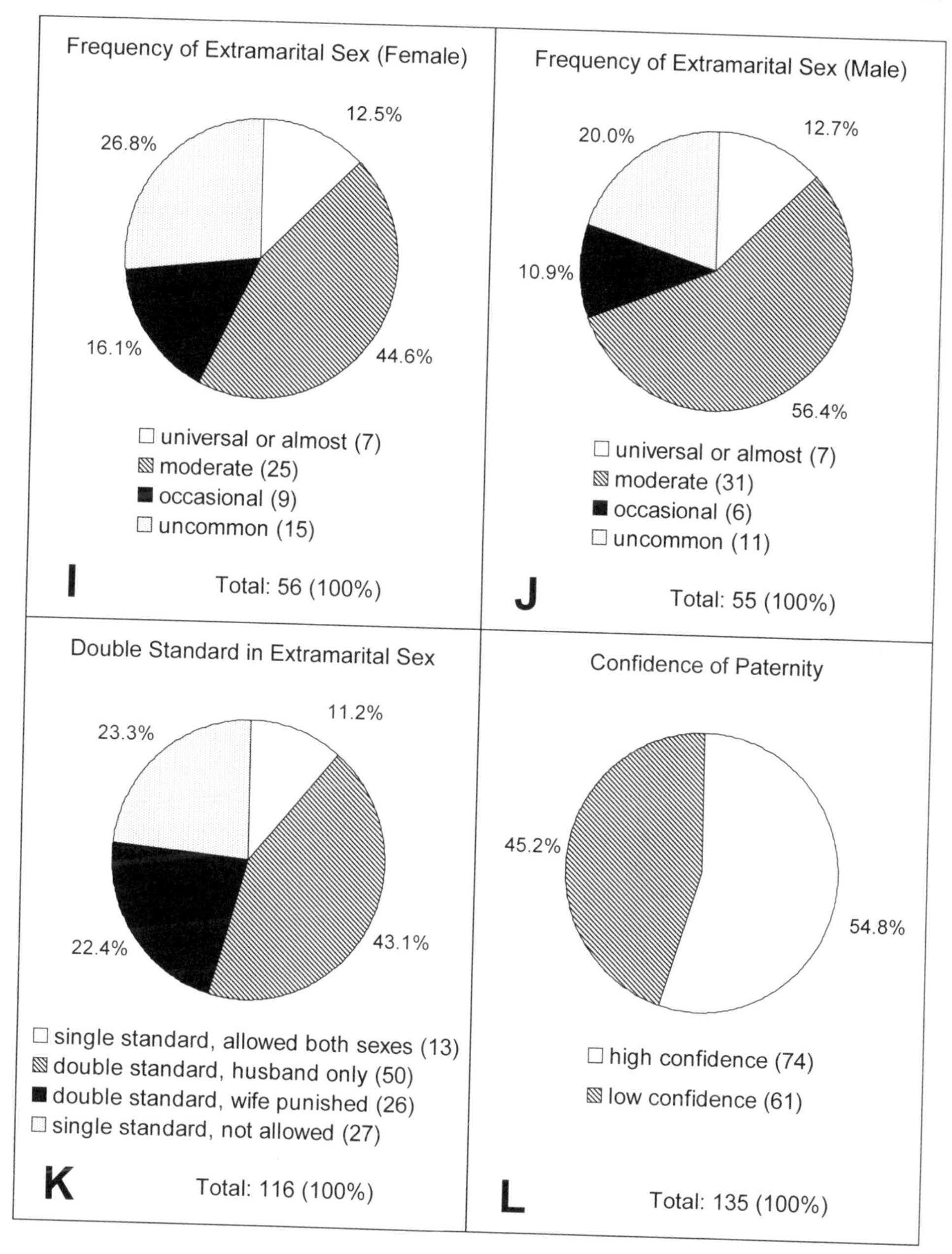


**Figure 1. (cont)** Cross-cultural indices of human mating systems, and sexual practices and attitudes.

## B. Rape

Rape is forced copulation of a female by a male (Thornhill and Thornhill 1983). Rape by an individual male (with ejaculation into the victim's reproductive tract) of a pair-bonded female will usually place the rapist's sperm in competition with that of her principal mate. "Gang rape," *i.e.*, sequential copulations forced on a female by several males in succession, probably creates the highest conceivable levels of ejaculate competition.

Broude and Green (1976) found rape to be common in 41.1%, uncommon in 35.3%, and absent in 23.5% of indexed societies (Fig. IE). Sanday (1981) broadly classified 95 societies, finding 45 to be "rape-free" and 17 to be "rape-prone." She concluded that rape was a cultural phenomenon; however, her criteria for identifying "rape-free" societies and her conclusions have been challenged (see Shields and Shields 1983). More than 60,000 rapes are **reported** in the United States each year, yet it is estimated that this figure represents only 10% of the actual cases (Green 1980).

Frequency of rape is highest during wars. Rape has been a universal aspect of human conflict from tribal battles to world wars, and has occurred throughout history. Brownmiller (1975) reviews the history of rape and provides an exhaustive assemblage of modern anecdotes as well as some statistics. For example, during the 9-month West Pakistani occupation of Bangladesh in 1971, between 200,000 and 400,000 rapes took place.

Rape is widely distributed and occurs with sufficient frequency to be considered a significant context for human sperm competition. Although it is not obvious that cultural evolution has progressively reduced the incidence of rape through time, I assume (as apparently Alexander and Noonan [1979] do also) that it was probably a more important sperm competition context in early human history, and relatively much more important in human prehistory.

## C. Prostitution

Prostitution is indiscriminate sexual activity for profit. Of specific importance here is the ad hoc sexual intercourse by females with males in exchange for resources as an agreed-to precondition for the intercourse. Because prostitutes may copulate with many men in a single day, the practice has potential for creating very intense sperm competition.

Prostitution definitely occurs in 142 of ca. 300 societies coded in the Human Relations Area Files (HRAF) and is explicitly excluded in only 12 of the 300 (Fig. IG). Censuses of prostitutes in various societies are not available generally, but some estimates have been accumulated by Symanski (1981). For the United States in the last decade, estimates ranged from 100,000 to 550,000 (Winick and Kinsie 1972), and Symanski presumed 250,000 to 350,000 full. and part-time prostitutes a credible estimate in 1981. In 1957, Poland allegedly had 230,000 prostitutes, Budapest ca. 10,000, and Tokyo more than 130,000. In the late 1960s, Rome was said to have had 100,000 prostitutes. A recent study reported 50,000 registered prostitutes in West Germany, and estimated 150,000 to work illegally. The estimate of 80,000 working prostitutes in Addis

Ababa, Ethiopia, in 1974 represented ca. 25% of the city's adult female population (Dirasse 1978), and illustrates the extremely high levels of sperm competition due to prostitution under certain socioeconomic conditions. Marco Polo found over 20,000 women living as prostitutes in Peking, and "too many to estimate" in Hangchau in the 13th century (Parks 1929). Use by males of female prostitutes was routine and almost universal in ancient Greece and Rome (Bullough and Bullough 1978). Kinsey *et al.* (1948) found that 69% of white males surveyed had had experience with prostitutes, and 15% used prostitutes regularly.

Commercial sex has had a long and colorful history (see Bullough and Bullough 1978), and though it is currently practiced by relatively few females in most societies, I shall assume that some form of prostitution has been an important context for sperm competition in human prehistory.

## D. Courtship

In cultures that permit adolescent females to choose their primary mates, the time between puberty and marriage may represent a brief period of relatively dynamic sexual activity, and concomitant potential for sperm competition. This is because the unmarried adolescent girl may not yet be under the control of any one male who has a reproductive interest in her, and hence may be free to sample the population of prospective long-term mates. Unmarried !Kung couples, for example, often leave their village to have intercourse in the bush, and when individuals find partners they like based on these experiences, they may marry (Howell 1979). Adolescent promiscuity (under the guise of courtship by both sexes) is well-known and generally tolerated in Western culture, and the cross-cultural index (Fig. IC) reveals that 44.7% of indexed societies approve of premarital sex or tolerate it without punishment. Another 25.5% disapprove only mildly and administer token punishment to violators, for a cumulative total of 70.2% lenient societies. Female premarital sex occurs in at least 79.8% of indexed societies (Fig. IB). Males (in a revealing discrepancy) from at least 89.9% of indexed societies are reported to engage in premarital sex (Fig. IC).

Some U.S. schoolgirls may use multi-male mating as a primary mate acquisition strategy (N. Chagnon, pers. comm.; see section V.B). An adolescent female may copulate repeatedly with several youths until impregnated, then name as father the favored phenotype (or best potential provider) from among the lot. Malinowski (1929) reported that Trobriand Island females were free to choose their lovers as an adolescent privilege but after marriage, female adultery was an offense that could be punished by death.

I contend that the period of female courtship is a common context for relatively intense human sperm competition, but both its historic and later prehistoric importance may be limited by its short duration in each female's reproductive life.

## E. Facultative Polyandry

The typical human female has one principal mate from whom she and her children receive care, protection, and material resources. Marriage (socially recognized bonding of a female to her principal mate) is a cross-cultural universal that applies virtually to all members of society (Daly and Wilson 1978, Hawthorn 1970). The male benefits in this arrangement by having routine sexual access to the female, and thus a consistent competitive advantage over that of any facultative mates, for his sperm in the fertilization of her ova. Overt long-term polyandry is extremely rare (Fig. IA); it seems to occur only under conditions of limited resources and most usually involves brothers sharing one wife (Aiyappan 1935, Kurland 1979, Beall and Goldstein 1981, Gowaty 1981, Hughes 1982). The universal convention of female monogamous pair-bonding is almost certainly an early hominid adaptation (Lovejoy 1981) that has served to limit the occurrence of sperm competition.

When a female voluntarily copulates with a male or males other than her primary mate, she is practicing facultative polyandry. Facultative polyandry may place ejaculates of extrabond mates at competition with that of the primary mate and/or cause competition among sperm of extrabond mates.

This context for human sperm competition almost certainly exceeds all others in importance. Female extramarital sex occurs in 73.2% of indexed societies and is common in 57.1% (Fig II). The cross-cultural occurrence of male and female extra-marital sex is surprisingly similar (Fig. II). Gaulin and Schlegel (1980) found low levels of paternity assurance in nearly half ($N = 61$) of the 135 societies they examined. Presumably, sperm competition is especially intense in these societies. Kinsey *et al.* (1953) found ca. 26% of U.S. adult females had engaged in extramarital intercourse, and over 50% of respondents in the *Cosmopolitan* survey (Wolfe 1981) indicated that they had had extramarital liaisons. Although similar data are not available for other cultures, it seems safe to assume that female extrabond sex occurs in all but the most restrictive societies, and that it is common.

## III. POTENTIAL FEMALE BENEFITS FROM FACULTATIVE POLYANDRY

Why should a female mate with males other than a primary mate? Parker, Knowlton and Greenwell, and others in this volume have enumerated the benefits that could accrue to a female of any species by multiple mating. These include: (1) good genes, (2) sons' effects (both somatic and gametic), (3) genetic diversity, (4) fertility backup, (5) material resources, and (6) protection for self and offspring. To this might be added another, perhaps uniquely human, benefit: enhancement of social status (Symons 1979, Barash 1982, Dickeman 1978). Proximate psychophysiological benefits, such as comfort and sexual pleasure, are not considered here except to the extent that they may represent mechanisms selected because they facilitate evolutionarily important results.

The good genes, sons' effect, and genetic diversity benefits have in common that all require extrabond mating to supply genetic material somehow superior to that available exclusively from a female's primary mate. All of these possible benefits are controversial and assailable on theoretical grounds.

### A. Good Genes

Most societies support polygyny (Fig. IA), but there are always constraints on the number of wives that even a "very high quality male" may effectively maintain. Consequently, some females will always have to settle for primary mates that are of apparently "inferior quality." This problem is intensified in cultures that prohibit polygyny.

A female who perceives her primary male to be genetically inferior to other males with whom she can casually mate may employ a mixed strategy, *i.e.*, accept resources from her primary mate and opportunistically copulate with an available superior male or males. The successful mixed-strategist female may deceive (cuckold) her primary mate into supporting a child or children fathered by the genetically superior male or males. This assumes that human females are capable of discriminating genetically-based differences among potential mates and deciding which phenotypes are superior.

The extent to which female mate choice may operate is a topic of current interest and no little controversy (see Bateson 1983; Parker 1983, this volume). Female intersexual selection is challenged on the grounds that if it occurred, it would quickly eliminate all of the variability in the male population. Others doubt that variation in human male status or ability to secure material resources has a genetic component.

The first of these objections would seem more credible if human females based mate-choice decisions on one or a very few simple aspects of the male phenotype such as physiognomy. However, physical appearance seems to rank relatively low among the characteristics females evaluate (Symons 1979). It is difficult to address the second objection, but I submit that there are many good human male phenotypes (and genes) when viewed against the backdrop of human culture and its occupational specializations. I do not doubt human females' abilities to discern especially good gene assemblages, but the task must be intellectually challenging. Processing mate-choice decisions was perhaps an important prehistoric challenge to human female intelligence.

In many cultures, however, choice of the first primary mate for a girl is made entirely by the parents and often exclusively by the daughter's male parent (Levi-Strauss 1969, Daly and Wilson 1978, Symons 1979). The parents' interests may, but probably will not be coincident with the daughter's. Thus a parent-offspring' conflict (see Trivers 1974) is created that may result in extramarital adventures by dissatisfied daughters.

Fickleness (mate choice ambivalence) is a characteristic that young males in Western culture attribute to the objects of their courtship. That human females have difficulty selecting primary mates when allowed the option is not surprising given the importance of the decision and the bewildering array of characteristics that must be comparatively evaluated. Female equivocation after a marriage is equally comprehensible, especially if the wife did not participate in the mate-choice decision or if her choice seems to have been a mistake. Such confusion may lead to facultative polyandry. Symons (1979) points out that polyandrous behavior may provide opportunities for the female to change principal mates.

### B. Sons' Effect

Sons' effect assumes that male children fathered by a particularly charming man may inherit paternal charm and thus a competitive advantage in seduction (see Weatherhead and Robertson 1979). likewise, the sperm production and sperm delivery systems of male children produced by paternal sperm with a competitive edge may deliver sperm exceptionally good at getting eggs in contests with gametes from other males. Recently, some theorists have turned their attention to the possibility of selection for daughters' effect (R. L. Trivers, pers. comm.).

### C. Genetic Diversity

Maintenance of genetic diversity is the evolutionary hedge against an unpredictably changing environment, and dynamic environments are responsible for retention of sexual reproduction in most animal populations (Williams 1966, 1975; Maynard Smith 1978). Human environments are dynamic and unpredictable. A small technological or cultural innovation, or a political event may dramatically alter human environment within a single generation, thus causing selection to shift favor from one phenotype to another or to suddenly reward a previously ignored or rejected behavioral morph. A recombinant perfect for computer programming may not have produced a good Pleistocene hunter, yet a modern person endowed with this set may secure large rewards, and possibly enhanced fitness. On a shorter time scale, male aggressivity may be considered in the contexts of war or peace. The highly aggressive male may be a better warrior than a more pacific type and thus win material rewards and differential reproductive privileges from a society grateful for his services during periods of conflict. In times of peace and social stability, with no acceptable outlet for his aggressivity, the warrior phenotype may commit criminal acts and be incarcerated and thus deprived of the opportunity to reproduce. Human females may benefit by diversifying the paternity of their children, especially during times of social instability when the future is relatively unpredictable.

### D. Fertility Backup

Failure to conceive may be the most compelling biological reason for facultative polyandry. It is estimated that 25% of couples in the United States have fertility problems, and in up to 35% of these, fault is with the male alone (Stangel 1979). A variety of factors may affect male fertility, including general health, diet, disease, stress, injury, frequency of ejaculation, genetic abnormalities or incompatibilities, and specific vaginal antibiosis. Male infertility may be temporary or permanent, depending on causes.

Failure of a female to conceive can be attributed to one of four general faults: (1) female sterility or reduced fertility, (2) male sterility or reduced fertility, (3) mutual sterility or reduced fertility, and (4) gametic incompatibility. In three of these four cases a female could benefit reproductively from extrabond mating. If extrabond mating is practiced, sperm competition will occur in all but the cases of absolute female or primary mate sterility.

Wives of polygynous males in West Africa are less fecund than those in monogamous marriages (Musham 1956; Dorjahn 1958, 1959; Isaac 1979). This may be attributable to the polygynous husband's rotational patterns among wives, or to reduced active sperm counts per ejaculate due to increased coital frequency. Whatever the cause, it makes facultative polyandry of greater potential benefit to the polygynously-married wife than to the monogamously-married one. It is possible that the effect of polygyny on female fecundity in the subject cultures may in fact be mitigated by facultative polyandry. Dorjahn (1959) notes that polygynously-married wives of the Temne are freer from supervision and are therefore more adulterous than monogamously-married females. Little (1967) implies this situation for the Mende of Sierra Leone as well.

### E. Material Resources

Prostitution is one extreme in a continuum of practices that capitalize on sex as a female service (see Burley and Symanski 1981). It probably has its origin in courtship gifting. Malinowski (1929) observed that Trobriand men regularly give their mistresses small presents; if nothing is offered, the mistress may refuse intercourse. It is said by the Bush Negroes: "The men who have the most success with women are always short of money" (Hurault 1961). Rich Hopi men are most successful in their "love affairs," presumably because Hopi women are in the habit of accepting money for providing sexual access (Titiev 1972). Among the Birom of northern Nigeria, it is customary for married women to have "lovers." The lover pays the husband for sexual access to his wife, and has commitments to the married mistress (Smedley 1976, in Daly and Wilson 1978). Pygmy men pay parents for the privilege of having affairs with their daughters (Turnbull 1965) Truk wives may be loaned to other men for material gain by their husbands (Kramer 1932). In a number of cultures, women not classified as prostitutes acknowledge that they use sex to secure supplemental income (Daly and Wilson 1978).

Nuptial gifting is common in many animal groups (Alcock 1984) and is probably a universal aspect of human courtship with its origins in distant prehistory (see section VIII). The first gifts were almost certainly high-quality food, probably meat. A good male hunter could give females excess meat in exchange for sex or sexual obligations. Females could likewise venture sex on a good prospect against the possibility of some future return in material resources. Males are more likely to share resources with females who cultivate investment by providing sex than randomly with all females. Males also may be inclined to provide resources to a married consort's children on the chance that some may be his. Female behavior in this regard represents a hedge against the possibility that a principal mate will be unable or unwilling to supply sufficient resources. In virtually all societies, females are economically dependent on males, and the extent to which a male is willing to support a female mate is a function of his perception of her reproductive value. Women of low reproductive value and high liability, such as nonvirgin girls, adulteresses, women of "low moral character," women past their reproductive prime, divorced or widowed women with children, and those afflicted with disease or deformity may be shunned by potential support providers, with the result that their only recourse may be to exchange sex directly and indiscriminately for economic return (Burley and Symanski 1981). The benefit from prostitution, for some females in many cultures, may simply be survival.

### F. Protection

Human males typically provide protection to their mates and children. Such protection may be delivered against all forms of exploitation (inter- and intra- specific) from rape to predation. This is an essential benefit to females from the pair bond, and female mate-choice decisions must be based in part on a male's perceived ability to protect.

A primary mate cannot always be available to defend his wife and children and, in his absence, it may be advantageous for a female to consort with another male for the protection he may offer. This arrangement could provide a female the advantage of secondary mate choice and simultaneously offer some protection against loss of choice by rape. Absence of the primary mate may create the opportunity and the need for extrabond mating. As in the case of resource sharing, a male may be inclined to protect the children of a married lover on the chance that his genes are represented among them.

Infanticide (see Hrdy 1979) occurs among some primates, including our closest extant relatives, the chimpanzees (*Pan*) and gorillas (*Gorilla*). In some cases, infanticide is apparently a male reproductive strategy that eliminates the dependent offspring of a competitor and reproductively releases the mother from lactational amenorrhea to the perpetrator (Hrdy 1979). Gorilla and chimpanzee females may reduce the probability of having offspring become victims of infanticide by mating with different males (see section VI). Among humans, infanticide is most commonly a parental manipulation tactic based on resource scarcity or sex preferences (Alexander 1974). However, there are occasional examples in several cultures of men killing children fathered by other men (Hrdy 1977), so there is at least the possibility that human females might likewise reduce infanticide under certain circumstances by obscuring paternity with multi-male mating.

### G. Status Enhancement

There is little question that human females prefer to marry high-status males (see van den Berghe and Barash 1977, Symons 1979, Trivers and Willard 1973), and in some cultures the tendency has become institutionalized (Dickeman 1979). Symons (1979) asserts that if no material compensation is given, selection can be expected to favor the female desire for high-status sex partners (as distinct from husbands) only if the status has a genetic basis and the female succeeds in conceiving by the high-status mate. I have addressed this possibility in section III.A, but another possible benefit is enhancement of the female's social position by association with a high-status male exclusive of any material or genic benefit he might supply. Liaison with a high-status but noncontributing male may permit the female temporary access to and possible opportunity in a higher social stratum (*i.e.* hypergamy). It may at least elevate her position in her own stratum. A hint of this effect may be found in attempts at high status imitation by occupants of lower social strata (Symons 1979).

## IV. POSSIBLE FEMALE COSTS OF FACULTATIVE POLYANDRY

My objective in the previous section has been to suggest benefits to females from facultative polyandry. These benefits may be considerable, but they may be offset by risks and costs.

Male sexual jealousy (see section VII.A.7) may result in punishment for female adultery that may include loss of resources, physical injury, or even death. Females must weigh the risks and consequences of discovery against the potential benefits to be derived from extrabond coitus. This model predicts that a female mated to a physically attractive, high status, wealthy, fertile, nurturant primary mate (if such exists) in times of societal stability, is not likely to engage in facultative polyandry. Also, the wife of a particularly brutal man, or of any man in a particularly brutal society, may reject facultative polyandry.

## V. EVIDENCE FOR HUMAN SPERM COMPETITION

There are hard data (albeit few) on sperm competition in humans. These come from (1) judicial proceedings of Western countries, supported by forensic genetics studies performed to exclude paternity; and (2) human population genetics studies conducted by geneticists and anthropologists. It is in the nature of these data that sperm competition is revealed only where the genes of contestant sperm other than those of putative fathers are expressed and can be detected. Therefore these data are conservative.

### A. Paternity Law and Forensic Genetics

The fact of legally contested paternity itself provides an indirect measure of probable sperm competition in Western culture. For example, courts in West Germany hear over 5,000 paternity cases each year (Spielmann and Kuhnl 1980); 4,200 paternity suits were filed in New York in 1959 (Anon, 1982); and 12,000 paternity suits were filed in Cuyahoga County, Ohio, between 1945 and 1959 (Whitlatch and Masters 1962).

A legal infrastructure including statutes, legal theory, legal specialists, and supporting forensic genetics laboratories, has grown of the need to adjudicate thousands of disputed paternity cases in Europe and the United States (Ellman and Kaye, pers. comm.). Use of blood test evidence in paternity actions did not begin in the United States until the 1930s. The first tests were based on the conventional ABO red blood cell antigens, and were relatively insensitive. At best the tests could only exclude ca. 14% of putative fathers who were not the biological fathers (Ellman and Kaye, pers. comm.). Later combination of ABO typing with other blood antigen systems succeeded in excluding about two-thirds of all falsely-accused defendants (Mendelson 1982) and, more recently, human leukocyte antigen (HLA) tests have produced exclusion probabilities of 95-99% (Terasaki 1978, Terasaki *et al.* 1978, Hummel 1979).

Presumably, few putative fathers would protest unless they had reason to suspect they were being cuckolded. It has been estimated, based on post-trial test results, that in ca. 40% of cases heard by the New York courts system, the defendant is innocent of paternity (Anon. 1982). However, in only about 10% of New York's paternity cases have the accused requested blood tests that may possibly exclude them; the reason seems to be the high cost of the tests which must be borne by the defendants. Sussman (in Anon. 1982) applied blood grouping tests *ex post judicum* to 67 cases that had been resolved in favor of the plaintiff. In each case the defendant had conceded paternity, yet Sussman's tests revealed that 18% of these men could not be the fathers of the offspring for whom they had accepted legal responsibility.

The defendant in most paternity cases is the primary or most conspicuous mate of the plaintiff and therefore may not credibly deny coitus with the mother around the time of conception. Defense and justification for paternity testing are therefore based on "multiple access" to (*excepto plurium concubentium*) or "promiscuity" (*excepto plurium generalis*) of the mother (Sass 1977). It is significant that prior to the availability of reliable paternity exclusion testing, the law in at least some judicial systems seems to have recognized the statistical advantage enjoyed by sperm of the principal mate, relative to that of the female's casual sex partners, in deciding paternity cases (Sass 1977).

The fertilization of separate ova in a single female by different males and the subsequent plural birth of half siblings (superfecundation) provides indisputable evidence of human sperm competition. More than a dozen cases of superfecundation have been reported since the early 1800s and several recent ones have attracted coverage by the popular media (Archer 1810, Sorgo 1973, Terasaki *et al.* 1978, Spielmann and Kühnl 1980, Associated Press 1983). The most celebrated and rigorously investigated cases are those discovered by Terasaki *et al.* (1978) and by Spielmann and Kühnl (1980). The latter case, reported in West Germany, was characterized by a conspicuous genetic marker: One of the fathers was a black U.S. soldier, the other a caucasian German businessman (Shearer 1978).

## B. Population Studies

Only three population genetics studies are known to illuminate levels of cuckoldry in humans, and none was done specifically to investigate confidence of paternity. Trivers (1972) reported a personal communication from the anthropologist Henry Harpending whose preliminary analysis of biochemical data on the !Kung indicated that about 2% of the children in that society were not the progeny of men to whom they were attributed. More recently, however, Harpending (pers. comm., and Howell 1979) has indicated that these paternity exclusions were statistically no more frequent than maternity exclusions, and should therefore be attributed to a small amount of error in labeling of samples. It seems likely, however, that the close conformity of actual to putative paternity in the !Kung is not attributable to fidelity, but rather to the fact that affairs usually lead to rapid divorce and remarriage of the participants (Howell 1979).

Neel and Weiss (1975) performed blood typing tests on 132 Yanomama children and their putative parents and found eight paternity exclusions. Considering that the typings would allow about one-third of true exclusions to go undetected, the researchers estimate that there would be four instances of undetectable nonpaternity in the 132 children. This yields an estimate that roughly 9% of children are fathered by males other than putative fathers.

Finally, an extensive study conducted by the University of Michigan Department of Human Genetics on a rural midwestern U.S. community allegedly revealed a ca. 10% discrepancy from putative paternity. This discrepancy may have been confined largely to first births, and was apparently primarily attributable to girls that "got pregnant" and then chose a husband from among the possible fathers (N. Chagnon, pers. comm.).

These meager population data are clearly insufficient to derive any generalizations, but it is remarkable that the discrepancy from presumed paternity in the !Kung and Yanomama were low, given the reported high levels of pre- and extra- marital sexual relations that characterize each culture (Howell 1979, Chagnon 1979). The results of these studies suggest that sperm competition adaptations, including marriage, successfully assure paternity for principal mates.

Recent advances in biotechnology may stimulate interest in and facilitate human population studies of this kind. Future investigators, whatever their primary objective, might well consider questions of paternity in their survey designs.

## VI. COMPARATIVE HOMINOID REPRODUCTIVE AND SOCIAL BIOLOGY

Although there is legitimate controversy on the exact relationship of humans and the great apes, it is generally agreed that the balance of evidence places *Pan* as our closest living relative, followed by *Gorilla* and *Pongo* (Pilbeam 1984). Schwartz (1984) offers an alternative, though weakly supported, view that sees the fossil *Sivapithecus*, *Pongo*, and *Homo sapiens* in the same clade, and envisions an earlier divergence of Pan, with Gorilla arising from the chimpanzee line.

Of interest here are not disputed relative positions of taxa, but rather the undisputed certainty that we are extremely close to three species having diverse ecologies, social systems, and sexually selected characteristics, and that the study of these relatives offers possibility for insights into the evolution of human characteristics. Short (1976, 1977, 1979, 1981) has contributed significantly by applying the comparative method (see Alcock 1984; Thornhill, this volume) to extant hominoids. Following Short, and in collaboration with him, Harcourt *et al.* (1981*a*) and Harvey and Harcourt (this volume) have used clever statistics and ingenious design to test predictions about the operation of selection on testis weight in the primates. Results of these studies are generally consistent with predictions based on species' mating systems.

I rely a great deal on the aforementioned work for this discussion. Tables I and II encapsulate anatomical and physiological aspects of male and female hominoid reproductive biology, and Table III assembles reproductive behavioral attributes for each species. Fig. 2 graphically represents sexual dimorphism and relative size of reproductive anatomical features in humans and the apes. These features are discussed below.

### A. Orangutan

Apparently because its food is randomly and sparsely distributed in space, the orangutan has evolved a solitary lifestyle. During estrus, orangutans form monogamous consortships, and pairs copulate about every two days over a period of 3–8 days. Females do not advertise estrus until a mate has been selected. Female mate choice is apparently paramount for, despite their much larger size, male orangutans cannot control females in their arboreal habitat. Forced copulation is not uncommon in free-living orangutans, but bonded females are successfully defended by their mates. So sparse is their distribution that a consorting pair may never contact another male during the average 5.4–day period of estrus. So there may be little opportunity for polyandrous mating even if bonded females were so inclined. And, since male orangutans make no parental investment, contribute nothing to the female (except limited defense), and are not reported to kill infants, there is little reason for a bonded female orangutan to mate with other males if her current consort is of high quality. This is a male dominance system, with insignificant incidence of sperm competition. The genital attributes of the species are consistent with expectation for negligible selection by ejaculate competition. The male has an extremely short unspecialized penis and very small nonscrotal testes that contribute low sperm concentration to low volume ejaculates. As is typical of male dominance polygyny, the male orangutan is much larger than the female (Fig. 2).

## B. Gorilla

Gorillas live in small bisexual groups that are dominated by an alpha (usually silverbacked) male. Within each group, the dominant male is responsible for all or nearly all of the conceptions (Harcourt *et al.* 1981*b*). Intragroup conflict is minimal. Female gorillas show a slight swelling of the external genitalia during estrus. Estrous females solicit copulations from the dominant male. The dominant male prevents subordinates from mating with fertile females with little violence, and allows them to mate with subadult and pregnant females. It is significant that immature and pregnant females submit to copulation by subordinate males and that alpha males allow it. This behavior may be adaptive in the context of the potential for adult male-perpetrated infanticide known to occur in gorillas (Fossey 1974, 1976; see section III.F, and Hrdy 1979, 1981). The absolute dominance of a single male within each gorilla group seems to militate against any significant sperm competition. Consequently, male gorillas, though much larger than females, have an unspecialized penis of about the same relative length as that of the orangutan and, like the orangutan, gorilla testes are small and nonscrotal. Mean ejaculate volume and sperm concentration for *Gorilla* are lowest of the hominoids (Table I). Copulation occurs about once every 2.5 hours during daylight for from 1-3 days. A dominant male gorilla may go for months or even as long as a year without mating with an estrous female. Short (1979) characterizes gorilla sexual behavior as ". . . a rare treat for these amiable vegetarians."

## C. Chimpanzee

Chimpanzees are reproductively extraordinary among the great apes. They are not strikingly dimorphic for size as are the other two species (Fig. 2), but male chimpanzees have enormous scrotal testes, proportionately about 5 and 10 times larger than *Pongo* and *Gorilla*, respectively, and a specialized penis more than twice as long as that of the much larger gorilla. Short (1981) has calculated that the testes of an average chimpanzee can sustain sperm production at a level that will produce at least four full-strength ejaculates/day, each containing several times the number of sperm in an average gorilla or orangutan ejaculate.

Female chimpanzees in estrus have greatly swollen external genitalia that advertise their receptivity. This display is visually discernible at a distance, and promotes the promiscuous mating typical of these animals. "Opportunistic mating" (Tutin and McGinnis 1981) may result in a female copulating as many as 50 times in one day with a dozen different males (Tutin 1975, in Hrdy 1979). This behavior would result in the highest levels of sperm competition of any hominoid species. Every male member of the group in turn is permitted sexual access by the estrous female, and no fights occur over coital privilege. Opportunistic mating is noncompetitive and even cooperative. It seems to promote group solidarity with probable defensive advantages for all participants. And, as noted for *Gorilla*, multimale mating may result in some future advantage for females in the differential survival of offspring when infanticide by adult males is a possibility (see Hrdy 1979, 1981 for reviews and discussion). Chimpanzee females are in estrus for about 10 days each month for 2-3 years prior to first pregnancy. Sexual swelling and

**Table I.** Male Reproductive Anatomy/Physiology in Humans and the Great Apes

| Characteristic | Human | Chimpanzee | Gorilla | Orangutan |
|---|---|---|---|---|
| Testis weight, g | 40.5 | 118.8 | 29.6 | 35.3 |
| % body weight | 0.079 | 0.300 | 0.031 | 0.063 |
| Scrotum[a] | highly developed, pendulous, hairless | moderately developed, pendulous, hairless, unpigmented skin | barely perceptible post-penial bulge, hair-covered | barely perceptible post-penial bulge, hairless skin, but concealed by hair |
| Penis length and other features[b] | 13 cm, thick, exposed when flaccid, conspicuous, no os penis, glans penis present | 8 cm, narrow, filiform, pointed at tip, concealed when flaccid, os penis present, no glans penis | 3cm, inconspicuous, conical, concealed when flaccid, os penis present, glans penis present | 4cm, inconspicuous, concealed when flaccid, os penis present, glans penis present |
| Seminal vesicles[b] | moderately large | extremely large | small | moderately large |
| Mean ejaculate volume[d] ml (range) | 2.5 (0.1–12.0) | 1.1 (0.1–2.5) | 0.4 (0.2–0.8) | 1.2 (0.2–3.6) |
| Mean sperm density[d] no./ml $\times 10^6 \times$ vol. of ejaculate (range) | 70 (0.1–600) | 548 (54–2750) | 162 (29–375) | 76 (10–165) |
| Mean sperm number / ejaculate[e] no./ml $\times 10^6$ x vol. of ejaculate (range) | 175 (0.01–7200) | 603 (5.4–6875) | 65 (5.8–300) | 91 (2–594) |
| Spermatozoa, qualitative characteristics[d] | pleiomorph for size, F- body visible on Y chromosome | F-body visible on autosomes | pleiomorphic for size, F-bodies visible on autosome and Y | F-body not present |
| Mean length, microns | 58.4 | 57.4 | 61.4 | 66.6 |
| Head length, microns | 3.5 | 2.9 | 7.1 | 3.8 |

[a]Harvey & Harcourt (1984). [b]Short (1979). [c]Graham and Bradley (1972). [d]Martin and Gould (1981). [e]Calculated.

**Table II.** Female Reproductive Anatomy/Physiology in Humans and the Great Apes

| Characteristic | Human | Chimpanzee | Gorilla | Orangutan |
|---|---|---|---|---|
| Ovary size, % body weight[a] | 0.014 | 0.010 | 0.012 | 0.006 |
| Vagina length, cm | ? | < 12.5[b] | ca. 70-80[b] | ? |
| Hymen | present[c] | absent, no evidence of homologue[b] | absent, no evidence of homologue[b] | absent, no evidence of homologue[b] |
| Breasts[a] | highly developed, pendulous, in nonlactating and lactating females from puberty for life | developed at the end of first pregnancy for duration of lactation | developed at the end of first pregnancy for duration of lactation | developed at the end of first pregnancy for duration of lactation |
| Sexual sign[a] | no sexual sign of any kind, cryptic ovulation | pronounced swelling of labia and circum-anal region with onset prior to menarche, lasting into pregnancy | slight perineal tumescence at time of estrus | perineal swelling absent during estrus, but present during pregnancy |
| Estrus period, days (range) | mean 28 (24–35)[d] continuously receptive[a] | mean 9.6 (7–17)[e] | median 1 (1–3)[f] | mean 5.4 (3–8)[g] |
| Menstrual cycle, days (range) | mean 28 (24–35)[d] | mean 37.3 (22–187)[h] | median 49 (36–72)[h] | median 30.5 (26–32)[h] |
| Lactational amenorrhea, months | 49.2[i] | 68.4[j] | 45.6[j] | ? |
| Orgasm | common[k] | not reported | reported[l] | not reported |

[a]Short (1979, 1981). [b]K. G. Gould (pers. comm.) [c]*Gray's Anatomy*. [d]Tortora and Anagnostakos (1984). [e]Tutin and McGinnis (1981). [f]Harcourt *et al.* (1981*b*). [g]Galdikas (1981). [h]Graham (1981). [i]Short (1976). [j]Nadler *et al.* (1981). [k]Fisher (1973). [l]Mitchell (1979).

**Table III.** Social and Sexual Behavior of Humans and Great Apes

| Characteristic | Human | Chimpanzee | Gorilla | Orangutan |
|---|---|---|---|---|
| Social system | highly social, family and larger social units[a] | moderate size, bisexual unit-group[b] | small bisexual groups with a single dominant "silverback" male[b] | solitary[b] |
| Mating system (male view) | monogamous or polygynous and opportunistically promiscuous[c] | promiscous and/or sequentially monogamous in consortships[d] | polygynous[e] | polygynous[f] |
| Mating system (female view) | monogamous or facultatively polyandrous; infrequently promiscuous[c] | promiscous and/or sequentially monogamous in consortships[d] | monogamous[e] | monogamous[f] |
| Frequency of copulation | 1-3.5 copulations/week, female continuously receptive[g] | mean 0.52 copulations/male/hour, female up to 50 copulations/day[d] | median 0.4 copulations/hour only when female in heat[e] | estimated 0.5/day during consortship lasting 5 days[h] |
| Courtship | extended courtship usually lasting several days at minimum, each coitus preceeded by a bout of "foreplay"[g] | male penile display and invitation directed to estrous females, duration < 1 min[d] | estrous female initiates by soliciting dominant (silverback) male[e] | consortship initiated by dominant male or by female, female choice exercised, short courtship[f] |
| Duration of copulation | variable, but usually extended period of foreplay and median 10 minute coitus (range 2 > 60 min)[i] | mean 7 sec (range 3–15 sec)[d] | median 96 sec[e] | mean 10.8 min (range 3–28 min)[f] |

**Table III (cont).** Social and Sexual Behavior of Humans and Great Apes

| Characteristic | Human | Chimpanzee | Gorilla | Orangutan |
|---|---|---|---|---|
| Copulatory position (female/male) | variable, but primarily ventral/ventral[j] | dorsal/ventral[d] | dorsal/ventral[e] | variable, but usually ventral/ventral[f] |
| Copulation, place and situation | variable, typically in sleeping place, privacy usually sort[j] | arboreal or on the ground[d] | on the ground[e] | invariably arboreal[f] |
| Copulation, time of day | variable, but usually nocturnal for pair-bonded individuals[j] | diurnal[d] | diurnal[e] | diurnal[f] |
| Rape (forced copulation) | common[k] | not reported | not reported | occasional |
| Male masturbation | common[g] | not reported in wild, common in captivity | not reported in wild | occasionally observed in juveniles[l] |
| Female masturbation | common[g] | not reported in wild | not reported in wild | occasionally observed in juveniles[m] |

[a]Wilson (1975). [b]Hamburg and McGown (1979). [c]Various sources (see introduction). [d]Tutin and McGinnis (1981). [e]Harcourt *et al.* (1981*b*). [f]Galdikas (1981). [g]Kinsey *et al.* (1948), Hunt (1974), Pietropinto and Simenauer (1977). [h]Harcourt (1981). [i]Age-dependent (male); Hunt (1974). [j]Alexander and Noonan (1979); Cross-cultural Index; Schlegel (pers. comm.). [k]Brownmiller (1975, Fig 1E). [l]Mitchell (1979). [m]Martin and Gould (1981).

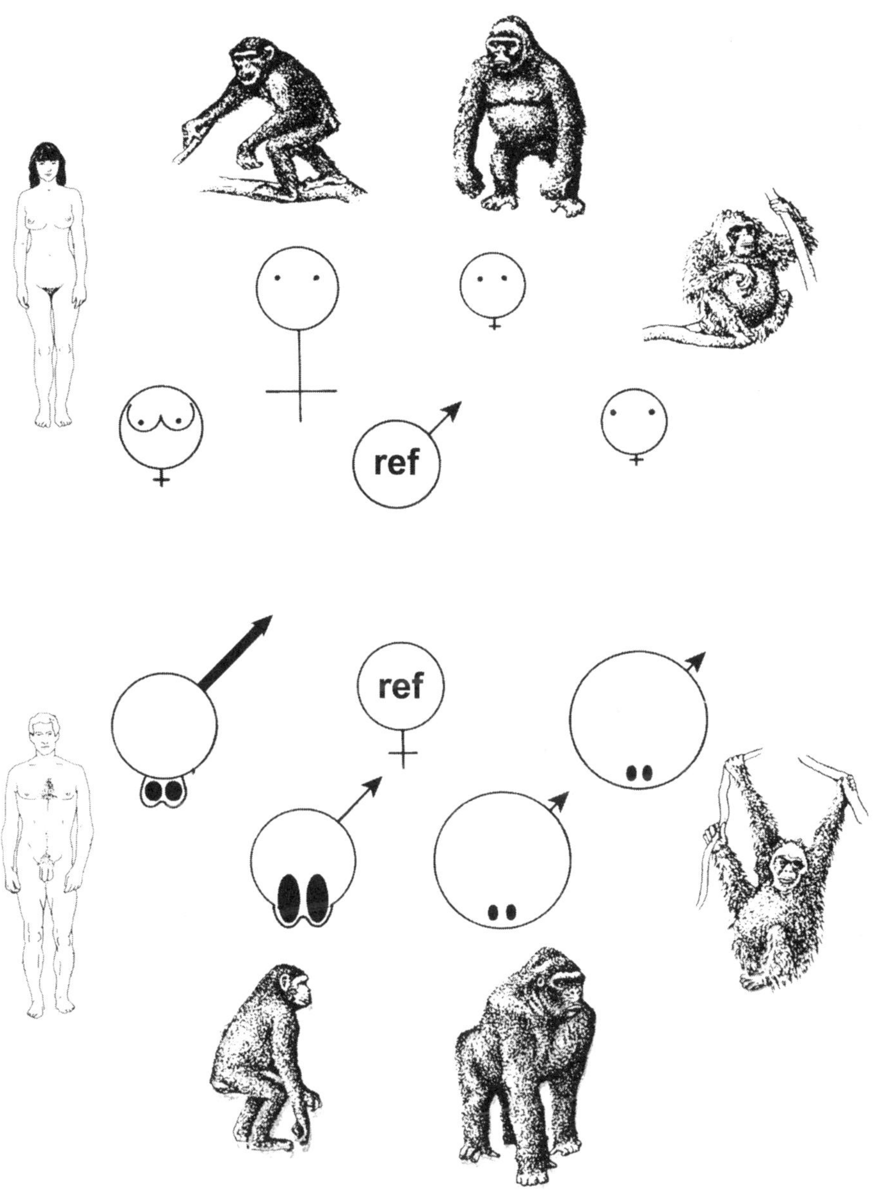


**Figure 2.** Diagrammatic representation of size dimorphism and relative sizes and features of sexual anatomy on humans and the great apes. Clockwise: human female, chimpanzee female, gorilla female, orangutan female, orangutan male, gorilla male, chimpanzee male, and human male. Figures labeled "ref" serve as size references for the figures of opposite sex that surround them. Adapted from Short (1977) and Halliday (1980).

receptivity begin before the menarche, and continue into the early months of pregnancy (Short 1979). These observations support the idea that female chimpanzees may use coitus for purposes other than obtaining sperm. I believe these conditions foreshadow cryptic ovulation and continuous receptivity in human females.

Chimpanzees have two mating strategies in addition to opportunistic mating (Tutin and McGinnis 1981). These are "possessiveness" and "consortship." Possessiveness is a short-term relationship initiated by a male in which he persistently attends to a particular female and restricts her copulations with other subordinate males. Such behavior confers an obvious competitive advantage on the sperm of the possessing male. Consortship contains the elements of possessiveness, but also involves the consort pair removing itself from the rest of the community. The male initiates consortship, but requires cooperation of the female. The advantage to the male is almost exclusive sexual access to "his" female, and greatly reduced sperm competition during the peak of her estrus. The female advantage is that of mate choice, shared premium food, and perhaps future special defense of her young. There are disadvantages as well. In isolation, both sexes are vulnerable to dangerous encounters with members of other communities. These encounters can result in death of the male and of the female's offspring. Also, both partners can expect hostile reunions with their own group when the consortship is terminated. Finally, the male who becomes involved in consortship gives up opportunistic matings with other females during the consortship.

## D. Humans

Fig. 2 shows similarities of *Pan* and *Homo*. Males of both species have large specialized penises and large testes. Human testes, though absolutely and proportionately smaller than those of the chimpanzee, are about twice as large as those of *Gorilla* and *Pongo*; chimpanzee and human males both have testes in scrota, in contrast to the peritoneal testes of the other two species. Finally, both human and chimpanzee males are only slightly larger than the females, while gorillas and orangutans are strikingly dimorphic.

Female chimpanzees and humans each have conspicuous, but different, anatomical features that distinguish them from unadorned orangutan and gorilla females. In the chimpanzee, the female feature of genital swelling is not present in the orangutan and is only slightly developed in the gorilla. Genital swelling is absent in human females, but they have pendulous breasts.

Behaviorally, chimpanzees and humans seem to share a preoccupation with sex compared to the relative indifference seen in thee gorilla and orangutan. Gorilla males not infrequently reject the advances of an estrous female, while chimpanzee males actively and frequently seek to copulate with any available and willing female. This tendency is also seen in human males, though usually mitigated by society. Although we do not know what apes think about, human males in the age class 12–19 report thinking about sex on average once every 5 minutes (Shanor 1978). Fortunately for the progress of civilization, this rate declines to from once every hour to only several times a day from age 50 on.

## VII. HUMAN ATTRIBUTES POSSIBLY EVOLVED IN THE CONTEXT OF SPERM COMPETITION

If sperm competition has been a significant factor in human evolution, we would expect human males to possess attributes to maximize the probability of fertilization by sperm from their every ejaculate, at the expense of those in competing ejaculates. Females, on the other hand, should have evolved characteristics to provide maximum control over the success in fertilization of sperm from competing ejaculates, and mechanisms that resist male control over their reproduction. This section explores human attributes that may have evolved by sperm competition. Where possible, I have tested characteristics by the comparative method, but for many intriguing elements, insufficient or inappropriate data make it impossible to do more than mention possible adaptation.

### A. Male Attributes

#### *1. Testes, Accessory Glands, and Ducts*

Human males have proportionately large testes relative to the great apes (Table I, Fig. 2). Spermatogenic capacity in the hominoids is a function of testicular weight. Short (1981) suggests that the size of the human testis (like that of the chimpanzee) must be related to high rates of copulation (relative to *Gorilla* and *Pongo*; see Table I). However, he does not mention sperm competition as a possible factor in the evolution of testis size.

Each gorilla ejaculate contains ca. 65 x $10^6$ sperm (Table I). Given a testes weight of ca. 30 g, daily sperm production for *Gorilla* is calculated to be 680 x $10^6$, which allows about 10 daily ejaculates with undiminished sperm densities (see Short 1981, for assumptions). The same calculations applied to the chimpanzee allow about 4.5 ejaculations per day with constant sperm counts, but each chimpanzee ejaculate contains almost 10 times the sperm found in a gorilla ejaculate. At ejaculation rates of greater than 3.5 per week, human sperm count begins to decrease (Freund 1962, 1963). At an ejaculation rate of 8.6 per week, human sperm counts are reduced to approximately that of *Gorilla*, but the average human ejaculate (at the 3.5/week rate) contains ca. 2.7 times the gorilla sperm count per ejaculate (see Table III). In other words, a gorilla with relatively small testes can copulate without reducing his sperm count, at over twice the frequency of a chimpanzee, and 20 times more frequently than a human male, although its absolute sperm production is much lower than that of the other two species. If these estimates are correct, selection for frequent copulation per se will not explain testis weight in hominoids. An alternative hypothesis is that increased testis weight evolved by sperm competition to produce ejaculates with larger numbers of spermatozoa (see Parker, this volume, and section VII.A.4).

Most other aspects of the male reproductive system have been studied in man, but not in the great apes. Data on the seminal vesicles of all four species are presented in Table I (from Short 1979). The size of the seminal vesicles follows the pattern for other genital organs, except in humans and the orangutan they are similar in size. Seminal vesicles in man contribute about 60% of semen volume. This secretion is alkaline and contains fructose; it apparently functions to nourish ejaculated sperm and raise the pH of the vagina. It probably affects sperm longevity and vagility. The prostate gland and the bulbourethral glands are well developed in both chimpanzees and humans, but apparently not in *Gorilla* or *Pongo* (Graham and Bradley 1972, Martin and Gould 1981). The epididymis is clearly exposed to selection in the context of sperm competition because of the sperm storage function of its caudal portion (see section VII.A.2), but there are no comparable morphometric data on this organ for the great apes.

Finally, the thickness of the muscle surrounding the lumen of the vas deferens is greater in relation to the lumen diameter than in any other tubular structure in man (Turner and Howards 1977). These muscles control the force of ejaculation and would therefore be subject to selection by sperm competition. It would be instructive to compare the thickness of vas deferens muscular layers among man and the great apes. Humans and chimpanzees should have a greater muscle to lumen ratio for this organ than *Pongo* and *Gorilla* if the theory is correct.

### *2. The Scrotum*

The scrotum, a thin pouch of skin and subcutaneous tissue that contains the testes of sexually mature males, is found in many eutherian mammals. It is a curious structure. Why should males of some mammal species and not others have been selected to move their testes from the relative protection of the abdominal cavity to a position of seemingly extreme vulnerability? This question is especially intriguing when applied to the great apes. Fig. 2 illustrates the relative positions of testes in the male orangutan, gorilla, chimpanzee, and human. The orangutan has a postpenial bulge of bare black skin under which the testes reside. A similar arrangement is found in the gorilla, but the postpenial bulge is covered with hair. The chimpanzee has a true pendulous scrotum that is hairless and unpigmented. It is exceeded in its pendulosity by the scrotum of human males.

Hypotheses advanced to explain the scrotum range from the mechanistic view of Woodland (1903) that the weight of maturing testes wears a hole in the lower abdominal wall, to the suggestion that the scrotum has evolved as an epigamic display (Portmann 1952). Since the scrotum is found only in mammals, and mammals are homeotherms, it has been proposed that the scrotum may function to maintain testicular temperatures lower than that of the body cavity, and this effect has indeed been demonstrated. Scrotal testicular temperature is typically lower (from 2–6°C, depending on species) than the abdominal cavity (Carrick and Setchell 1977).

Bedford (1977), in an insightful paper, takes issue with all previously proposed hypotheses and garners impressive support for an alternative explanation, *i.e.* that testicular migration to the scrotal position has been driven by progressive selective advantage in epididymal sperm storage. Fig. 3A–C depicts a possible evolutionary. progression. It shows migration of the sperm-storage region of the epididymis preceding descent of the testes into the scrotum. The reverse is illustrated by a nonexistent arrangement depicted in Fig. 3F. Scrotal topographies in a variety of species over diverse taxa are apparently patterned to ensure cooling of the epididymal sperm-storage area rather than the testes. Typically, the pattern involves abrupt hairlessness at the lower border of the scrotum adjacent to the cauda epididymis where mature spermatozoa are stored. Experiments with rabbits and rats, both scrotal mammals, in which the epididymis was surgically reflected into the abdomen, produced no effect on sperm maturation but dramatically altered the longevity of stored sperm. Rat sperm in the scrotal cauda epididymis remained motile for up to 42 days and viable for 21 days. In the rabbit, viability of scrotal stored sperm was 35 days. Sperm retained by ligatures in abdominal epididymis remained viable for only 3–4 days in rats, and 8–10 days in rabbits.

The sperm storage hypothesis to explain the scrotum is intuitively satisfying. Mammalian sperm that compete would be selected to possess metabolic enzymes that operate most efficiently at female deep body temperatures. However, sperm that contain such enzymes are likely to store much better at lower temperatures because of enzyme deactivation and conservation of metabolic substrate. Conflicts between storage and operational efficiencies of sperm in a system subject to selection by sperm competition could have driven the migration of the sperm storage organ to a scrotal position. Sperm of mammals with negligible sperm competition may not evolve high temperature adapted enzymes, therefore causing no selection for migration of the epididymis into a scrotum.

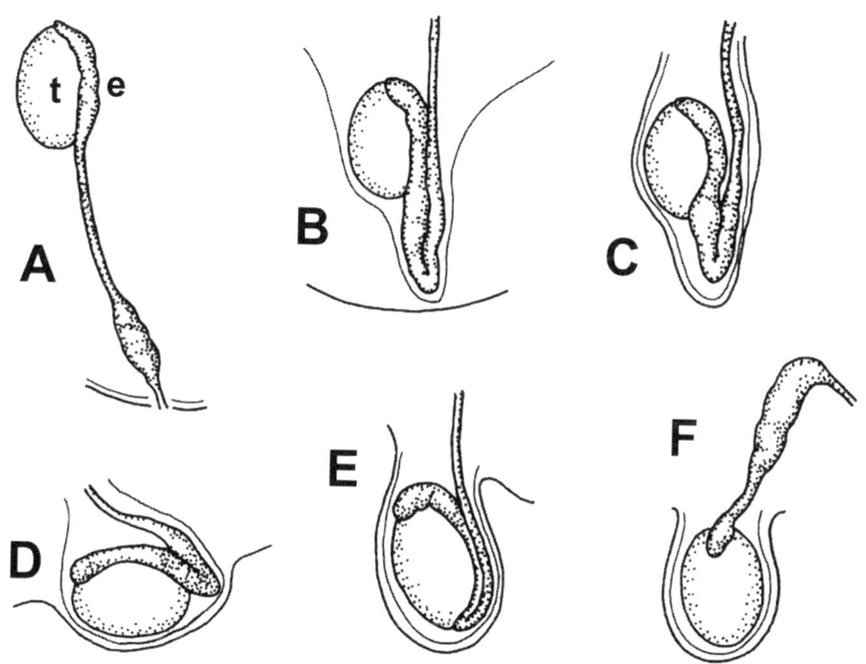


**Figure 3.** A–C and D. Diagrammatic examples of eutherian mammal testes showing a possible evolutionary progression to the scrotal condition, led by migration of the epididymis. E. Human. F. Hypothetical arrangement expected if testis function were the only important element in the evolution of the scrotum. Adapted from Bedford (1977).

Bedford (1977) suggests that the scrotal state should be found in mammal species where "the male must be able to produce fertile ejaculates repeatedly over a relatively short period." He emphasizes polygyny as a mating system that would create these demands on the male system, but fails to identify competitive mating and sperm competition as a context. I propose that sperm competition drove the evolution of the scrotum in *Homo* and *Pan*, and note that this hypothesis would also account for the lack of a scrotum in *Pongo* and *Gorilla*, species that are polygynous, but without intense sperm competition.

*3. The Penis*

The size of the human male penis is extraordinary relative to other hominoids. It is nearly twice as long and over twice the diameter of the chimpanzee penis (Table I, Fig. 2). The mean length of the adult human penis is 13 cm (range = 11–15 cm). Penis lengths measure 3 cm, 4 cm, and 8 cm in the gorilla, orangutan, and chimpanzee, respectively (Table I, and Short 1980).

Short (1980, 1981) and Halliday (1980) review speculation on why so large an intromittent organ has evolved in humans. Speculation as to function include: aggressive display (intrasexual selection), attractive display (epigamic selection), and facilitation of a variety of copulatory positions to enhance female sexual stimulation during coitus. To these proposed functions, I add what would seem to be the most obvious: delivery of ejaculate as close as possible to ova.

The display theses are not compelling. Penis size is inherited independently of other aspects of human male physiognomy (Schonfield 1943). So the idea that the human penis is used in aggression raises the improbable scene of a large muscular man with a small penis cowering at the penile display of a “well-endowed” shrimp. There are no anthropological data to suggest that the human penis is used in aggressive display between males, and no evidence that human males experience erection in agonistic encounters.

It seems equally unlikely that the large human penis evolved by epigamic selection for display, or because larger penises enhance female enjoyment of intercourse. In a survey conducted by the *Village Voice* (Diagram Group 1981), women were asked to select the anatomical features they admired in men. Only 2% indicated an interest in penises (buttocks ranked highest at 39%). An indirect measure of female indifference to penises is the failure of U.S. magazines featuring photographs of nude males to secure a substantial female readership. This contrasts dramatically with the male market for photographs of nude females (featuring genitalia) that supports a dozen or so periodicals in the United States alone (see Symons 1979). The work of Masters and Johnson (1966) and results of recent surveys (Hite 1976, Wolfe 1981) have shown that penis size has little to do with female enjoyment of coitus and the achievement of female orgasm. The idea that the large human penis evolved to permit a variety of copulatory positions is tenuous at best (see Table III). Orangutan males have penises that average 4 cm, and they regularly copulate in the ventral/ventral position. Chimpanzees have 8 cm penises that would probably permit ventral/ventral coitus, but they always adopt the dorsal/ventral position. The ventral/ventral position is cross-cultural in humans and apparently can be achieved by males having penises through the full range of sizes (Masters and Johnson 1966). Finally, Symons (1979) observed that if females had the tendency to be sexually aroused by the sight of male genitalia, men would be able to obtain copulations by genital display, as is apparently the case with chimpanzees; however, human females seem to be singularly unimpressed and decidedly unaroused by "flashers."

An excessively large penis produced by epigamic selection would be no less effective in delivering sperm than one of optimal length, but a substantially shorter (than optimal) penis would obviously place its owner's ejaculates at a disadvantage in competition with those deposited by a longer organ. Therefore it may be possible to test some predictions of the ejaculate delivery/sperm competition hypothesis for human penis length and distinguish this function from one of display.

If human penis length has been selected to deliver sperm as close as possible to ova descending the female tract, then length should not on average exceed the mean depth of the vaginal column (of the sexually aroused female) from labia to cervical os, and there should be racial differences in vaginal column and penis length consistent with this prediction.

If penis length has been selected for display, then the organ may on average be longer or shorter than the vagina, and its length or proportional length perhaps randomly distributed among races, depending on the intensity of epigamic selection from race to race. Epigamic selection intensity might be greatest in the tropics where sparse clothing would permit the community of females to regularly view male genitalia. However, in this regard, Short (1980) notes that penis sheaths were worn out of modesty and decorum rather than for display by the Telefomin of New Guinea. Unfortunately, there are no genitalic morphometric data that would permit interracial comparisons.

Short (1979) also argues that the human penis probably has a display function because the flaccid organ is exposed outside of the body cavity. An alternative hypothesis is that in attaining its present length by sperm competition, it lost the os penis (penis bone) and the protection of an abdominal prepuce.

The chimpanzee penis is pointed at the apex and differs from that of the other apes and man in lacking a glans. The specialized apex may permit the penis tip to penetrate the cervical os; gross anatomy of the female tract suggests that this is possible (Gould and Martin 1981). Although there are no data on chimpanzee vaginal length, K. G. Gould (pers. comm.) uses a 12.5 cm speculum to dilate chimpanzee vaginas for artificial insemination. This is longer than the 8 cm mean penis length in these animals, but if the cervix dips a couple of centimeters into the vagina, and the speculum is not inserted completely, the fit could be made.

In the absence of selection by sperm competition or female preference, penis length should stabilize at that which consistently transfers an effective ejaculate, and would likely be less than optimal length because of compromising selection. *Gorilla* and *Pongo* are candidates for testing this prediction.

### *4. Semen*

Seminal fluid is not essential for effective fertilization. Spermatozoa taken directly from the vas deferens will fertilize ova (Johnson and Everitt 1980). So, although many constituents of human seminal plasma have been identified, and some of the more obvious functions for major constituents established, there is yet much left to be learned. Human reproductive physiologists apparently have not considered the possibility that accessory gland secretions may function in the context of sperm competition. In this regard, prostaglandins, so named because they were originally discovered in extracts from the human prostate gland, are known to cause strong uterine smooth muscle contractions (Eliasson and Posse 1965, Pickles 1967). Much work has been done to identify the prostaglandin content and secretions of the reproductive systems of humans. The concentrations of "E" prostaglandins (PGE) in human semen have been found to be about 100 times greater than found in all other reproductive tissues (Bergstrom 1973). Goldberg and Ramwell (1977) observe that "PG's in the male tract occupy a somewhat existential niche in biology: although everyone admits that they exist, no one knows why." Bygdeman *et al.* (1970) found that 40% of infertile men whose infertility could not be attributed to another cause had significantly lower PGE levels than fertile men. I propose that these compounds are produced and incorporated into ejaculatory plasma as a means for controlling the female reproductive tract by causing uterine contractions that move the sperm in an ejaculate toward the ovum (see section VII.B.4). This intriguing possibility is reminiscent of factors found in *Drosophila* ejaculates that impose some control on female physiology and behavior (see Gromko *et al.*, this volume). If this is a function, presence and concentration of prostaglandins would be sensitive to selection by sperm competition. In this regard, it would be useful to compare prostaglandin titers among the ejaculates of the apes and man.

A large seminal plasma volume might secure some competitive advantage for the sperm it contains by providing optimal environment and nutrients for spermatozoa, and/or by debilitating alien sperm. Human males have not only the largest mean ejaculate volume, over twice as large as that of any of the great apes, but also volumes that range to 12 ml, more than four times the volume of any ape species.

Martin and Gould (1981) compare the morphometrics and gross anatomical features of spermatozoa among the apes and man (see Table I) and reveal similarities among them. No currently available data suggest the differential operation of selection on the anatomy of any species of hominoid sperm. Future comparative ultrastructural studies may, however, reveal unique adaptations. Acrosomal chemistry may be selected for egg penetration efficiencies in the final context among competing spermatozoa (see section VII.B.5).

We know from artificial insemination technology that a small fraction of sperm contained in an ejaculate can achieve fertilization when accurately placed at the proper time. So why do ejaculates contain so many sperm? Parker (this volume) shows that high sperm densities can evolve by sperm competition (the strategy being to overwhelm opponent sperm by sheer numbers). The comparative work of Short (1979) on humans and the great apes provides results that fit this model. Chimpanzees and humans have much higher levels of sperm competition relative to the other two species and, predictably, produce ejaculates that contain many more sperm than those of *Gorilla* and *Pongo* (Table I).

Foote (1973) notes that the quality (sperm density) of ejaculates obtained from bulls for artificial insemination is dependent on providing proper "teasers" to sexually stimulate the animal prior to ejaculation. He speculates that this factor may also be of "considerable importance" in humans. Is it possible that the human male is able to adjust sperm density levels to meet anticipated competitive challenges to his ejaculate? There are no data on this, but the well-known Coolidge effect (sexual habituation to an old mate, and arousal in response to the presence of a potential new mate; see Symons 1979 for full discussion) certainly demonstrates the psychosexual power of a novel stimulus, and portends the possibility of other physiological effects.

Masturbation seems a bizarre, maladaptive behavior because it wastes sperm, but the practice is almost universal among U.S. males and is particularly common in adolescents (Kinsey *et al.* 1948, Sorensen 1973). Masturbation is also commonly observed in some captive primates. It has been said that "primates masturbate, because they can," but if autoeroticism were maladaptive, the tendency should have been repressed by natural selection. A clue to a possible function is found in the fact that males who do not ejaculate by coitus or masturbation for some period of time experience nocturnal emissions, *i.e.* spontaneous seminal discharge during sleep. This suggests that stored sperm and/or accessory gland products may have a definite "shelf life" after which they are best discarded and replaced with new in order to stay competitive. Another possibility that has not been properly investigated is that frequent ejaculation may activate some feedback mechanism (*e.g.*, testosterone/inhibin) to stimulate higher levels of spermatogenesis and accessory gland secretion.

No studies are available that make comparisons of human sperm motility, longevity, time to capacitation, or other measure of performance that would reveal mixed strategies within an ejaculate, or specializations between ejaculates. Although there has been a great deal of work done on the causes of male infertility, I find no data that compare performance of sperm among ejaculates of fully fertile men. No published observations of sperm in mixed human ejaculates exist that might reveal incapacitation or selective spermaticidal effects as competitive tactics. Finally, none of these parameters has been studied in the great apes, so there is no basis for interspecific comparison.

### 5. *Mating Patterns*

Human males employ a variety of highly variable reproductive tactics, each mix depending on the opportunities to place sperm in competition and the need to defend against competing sperm. Symons (1979) devotes a chapter in his book to the human male desire for sexual variety, which he introduces with the following reproductive arithmetic: "A male hunter/gatherer with one wife (who will probably produce four or five children during her lifetime) may increase his reproductive success an enormous 20 to 25% if he sires a single child by another woman during his lifetime," Small wonder that selection has favored sex-crazed human males. The most common human male reproductive strategy involves attempts to monopolize sexual access to one or more women through pair-bonding. Maintenance of wives requires a regular outlay of resources, and failure of the male to provide adequately may result in his loss of sexual exclusivity and even possible loss of a wife.

As will be discussed (section VIII), marriage will on average produce higher male reproductive returns than any mix of other tactics that could be pursued at the expense of equivalent resources. Thus whatever mix of extramarital tactics a human male employs should be free or cost less than required to support another wife or equivalent (*i.e.*, mistress/concubine). Prostitution, for example, is an institution that provides males the opportunity to make small cost highly speculative reproductive investments. If marriage is a blue chip stock, then intercourse with a prostitute is the reproductive equivalent of purchasing a lottery ticket. A very few purchasers win big.

Shields and Shields (1983) and Thornhill and Thornhill (1983) have meticulously searched and analyzed the available data on rape with the purpose of seeking an explanation for its ultimate cause. Both teams found that rape has reproductive value, and is therefore subject to the forces of natural selection. There is no question that rape results in pregnancies. Best estimates are that 25,000 women became pregnant as a result of the hundreds of thousands of rapes that occurred during the West Pakistani occupation of Bangladesh (Brownmiller 1975), and several gynecologists recorded hundreds of conceptions following the many rapes that occurred in Germany in the three postwar years of 1945–1947 (Jöchle 1973).

Shields and Shields (1983) concluded that ultimately, rape should be expected to occur when its potential benefit (in production of extra offspring) exceeds its potential costs (in energy expended and risk taken). They agree with the socio-cultural view that as male hostility increases so likewise does the probability of rape. They contend that the external stimulus for rape is a sufficiently vulnerable female.

Shields and Shields (1983) suggest that warfare is a context that may precipitate rape by almost any man because the risks of detection and punishment are low, and hostilities are high. They find substantial anecdotal support (in Brownmiller 1975) for their contention that the extremely high frequency of rape during war is a function of more males raping, rather than simply some males raping more. Thornhill and Thornhill (1983) propose and supply much data in support of the position that rape is an evolved facultative reproductive strategy employed primarily by men who are unable to compete effectively for the status and resources necessary to attract and mate with desirable females. They further demonstrate that rape may be incorporated into a repertoire of other patterns, including low committal pairbonding and/or directing resources to sister's offspring (the avunculate).

There is apparently some evidence for unusually high rates of conception following rape (Fox *et al.* 1970; Jöchle 1973, 1975). Two theories attempt to explain this phenomenon. Fox *et al.* (1970) suggest that the release of adrenalin caused by fear and anger may induce contractions in the uterus that facilitate uptake of semen. Jöchle (1973, 1975) proposed that rape may induce ovulation in human females. Both the incidence of conception following rape, and these theses bear further investigation, especially in consideration of their apparent maladaptiveness for females.

Another reason rape may succeed as a reproductive strategy is that it often eliminates subsequent competition from sperm of the victim's primary mate. Genitalic and somatic injury and the psychological trauma that usually accompany rape minimize the possibility that the victim will soon be receptive to coitus (see Thornhill and Thornhill 1983). This leaves the rapist's sperm uncontested in the victim's reproductive tract. Furthermore, the rape victim's principal mate will very likely reject her or refuse normal sexual activity with her following sexual assault (Thornhill and Thornhill 1983). A substantial number of cultures take a surprisingly relaxed view toward rape (Fig. IF). This may be due to low incidence of rape, or an exceptionally low reproductive payoff potential for forced copulation. In any case, the study of these societies may offer significant new insights into cultural causes of rape.

*6. Coitus*

Aspects of copulation itself (*i.e.*, position, duration, time of day, and frequency) can be male-controlled and may therefore be optimized by natural selection and adjusted for circumstances to achieve maximum probability of fertilization.

Humans are able to copulate in a variety of positions, but most are variations on two basic positions, dorsal/ventral and ventral/ventral. Most primates, and other mammals for that matter, copulate in the dorsal/ventral position, but humans use primarily the ventral/ventral ("missionary") position. It is reported that females prefer this position because it produces maximum clitoral stimulation and probability of orgasm (Masters and Johnson 1966, Beach 1973, Fisher 1973). It is probably in the male's interest to have his mates experience orgasm, but it is not always in the female interest (see section VII.B.4). The ventral/ventral position does not, however, result in ideal placement of the ejaculate. Ejaculation in the ventral/ ventral position produces a semen pool below the cervix (Masters and Johnson 1966), and if the female remains supine following coitus, most of the seminal reservoir is unavailable to the uterus. In contrast, the dorsal/ventral position leads to appression of the male glans to the cervix at full penetration such that semen may be ejaculated directly through the cervical os into the uterus, producing an obvious advantage over vaginal ejaculation. In addition, the ejaculate would be aided by gravity in the dorsal/ventral position. Consequently, this position should be the unmitigated preference of males, but I find no useful data on male preferences to test this prediction. In a similar vein, males should prefer to ejaculate during coitus rather than by any other stimulation. Melanesian island couples of "East Bay" routinely engage in extended mutual heterosexual masturbation, but coitus always takes place just before male orgasm (Davenport 1977).

Human females require about 8 minutes of coitus on average to achieve orgasm with a range of from 1 to 30 minutes, and women may achieve multiple orgasms during extended coitus (Masters and Johnson 1966, Fisher 1973). If female orgasm facilitates uptake of semen and fertilization (as suggested below), or if it is important in mate retention, then males should be selected to choose coital positions and sustain copulation for a minimum period that would allow the typical female to achieve orgasm. As noted, humans usually copulate in the ventral/ventral position and average duration of coitus in humans is about 10 minutes, so the prediction is supported (Table III).

The preference of human males to copulate with their mates at night in the sleeping place (Table III) secures a competitive advantage for primary mates in two ways. First, "sleeping together" permits the investing male to passively guard his mate for several hours after coitus. Second, the male's ejaculate is likely to be retained longer in the sleeping female's reproductive tract.

Human males usually initiate coitus and, therefore, generally have control over its frequency. Coital frequency is highly variable for humans, but typically occurs from 1–3.5 times/week (Table III). It might be expected that the frequency of male-initiated copulation should relate to the perceived potential threat of sperm competition in a relationship. For example, copulation frequency should be low in an old and stable relationship (a corollary to the Coolidge effect), and high in a young or threatened relationship. Clinical psychologists may possess some data useful for testing this prediction.

*7. Male Sexual Jealousy and Paternity Assurance*

Male sexual jealousy is a behavioral/motivational complex that causes the "dogged inclination" of men to possess and control women, and to use violence or threat of violence to achieve sexual exclusivity and control (Daly *et al.* 1982). The emotion, by defending males' exclusive sexual access to their wives, functions to assure paternity in a system that threatens sperm competition. The proximate manifestations of male jealousy include male precautionary attitudes and methods designed to prevent the occurrence of sperm competition, violence directed toward unfaithful wives and their facultative mates, and universal societal rules and laws that buttress and facilitate these male attitudinal and behavioral predispositions.

Male anticuckoldry tactics have been reviewed by Dickeman (1979, 1981). These include the more conventional practices of virginity tests and guarantees, veiling, chaperoning, and claustration. Also of note are the bizarre and brutal practices of footbinding to prevent females from "straying," clitorectomy to preclude female orgasm and thereby diminish feminine interest in extramarital sex, and infibulation (suturing shut the labia majora) of girls to prevent infelicitous or untimely intercourse. Although footbinding is no longer practiced in China, Hosken (1979) estimates that 65 million living African women have undergone some sort of genital mutilation.

Human males are typically unpleasant about their wives' infidelity. Daly *et al.* (1982) point out that most wife beating is the result of the husband's suspicion or knowledge of his wife's sexual infidelity. Also, male jealousy has ranked high as a motive for domestic homicide. Up to one-third of the wife killing in the United States can be attributed to male sexual jealousy (see Daly *et al.* 1982), and this cause is important in other countries and cultures. For example, a Canadian study credited 85.3% of spouse killings to "sexual matters" (Chimbos 1978); Lobban (1972) discovered male jealousy was the leading motive category for all homicides in the Sudan; Tanner (1970) in a Ugandan study found killings due to male sexual jealousy followed only those committed during robbery and property disputes: 45% of all murdered wives in several cultures in British colonial Africa (Bohannan 1960) involved sexual matters (primarily adultery); and studies of homicide among aboriginal people in India revealed a high proportion attributable to adultery and related sexual problems (see Daly *et al.* 1982).

The double standard is cross-cultural (Fig. IK). Over 65% of indexed societies have rules against female extramarital sex but allow male affairs. Twenty-three percent prohibit extramarital liaisons by both sexes, but the prohibition on males in most of these seems designed more to create the illusion of equity than to enforce it (*i.e.*, females are punished for their indiscretions and males are not, or if so, less severely). Daly *et al.* (1982) studied adultery law and found a remarkable consistency of concept: "Sexual intercourse between a married woman and a man other than her husband is an offense." Typically, the offense is viewed as a property violation. The husband is the victim and, depending on the culture to which he belongs, is entitled to extract compensation from the offending male and/or violent revenge on his wife and/or her facultative mate. Clearly, male-dominated politics have universally produced laws and social conventions that reflect and support male sexual jealousy and its ultimate function: to prevent sperm competition and thereby protect male investment in wives and offspring.

## B. Female Attributes

### *1. Cryptic Ovulation and Continuous Sexual Receptivity*

Reproductive interests of human males and females are often in conflict. Males attempt to achieve exclusive sexual access to wives with minimum expenditure. By contrast, selection favors females who extract maximum contribution from principal mates while maintaining the option of extramarital mating (see section III). Overt female tactics are not likely to have been successful, and because of males' superior physical strength and political power, would result in reduced female fitness. Thus, the evolution of more subtle female stratagems.

Human males could efficiently allocate their resources with relatively low expenditure per mate if they were provided unambiguous information about the probability of a female conceiving at any given time. I propose that natural selection has produced just the opposite effect. I argue (following others) that continuous sexual receptivity, cryptic ovulation, and some other feminine characteristics have evolved to obscure a human female's current reproductive value and confuse males as a countermeasure to male resource allocation and anticuckoldry strategies. These female adaptations enhance opportunities for facultative polyandry and thus promote human sperm competition.

Most animals have regular periods of mating when copulation can result in pregnancy. This periodicity is usually synchronized between the sexes, and is mediated by hormones released in response to a variety of environmental stimuli. Some tropical mammals reproduce continuously; in these species, female receptivity is coincident with cyclical ovulation and males are always fertile.

Heape (1900) observed that primates deviate from the typical mammalian pattern in that males occasionally copulate with infertile females. This behavior is widespread among the primates (Hrdy 1979, 1981) and is most highly developed in humans. Human females do not experience estrus, offer no conspicuous morphological or behavioral evidence of ovulation, and are more or less continuously receptive to coitus from the onset of puberty.

Cryptic ovulation and perennial sexual receptivity in humans has intrigued sociobiologists, and many have attempted to identify the ultimate causation of these phenomena (Etkin 1963, Pfeiffer 1972, Alexander and Noonan 1979, Hrdy 1979, Burley 1979, Benshoof and Thornhill 1979, Strassmann 1981, Daniels 1983). Most theorists suggest that cryptic ovulation and continuous female receptivity facilitate monogamous pairbonding through the mechanism of permanent sexual attractiveness (Etkin 1954, Morris 1967, Washburn and Lancaster 1971, Crook 1972, Campbell 1974, Washburn 1974, Cook 1975, Halliday 1980, Short 1981, Lovejoy 1981). These notions imply that diminution of or loss in signals about a female's reproductive condition and the extension of sexual receptivity have somehow trapped promiscuous males into monogamy, which is maintained by conjugal sexual contentment for the ultimate benefit of altricial offspring.

Comparative primate behavior studies do not support this hypothesis. Cryptic ovulation is unlikely to evolve in a promiscuous mating system. Consider, for example, the chimpanzee (Table III). Male chimpanzees prefer to mate with females who are on the verge of ovulating. High status males are likely to be most successful in realizing this preference. If the majority of females in such a system provide an honest sign of high reproductive value, such as labial and circum-anal swelling (Table II), then the best males will attend the most swollen females. Female chimpanzees at the peak of estrus (those most swollen) do get much more attention from males than do less swollen females (Tutin and McGinnis 1981). Therefore any female variant that de-emphasized swelling, as a first move toward concealed ovulation, would have had diminished attractiveness to the best males, and would therefore have been selected against. Another difficulty with the hypothesis comes from the fact that increased infant dependency could only evolve in response to increased parental investment and not the reverse. Greater offspring dependency unmet by nurturant parents would only have increased infant mortality and could not have evolved. Finally, natural selection favors individuals whose behavior patterns tend to maximize their reproductive success and not their sexual contentment, so males could not have been bound by the continuous sexual attractiveness of wives if the binding itself did not contribute to fitness.

Of the several theories on the original function of concealed ovulation and continuous receptivity, Benshoof and Thornhill's (1979) discussion best recognizes and avoids these pitfalls. They propose that unrevealed ovulation and continuous female receptivity evolved after female monogamy because any group-living female with inconspicuous estrus in a monogamous mating system would be in a better position to mate with a superior male without her being detected by her primary mate. Benshoof and Thornhill further suggest that ovulation is apparently concealed from the female because some degree of self-deception facilitates the deceit of mates.

There is some debate on whether ovulation is actually concealed from the female, or if the event is only protected from her conscious detection. "*Mittelschmertz*," a pain in the lower abdomen occurring midway through a female's intermenstrual interval is thought to be caused by irritation of the pelvic peritoneum by blood or other fluid escaping from the ovary at ovulation (Osol 1972). If *mittelschmertz* occurs regularly, it might somehow trigger complex adaptive behavioral or physiological processes in a female without her conscious involvement. This could permit females' selectivity of sperm without alerting males. Support for this idea comes from studies that reveal females copulate more frequently at midcycle than at other times (Udry and Morris 1968, Adams *et al.* 1978, and see Hray 1979).

Heightened female libido during the first few months of pregnancy (see Masters and Johnson 1966) may also contribute to female reproductive inscrutability, and facilitate cuckoldry under conditions of facultative polyandry. In this regard, Howell (1979) found that !Kung women do not reveal or acknowledge their pregnancy until it shows after about the third or fourth month.

### *2. Perennial Pendulous Breasts*

Following parturition, female mammals cease to cycle reproductively for variable periods of time generally coincident with the duration of lactation (Tortora and Anagnostakos 1984). Lactation is maintained by prolactin, a hormone produced by the adenohypophysis in response to suckling stimulation of the breasts. Active mammary gland tissues expand, causing breast enlargment and pendulosity. Non-human primates, including the great apes, develop pendulous breasts only at the onset of lactation, and their breasts remain pendulous only for the duration of lactation.

Human females are unique among the hominoids (and mammals) in that they develop pendulous breasts at puberty and retain them throughout life irrespective of lactation. Human pubertal breast development involves estrogen-induced hypertrophy of the adipose and stromal tissue of the mammary gland, rather than activity of the glandular epithelium itself. The quantity of contained adipose determines breast size, and size bears no relationship to milk production ability (Tortora and Anagnostakos 1984).

Short (1980), Halliday (1980), and others have suggested that human breasts evolved their perennially pendulous condition to maintain constant male sexual interest in the female and therefore facilitate pair bonding and paternal investment. Although males in Western cultures are undeniably fascinated by female breasts, it is not clear this preoccupation is a cross-cultural universal. Male African graduate students have expressed surprise and amusement at this focus of U.S. males. "We look at the whole woman," is their typical reaction.

Although cross-cultural studies on intersexual attraction have been conducted (*e.g.*. Rosenblatt 1974), the question of which specific female features and which qualities of these are admired by males in different cultures apparently has not been investigated. Ford and Beach (1951) noted that appearance is important cross-culturally, especially in males' assessment of females, but found that the criteria of physical attractiveness vary widely among cultures. The only regularly recurrent characteristics that could be identified were health and feminine plumpness. This latter characteristic, although alien to members of Western food-rich, diet-conscious cultures, must provide an important indication of female health and reproductive potential in a predominantly food-limited world. Plumpness must also have been an excellent predictor of female fecundity throughout most of our evolutionary history. High investment males may therefore be attracted by females having pendulous breasts because this feature conspicuously displays stored feminine resources, emphasizes the principal organs of maternal contribution, and may predict females' ability to contribute. However, a problem is apparent with all of these proposed "functions." None of the aforementioned seems likely to have **initiated** selection for perennially pendulous breasts in hominids.

Pendulous breasts should initially have been sexually repulsive, at least to high status males, because they would have signaled lactation, amenorrhea, anovulation, and therefore a female of little or no current reproductive value. Astute males certainly would tend to ignore lactating females in order to focus their attention on others with obvious current reproductive value. As explained later (section VIII. B), selection for perennial pendulous breasts prior to the evolution of extended human pair bonding would nave been especially problematical. In this regard, Tutin and McGinnis (1981) observed that female chimpanzees who resumed sexual cycles while still lactating mated infrequently with adult males. In contrast, a female who resumed sexual cycling (with flattened breasts) 2 months after the death of her 2.5-year-old son was observed to be extremely attractive to males.

So, if the evolution of perennial breast pendulosity was not initiated by epigamic selection, what forces were involved? I propose the same function as discussed above for the evolution of concealed ovulation and continuous sexual receptivity, *i.e.*, to render additional ambiguity of female reproductive value in an imperfect system of female monogamy imposed by male domination.

After the evolution of concealed ovulation and continuous receptivity by females, males still would have had two conspicuous indicators of low female reproductive value: the swollen abdomen of third trimester pregnancy, and the swollen breasts of lactation. Both signs indicate current low reproductive value, and neither of these functional conditions could be disguised. However, the informational content of enlarged breasts could be progressively obscured by gradually extending the period of enlargement. Presumably. lactating females were not as assiduously guarded by their principal mates, and a mechanism to extend the relaxation of male vigilance into a post-lactational ovulation would enable sexual liaisons with other males. Such facultative polyandry would have been advantageous for the social, material, and genetic benefits discussed previously (section III). Though potential facultative mates would perceive females with enlarged breasts to have low reproductive value, mating would occur and female benefits accrue as is seen in other primates (see section VI, and Hrdy 1979). In a primarily monogamous mating system, promiscuous mating opportunities for males are relatively rare and should therefore almost always be beneficial, even with low value females. So, the evolution of perennial pendulous breasts could have initially evolved to provide females greater opportunity for facultative polyandry. Male epigamic selection may have operated, perhaps on the size and form of breasts, but only after the general trait had been fixed by selection for its original function.

### *3. The Hymen*

Males in most societies place a high premium on virginity of prospective wives, and this preference has been institutionalized in some cultures (Dickeman 1978, 1981). Virginity tests typically involve examination of the female to determine that the hymen is intact, or of the nuptial bedding for blood stains-evidence of hymen rupture by first intercourse.

The hymen is a membranous partition partially blocking the orifice of the vagina. It may take on a variety of forms, but is typically circular or crescentic. Occasionally, it may be multiple, entirely lacking, or imperforate (Osol 1972). Although the structure may be ruptured by activities other than sexual intercourse, usually the first coitus breaks the hymen, causing female pain and bleeding. The hymen is one of the great unsolved mysteries of human anatomy. I know of no plausible hypothesis for any physiological function it may serve, and I know of no other organ in the animal kingdom evolved inevitably to be injured. However, the trauma caused females of certain bugs (see Thornhill and Alcock 1983) and poeciliid fishes (see Constantz, this volume) by male genitalia during copulation is analogous. If the hymen does have some as yet undiscovered physiological function, why should it cease to be important after the female's first copulation?

Dickeman (1978), among others, apparently believes that the vaginal hymen evolved by selection due to male preference for virgin wives in hypergamous systems. It is true that members of most societies are aware of the hymen and its significance as an indicator of female virginity. Therefore, in some cultures, a mutant female who lacked this "evidence" of virginity could be selected against. The question here, as with concealed ovulation and perennial pendulous breasts, is how did the characteristic originally increase in frequency? The trait in a new variant would have no instant value in the context of intersexual selection, and would seem disadvantageous since any structure subject to injury should be selected against unless it confers some compensatory advantage. If the hymen were present in females of closely related species, some primitive function might be discerned, but K. G. Gould (pers. comm.) of the Yerkes Primate Research Center finds no evidence for a structure homologous to the hymen in any of the great apes. I assume, then, that the structure originated in the hominid line and perhaps first arose in *Homo sapiens*.

I offer no hypothesis for the initial evolution of the hymen, but concede that the structure may have been maintained and even further developed in human females by intersexual selection in the context of male defense against sperm competition.

### *4. Female Orgasm*

It would be adaptive for facultatively polyandrous or promiscuous females to assist the sperm of preferred mates and handicap those of nonpreferred mates. In this way a female might exercise some control over the paternity of her offspring while pursuing the nonreproductive benefits of facultative polyandry. Orgasm might permit females to achieve both of these objectives.

Female orgasm has been investigated by Masters and Johnson (1966). They found that single or multiple orgasms may be achieved by either direct or indirect clitoral stimulation. Indirect stimulation of the clitoris sufficient to induce female orgasm can be accomplished most effectively during intercourse in the ventral/ventral position (Masters and Johnson 1966, and Beach 1973). At the peak of female orgasm, the outer part of the vagina contracts rhythmically from 3–15 times, the inner blind end of the vagina expands, and the uterus contracts rhythmically for several seconds.

These events are believed to be mediated by the hormone oxytocin released from the posterior pituitary into the circulation as an autonomic response to clitoral stimulation (Fox and Fox 1969; Bishop 1971, 1973). Until recently, this belief was based largely on studies of nonhuman animals and circumstantial evidence; however, recent experiments have revealed a consistent, significant elevation of blood serum oxytocin in female subjects coincident with their orgasms induced by masturbation (J. Davidson, pers. comm.).

Symons (1979) views female orgasm as an essentially nonadaptive, or even "dysfunctional" byproduct of mammalian bisexual potential. He state "... orgasm may be possible for female mammals because it is adaptive for males." Symons sees the capacity of human females to experience multiple orgasms as an incidental effect of their inability to ejaculate. These attitudes seem as naive as they are chauvinistic. If there is as much variation in the ability of human females to achieve orgasm as is noted by Symons in support of his thesis, variants must have existed upon which natural selection could have operated to remove this complicated "artifactual, possibly maladaptive trait" from female populations.

An alternative to Symon's view is that human female orgasm may facilitate the transport of semen from the vagina into the uterus, and thereby provide the female with some control over the use of ejaculates. Grafenberg (1950) and Masters and Johnson

(1966) conducted experiments on human females that addressed this possibility. However, they failed to demonstrate the entry of radio-opaque fluid from the cervical cap into the uterus after either coitus or clitoral stimulation. The latter experiment has been criticized by Fox and Fox (1967). In addition, Fox *et al.* (1970) used radio telemetry to demonstrate a positive intrauterine pressure of ca. +40 cm $H_2O$ during coitus and a negative intrauterine pressure of ca. –26 cm $H_2O$ immediately after female orgasm (see also Fox and Fox 1971).

These results confirm the uterus creates suction immediately following orgasm and support the idea that female orgasm may actively transport semen and thereby facilitate fertilization. If this conclusion is indeed correct, orgasm may provide females with some opportunity to control fertilization. For example, a female might achieve orgasm during, or masturbate to orgasm following coitus with a preferred mate. Conversely, she could avoid orgasm during or following coitus with a less desirable potential father.

The Mangaia islanders practice penile subincision, a curious form of male genital mutilization which entails slitting the urethra from its orifice for a variable distance toward the scrotum (Marshall 1971). Aboriginals appropriately refer to this process as mika, meaning "terrible rite" (Short 1980). Subincision almost certainly results in semen being shallowly discharged against the vaginal wall with the result that female orgasm may be required in order for fertilization to occur. It is perhaps no coincidence therefore that the Mangaian culture is also known for its elaborate instruction to pubertal males on coital methods designed to cause multiple orgasms in the female and to insure female orgasm simultaneous with male ejaculation (Marshall 1971).

Psychosexual anecdotes suggest that female orgasm during coitus is most easily accomplished if the female feels generally secure and is convinced of at least some degree of future support from her partner (Fisher 1973, Hite 1976). It is significant that call girls, prostitutes who are well paid and well treated by high status clientele, have orgasms as frequently as non-prostitutes. Streetwalkers, by contrast, have difficulty experiencing orgasm (Exner *et al.* 1977). These prostitutes are subject to harassment from both clients and procurers and are unable to exercise selectivity on clients who typically belong to the lowest socioeconomic strata. This pattern is consistent with the thesis that females may choose the fathers of their children after mating, or at least minimize the possibility of impregnation by undesirable males.

Finally, if female orgasm facilitates sperm transport and fertilization, its absence during and following rape might militate against conception by a rapist's sperm. Consistent with this idea, I find no reports of rape victims having experienced orgasm during the sexual assault. However, this issue is confounded by the possible existence of other mechanisms that may enhance the prospects for rapists' sperm (see section VII.A.5).

### *5. Female Reproductive Tract and Ova*

Females may also control sperm by means of the physical, chemical, and immunological characteristics of their reproductive tract secretions. The secretions of the female tract have been studied both to determine causes of infertility and in search of improved contraceptive technologies, but never from the viewpoint of sperm competition. Here, I trace the movement of sperm in the female tract and consider how and where the female system may aid or impede a particular ejaculate, or discriminate among sperm from different ejaculates.

Female emotional factors such as attraction, affection, and trust together with the investment in sexual foreplay rendered by a mate prior to coitus are known to influence the vaginal environment. Specifically, the quality and quantity of vaginal mucus produced by the vestibular glands are affected (Tortora and Anagnostakos 1984). Mucus lubricates the vagina to facilitate intercourse and perhaps the passage of sperm to the cervical os.

Insufficient lubrication prior to coitus, as in rape, may result in trauma to the vaginal wall with the discharge of blood into the vaginal cavity. Blood serum contains the spermatotoxic protein gamma globulin and often sperm antibodies as well (Franklin and Dukes 1964*a*,*b*), so females who experience vaginal trauma with bleeding during forced copulation may be defended to some extent against impregnation by constituents of their own blood, and by the lack of vaginal mucus to aid sperm movement.

The human vaginal environment is considered generally hostile toward sperm. Evidence for this is partially circumstantial in that only a very small fraction of ejaculated sperm ever reach the uterine tubes where fertilization usually occurs. Of the hundreds of millions of sperm contained in each ejaculate, only about 2,000 arrive in the vicinity of the descending ovum (Tortora and Anagnostakos 1984). Selection on the female vagina in the context of ejaculate competition may have favored the evolution of secretions that produce extraordinarily rigorous environments for sperm. This possibility might be explored by experiments to test the prediction that vaginal secretions in humans and the chimpanzee will be more hostile toward sperm than those of the gorilla and orangutan.

The cervix may be an important filter of sperm because of its key position in the route to fertilization (Porter and Finn 1977). All ascending sperm must negotiate the cervix and contact cervical mucus. During most of the female reproductive cycle, the high viscosity and elevated protein and leucocyte content of the cervical mucus impede the progress of sperm. At midcycle, there is a 10-fold increase in volume of cervical mucus produced. The midcycle mucus secreted beginning about 6.2 days prior to ovulation is a low viscosity, low protein, more hydrated product than the earlier cycle secretions. Midcycle mucus facilitates passage of sperm, and is therefore helpful to male gametes contained in ejaculates deposited during this time. Females may use subliminal cues to copulate with preferred mates during this time and avoid coitus with non-preferred mates. The change in cervical mucus is apparently so reliably associated with ovulation and easily recognizable by females, that it has been suggested as a means of predicting ovulation (see Porter and Finn 1977).

Franklin and Dukes (1964*a*,*b*) first reported a high incidence of sperm agglutinin in the blood serum of females with sterility of unknown etiology. A sperm agglutination factor believed to be transferred from serum is found in cervical mucus (Isojima 1973). Antibodies in cervical mucus have been reported at higher levels than in the blood (Eyquem *et al.* 1976, Parish *et al.* 1967). The highest levels of sperm agglutinin factor have been found in prostitutes, followed by married females, who in turn have higher levels than unmarried women. Conceivably, antibodies produced by married women may be specific to husbands' sperm. This raises the discomfiting possibility (for married men) that the gametes of facultative mates may enjoy a competitive advantage over those of husbands' handicapped by wives' immune systems. Significantly, "condom therapy" that shields wives from exposure to husbands' sperm for a period of several months, apparently halts production of antibodies and causes reduced female titers of sperm agglutinating and immobilizing factors (Kay 1977).

The uterine tube is the final leg in the journey of human sperm. Sperm are usually capacitated and activated (made biochemically ready for penetrating the ovum) in the tubal environment. Tubal secretions are known to be involved in these processes, although the molecular mechanisms are not well understood (Johnson and Everitt 1980). Oviducal fluid is produced most copiously around the time of ovulation and its composition at this time stimulates maximum spermatic metabolism (Blandau *et al.* 1977). The change in the quantity and characteristics of oviducal fluid is apparently under hormonal control.

Psychological factors are know to influence reproductive endocrine functions (Okamura *et al.* 1973), and psychosomatic sterility is an established syndrome in human females (Mori *et al.* 1973, Uemura and Suzuki 1973). Although the details of this syndrome are vague, it is known that emotional anxiety in psychologically normal females can influence oviducal motility and the constitution of tubal fluid (Asaoka *et al.* 1973). Since both tubal mobility and fluid composition have a critical influence on the progress and activation of sperm, it seems reasonable to suggest that females have the capability of exercising control over sperm in the uterine tubes.

The final barrier to fertilization is the zona pellucida of the ovum. In order for fertilization to occur, the acrosome of a spermatozoon must release its enzymes against the egg's gelatinous covering, and the enzymes must succeed in dissolving a hole so that penetration may occur. The ovum may be surrounded by many sperm attempting to penetrate, but only one will succeed and this may not necessarily be the sperm first to arrive at the membrane (Chang *et al.* 1977). Though no data specifically address the subject, it seems evident from descriptions of in vitro fertilization that egg membranes have properties that may favor penetration by some sperm over others within an ejaculate. The much higher variability among sperm from mixed competitive ejaculates would seem to permit greater opportunity for discrimination by the egg membrane.

## VIII. SPERM COMPETITION AND HUMAN EVOLUTION

Pilbeam (1984) recently provided an update on hominoid historical evolution that synthesizes the past 5 years' fossil and archaeological discoveries. I have abstracted this useful paper in the following paragraphs to provide a factual framework upon which I attempt to construct a scenario for the evolution of hominoid mating systems and to fit the elements presented in Section VII.

### A. Hominoid Evolution

It is not clear precisely when the hominid line diverged from the Old World monkeys, but current consensus places the event after middle Oligocene and before early Miocene (20–30 million years BP). The molecular "clock" and other evidence has the orangutan line branching about IS million years BP, followed by the gorilla line 9–10 million BP. Then the chimpanzee line diverged between 6–10 million years ago from the lineage that ultimately gave rise to the hominids. Experiments on DNA hybridization have revealed at least 98% identity in nonrepeated DNA in humans and the chimpanzee, indicating a remarkably close relationship (see Lovejoy 1981).

Bipedal hominids were present in eastern Africa by at least 3.5–4 million years BP. *Australopithecus africanus*, the first species reasonably well represented in the fossil record (primarily from Hadar, Ethiopia), was strongly dimorphic, with males from 50–100 percent larger than females. This species had chimpanzee-like facial features, but the overall skull was more gorilla-like. The canines, however, were low-crowned and without ape-like forward projection. Between 2–2.5 million years BP, another sexually dimorphic species, *A. boisei*, appeared in Africa. This species lived in an area of woodland and savanna, away from the forests. Ecologically, these early hominids shared more in common with modern *Gorilla* than with humans. Their dentition (*i.e.*, large cheek teeth capped with thick enamel) indicates a vegetable diet. Walker and Grime (see Pilbeam 1984) have studied the teeth of *Australopithecus* and concluded that all were apelike vegetarians, eating broadly similar diets that required a great deal of chewing. The *Australopithecus* diet probably varied from that of modern forest-dwelling apes only in that it may have contained more roots and tubers and less fruit. No stone tools have been found associated with fossils, so it is assumed that if tools were constructed, they must have been of perishable materials. Tool use is inferred from reduction of canines, bipedal locomotion, and skeletal evidence of manual dexterity.

*Homo habilis* was present at the same time as *A. boisei*, and persisted until about 1.75 million years ago. *H. habilis* vanished and was replaced by *H. erectus*, who had a larger brain than the australopithecines, but resembled members of the older genus. The appearance of *H. habilis* was coincident with the first archaeological sites that contained used or altered stone, as well as non-hominoid animal remains.

It is widely acknowledged that this period is marked by a significant shift in diet from plant to animal food. It is hazardous to suppose any particulars in the lifestyle of *H. habilis*. Perhaps it was a hunter-gatherer with home camps and division of labor with shared resources. Alternatively, the species may have been primarily vegetarian with some opportunistic scavenging of meat and a lower level of social organization. The physical evidence is sufficient to support either scenario, but inadequate to distinguish one from the other. *Homo habilis* survived for a few hundred thousand years and was replaced by *H. erectus*.

*Homo erectus* was a widely-distributed species that first appeared in Africa. At 1 million years BP, it was present in southeastern and eastern Asia, where it survived until at least 300,000 years ago. This species made flaked stone tools and lived in groups, some of which may have used fire. It resembled early *H. sapiens* in gross body morphology, and seems to have been both morphologically and culturally stable for about 1.5 million years. A gradual transition from *H. erectus* to our own species began about a half million years ago. The transition from "archaic" *H. sapiens* (including Neanderthal man) to "modern" *H. sapiens* apparently represents a continuum.

Neanderthals and their contemporaries were different from us both physically and behaviorally. Skeletal evidence suggests that they were much more robust and muscular than modem humans. Their teeth were larger than those of modem *H. sapiens*. Dental wear patterns indicate that Neanderthals used their teeth for non-feeding activities, probably chewing of hides to soften them, by analogy with the modem Eskimo. This implies the use of clothing.

*Homo sapiens sapiens* appeared sometime between 50,000 and 100,000 years BP, and was characterized by loss of robustness and changes in the female pelvis. The female skeletal changes are indicative of some change in pregnancy and births. These alterations may have included reduction in the gestation period from 11 months to the present 9 months and a concomitant increase in dependency of infants on parental care. The prediction is made on the basis of morphometric comparisons with the great apes and other mammals (see Pilbeam 1984).

## B. Hominid Mating Systems and Sperm Competition

There is no indication that the strongly dimorphic species *Australopithecus africanus* and *A. boisei* had a sexual division of labor that would necessitate the separation of males from females or give males control over high quality food resources not directly available to females. There is likewise no physical evidence that the australopithecines used tools or sophisticated weapons for hunting big game, and in fact nothing to suggest that they regularly ate meat. Consequently, their social organization was probably similar to that of extant gorillas, *i.e.*, small mixed sex groups, each with a dominant male who was in constant contact with and control of females. It may be reasonable to suggest that their mating pattern consisted of male dominance polygyny and female monogamy, as in the system of *Gorilla*. If this assessment is correct, sperm competition should not have been a significant evolutionary force in the australopithecines. Two factors, intergroup competition and the sharing of high quality food, probably played a major role in destabilizing the single male dominance mating system of these early ancestors. By analogy, the first of these factors seems to have been important in development of female promiscuity in chimpanzees, and may well have played a role in humans (see Alexander and Noonan 1979). Some fossil evidence exists to mark the historical onset of the second element.

*Homo habilis* probably hunted meat by scavenging. Hunting for meat first took the form of scavenging, which initiated a sexual division of labor. Scavenging the kills of large predators clearly involved more risk than gathering vegetable material, insect larvae, and the like. Early hominid scavengers would regularly encounter large predators, and no doubt had to compete directly with other species highly skilled in this specialization. The rigors and risks of scavenging would almost certainly have excluded pregnant females and those with infants.

Possession of meat by subordinate males would have been disruptive to the single male dominance system in two ways. First, it would have given subordinate males the power to lure females away from dominant males while giving females a strong incentive to mate with subordinates. Second, if a dominant male began to engage in scavenging, he would be separated from "his" females, thus increasing their opportunities for facultative polyandry. Females at this time were probably selected to deceive the dominant males in order to maximize their harvest of meat from subordinate males. These changes in turn diminished most of the male inter. sexual advantage for being large, and therefore selection for large male size was relaxed.

Sperm competition would have intensified, triggering a series of reciprocal evolutionary changes in the reproductive organs of males and females. First, males probably rapidly developed both enlarged penises and large scrotal testes. In response, females would have evolved more rigorous vaginal, uterine, and tubal environments. Males countered with the elaboration of protective and manipulative accessory gland secretions, while at the same time increasing the volume of their ejaculates to neutralize vaginal defenses and otherwise mitigate against female control of sperm. Females may now have begun to exert psychological control over their reproductive tracts and secretions.

Males were selected to identify and favor females of the highest reproductive value, in the exchange of meat for sex. As a consequence, females may initially have evolved honest signals that emphasized their actual high reproductive value, and later modified these signals to create the illusion of high value. Selection could have favored extension of female sexual receptivity before and after ovulation (as has occurred in chimpanzees), but it seems unlikely that female reproductive crypsis and perennial pendulous breasts could have been initiated at this stage because any tendency to crypsis in a female would have made her less desirable than signaling females with the result that she would have received less male attention and, hence, fewer male resources. Similarly, pendulous breasts would have signaled low reproductive value and would have resulted in reduced male interest.

The sex-for-meat strategy pursued by males may have been the driving force behind the transition from scavenging to hunting, a far more hazardous, but potentially more rewarding method of obtaining meat. Individual hunting with primitive weapons may not have been as efficient as scavenging, but cooperative hunting of big game developed into a hugely successful operation probably beginning with *H. erectus* about 1 million years ago.

Though a certain level of cooperation may have evolved earlier by selection arising from intergroup conflict, the benefits of cooperative hunting certainly resulted in even greater selection for behavior patterns and cultural adaptations tending to increase male solidarity and reduce overt intermale competition and its disruptive consequences. It is reasonable to assume that these adaptations would have included mechanisms to reduce conflict over mates. The form of these adaptations may have resembled those of the chimpanzee system of female promiscuity.

So, the early social contract would have guaranteed meat and sex for every contributing male. This obviously would have created very intense sperm competition that drove genital selection to remove all of the heritable variation in the male population. As a result all males had approximately equal chance to reproduce, but no male could have any confidence of his paternity for any offspring of any female. This situation would seem to be unstable, especially if the product of cooperative hunting was not equally distributed. Good hunters or hunt leaders probably demanded and were afforded an extra share of meat from each kill. This inequity would have created the opportunity to translate hunting/leadership skills into extra matings and hence potential differential reproductive success. However, variance in male reproductive success would have been reduced by female promiscuity. Consequently, the reproductive success of superior males with abundant resources to exchange for sex would be limited only by the time to copulate, and by ejaculate production capability. At this point, as in the chimpanzee analogy, high quality males would be selected to supplement promiscuous mating with temporary consortship. The relatively high level of paternity confidence afforded by consortship would have been of great advantage, especially to top-ranking males with the most abundant resources at their disposal for investment in offspring. The gradual shift in mating strategies from purely promiscuous to a mix that emphasized mate defense polygyny would not destabilize the primitive society. The now obligatory interdependence of males would make for stability and non-consorting males would continue to have sexual access to nonconsorting females.

Improved weaponry and hunting techniques would have increased the resource base so that progressively more males could afford to engage in consortships. This caused competition among females for males able to establish consortships, and selected for female inclination to consort rather than to mate promiscuously. The disappearance of female promiscuous mating may have begun selection for forced copulation as a low investment male reproductive option.

Consortship would have gradually given way to long-term pairbonding with progressively higher levels of male investment in mates and offspring. Increased male investment coincided with the increased levels of paternity assurance that resulted from improved male efficiencies in controlling their mates' reproductive activities. Selection strongly favored sexually jealous males who maintained tight control over their mates, and this progressively and substantially reduced the. levels of sperm competition. Females evolved cryptic ovulation, perennial pendulous breasts, and behavioral deceptions to militate against male control efficiencies, and to achieve reproductive and other advantages from facultative polyandry. Paradoxically, female opportunity to secure advantages by facultative polyandry was created and propagated by the ability of males to mix high and low investment strategies.

Long-term pair bonding and a willingness of males to make parental investment, as distinguished from mating effort (see Gwynne, this volume) probably enabled reduced gestation and increased infant dependency. As evidenced in the changes of the female pelvis that distinguish *Homo sapiens sapiens* from earlier forms of our species, paternal care and increased infant dependency may be relatively recent events in hominid evolution.

The potential for sperm competition apparently has remained sufficiently strong into recent prehistoric and historical times for there to be no relaxation of selection on what may be very old male genitalic and emotional/behavioral anti-cuckoldry adaptations. Historically, male-dominated societies have universally evolved politics that reinforce individual anticuckoldry adaptations and have instituted a variety of new practices that serve this function. Very recent social institutions evolved in the context of sperm competition may even have selected a female structure, the hymen, that is unique to humans.

## SUMMARY

Sperm competition occurs in humans and apparently has been a force in the evolution of certain human characteristics. It occurs in the contexts of communal sex, rape, prostitution, courtship, and (most commonly) facultative polyandry. Facultative polyandry permits a female to maximize genic benefits for her offspring, and to secure material resources and protection from males. It may also accomplish social benefits. Data on the incidence of human sperm competition are meager. Most come from forensic genetics studies conducted to exclude paternity, and very few from human population genetics studies. These data indicate the occurrence of sperm competition and the threat of cuckoldry, but reveal that long-term pairbonding (*i.e.*, marriage) achieves a high level of paternity assurance.

Among our relatives the great apes, chimpanzees have a social system and some sexual behavior patterns that most closely resemble our own. Relatively high levels of sperm competition occur in humans and chimpanzees, compared with the gorilla and the orangutan. Coincident with these differences are male genitalic similarities between humans and chimpanzees that include large specialized penises and large scrotal testes. Male gorillas and orangutans have small penises and nonscrotal testes. Human and chimpanzee females both have conspicuous sexual ornamentation (breasts in humans and genital swelling in chimpanzees), but female gorillas and orangutans lack conspicuous sexual displays.

A variety of human anatomical, physiological, behavioral, and cultural characteristics may have evolved in the context of sperm competition. Among male characteristics probably influenced by this force are the size and structure of the testes, accessory glands, ejaculatory duct, and penis. The human scrotum probably evolved by sperm competition favoring efficient storage of spermatozoa. Seminal plasma constituents and ejaculate production capability were almost certainly influenced by interejaculate competition. Male masturbation and nocturnal emission in lieu of intercourse may maintain highly competitive ejaculates. A variety of human male mating tactics have evolved to place sperm in competition and to defend against competing sperm. Male controlled aspects of copulation including coital position, time, duration, and frequency may be optimized in the context of sperm competition. Finally, male sexual jealousy encompasses a constellation of masculine attitudes and aggressive behavior patterns evolved to defend against sperm competition. These attitudes and behavior patterns are universally reinforced by cultural institutions.

Human female characteristics, including continuous sexual receptivity, cryptic ovulation, and perennial pendulous breasts, have apparently evolved as countermeasures to male resource allocation and anticuckoldry strategies. These female attributes enhance opportunities for facultative polyandry, promote sperm competition, and may permit females some control over paternity of offspring. The motility and secretions of the female reproductive tract are influenced by female emotional states, which offers females opportunity to control the fertilization of ova in a facultatively polyandrous mating system by selectively aiding sperm from preferred mates or handicapping sperm from undesirable mates. The hymen is a mysterious structure because of its obscure origin and its absence in the great apes, but it may have been maintained and even elaborated in human females by an institutionalized male preference to marry virgins in hypergamous societies.

Early hominids, *i.e.*, *Australopithecus* spp., probably formed mixed groups, each with a large dominant male. These species would then have had a male dominance polygyny/female monogamy mating system, and thus little or no sperm competition. I postulate that this system was destabilized by hunting (scavenging) and possession of meat by subordinate males beginning with *Homo habilis*. Cooperative hunting probably begun by *H. erectus* and set conditions that led to a promiscuous mating system and high levels of sperm competition. This in turn would have caused intense reciprocal selection on the genital organs of males and females such that all of the heritable variation was quickly removed from populations. Unequal distribution of the product of cooperative hunting probably caused some males to begin consortships which would have reduced sperm competition, improved individual paternity assurance, and begun to destabilize the promiscuous mating system. Improved paternity assurance permitted evolution of longer term pairbonding, paternal investment, and intensified male sexual jealousy. More efficient male control of females selected female reproductive crypsis in the form of continuous sexual receptivity, concealed ovulation, perennial pendulous breasts, and other deceptive tactics. Paternal investment enabled reduced gestation and increased infant dependency in *H. sapiens sapiens*.

Historically, sperm competition has apparently remained a sufficiently important force to permit no relaxation of selection on very old genitalic and male emotional/behavioral anticuckoldry adaptations. During recorded human history, male attitudes forged by the potential for sperm competition have come to be reflected in political, legal, and social institutions that repress human females.

## ACKNOWLEDGMENTS

I am grateful to the following people for assistance by way of useful discussions, papers, and personal communications: J. Alcock, J. Brewer, N. Burley, N. A. Chagnon, J. M. Davidson, M. Dickeman, L M. Ellman, K. G. Gould, H. Harpending, K. Hummel, D. Kaye, H. P. Krutsch, A. Schlegel, P. L Terasaki, R. Thornhill, R. L Trivers, and D. W. Zeh, J. Alcock, D. N. Byrne, T. Myles, W. L Nutting, F. G. Werner, J. Smith, and D. W. Zeh reviewed the manuscript and all provided useful suggestions. David Zeh deserves special recognition for carefully editing the editor and substantially improving the final product. Finally, I thank Jill Smith for acting as the sounding board for all of my ideas, for producing the figures, and for helping with all aspects of the project.

## REFERENCES

Adams, D. B, A. R. Gold, & A. D. Rise. 1978. Rise in female-initiated sexual activity at ovulation and its suppression by oral contraceptives. *New England J. Med.* **229**:1145–1150.

Aiyappan, A. 1935. Fraternal polyandry in Malabar. *Man India* **15**:108–118.

Alcock, J. 1984. *Animal Behaviour: An Evolutionary Approach*. Sinauer, Sunderland, MA.

Alexander, R. D. 1974. The evolution of social behavior. *Annu. Rev. Ecol. Syst.* **5**: 325–383.

Alexander, R. D. 1977. Natural selection and the analysis of human sociality. In *Changing Scenes in the Natural Sciences*: 1776–1976, C. E. Goulden (ed.), pp. 283–337. Philadelphia Acad. Nat. Sci. Special Publ 12.

Alexander, R. D. 1979. *Darwinism and Human Affairs*. Univ. Washington Press, Seattle, WA.

Alexander., R. D., & K. N. Noonan. 1979. Concealment of ovulation, parental care, and human social evolution. In *Evolutionary Biology and Human Social Behavior: An Anthropological Perspective*, N. A. Chagnon & W. G. Irons (eds.), pp. 436–453. Duxbury Press, North Scituate, MA.

Anonymous. 1982. Thirty-nine percent of paternity charges are phony, tests reveal. *Current Med. Attorneys* **29**:10.

Archer, J. 1810. Facts illustrating a disease peculiar to the female children of Negro slaves. *Med. Reposit.* **1**:319.

Asaoka, T., S. Iwabuchi, & H. Yamamoto. 1973. Influence of emotional anxiety on tubal factor in infertile women. In *Fertility and Sterility*, T. Hasegawa, M. Hayashi, F. J. G. Ebling, & I. W. Henderson (eds.), p. 970. Excerpta Medica, Amsterdam.

Athanasiou, R. 1973. A review of public attitudes on sexual issues. In *Contemporary Sexual Behavior*, J. Zubin & J. Money (eds.), pp. 361–390. Johns Hopkins Univ. Press, Baltimore, MD.
Barash, D. P. 1982. *Sociobiology and Behavior, 2nd ed.* Elsevier, New York.
Bateson, P. (ed.) 1983. *Mate Choice*. Cambridge Univ. Press, Cambridge.
Beach, F. A. 1973. Human sexuality and evolution. In *Advances in Behavioral Biology, Vol. II Reproductive Behavior*, W. Montagna and W. A. Sadler (eds.), pp. 333–365. Plenum Press, New York.
Beall, C. M., & M. S. Goldstein. 1981. Tibetan fraternal polyandry: A test of sociobiological theory. *Am. Anthropol.* **83**:5–12.
Bedford, J. M. 1977. Evolution of the scrotum: The epididymis as the prime mover? In *Reproduction and Evolution*, J. H. Calaby & C. H. Tyndale-Boscoc (eds.), pp. 171–182. Aust. Acad. Sci., Canberra.
Benshoof, L., & R. Thornhill. 1979. The evolution of monogamy and concealed ovulation in humans. *J. Soc. Biol. Struct.* **2**:95–106.
Bergström, S. 1973. Prostaglandins and human reproduction. In *Fertility and Sterility*, T. Hasegawa, M. Hayashi. F. J. G. Ebling, & I. W. Henderson (eds.), pp. 40–44. Excerpta Medica, Amsterdam.
Bishop, D. W. 1971. Sperm transport in the fallopian tube. In *Pathways to Conception: The Role of the Cervix and the Oviduct*, A. I. Sherman (ed.), pp. 99–109. Charles Thomas Publ., Springfield, IL.
Bishop, D. W. 1973. Biology of spermatozoa. In *Sex and Internal Secretions, Vol. 2*, W. C. Young & G. W. Corncr (cds.), pp. 707–796. Robert E. Krieger Publ., Huntington, NY.
Blandau, R. J., B. Brackett, R. M. Brenner, J. L Boling, S. H. Broderson, C. Hammer, & L. Mastroianni. 1977. The oviduct. In *Frontiers in Reproduction and Fertility Control*, R. O. Greep & M. A. Koblinski (eds.), pp. 132–145. MIT Press, Cambridge, MA.
Bohannan, P. 1960. *African Homicide and Suicide*. Princeton Univ. Press, Princeton, NJ.
Broude, G. E., & S. J. Greene. 1976. Crosscultural codes on twenty sexual attitudes and practices. *Ethnology* **15**:410–429.
Brownmiller, S. 1975. *Against Our Will: Men, Women, and Rape*. Simon and Shuster, New York.
Bullough, V., & B. Bullough. 1978. *Prostitution, An Illustrated Social History*. Crown Publ., Inc., New York.
Burley, N. 1979. The evolution of concealed ovulation. *Am. Nat.* **114**:835–858.
Burley, N., & R. Symanski. 1981. Women without: An evolutionary and cross-cultural perspective on prostitution. In *The Immoral Landscape: Female Prostitution in Western Societies*, pp. 239–273. Butterworth, Toronto.
Bygdeman, M., B. Fredricsson, K. Svanborg, & B. Samuelsson. 1970. The relation between fertility and prostaglandin content of seminal fluid in man. *Fertil. Steril.* **21**:622–629.
Campbell, B. G. 1974. *Human Evolution, 2nd ed.* Aldine, Chicago.
Carrick, F. N., & B. P. Setchell. 1977. The evolution of the scrotum. In *Reproduction and Evolution*, J. H. Calaby and C. H. Tyndaie-Boscoe (eds.), pp. 165–170. Aust. Acad. Sci., Canberra.
Chagnon, N. A. 1979. Mate competition favoring close kin, and village fissioning among the Yanomamo Indians. In *Evolutionary Biology and Human Social Behavior, An Anthropological Perspective*, N. A. Chagnon and W. G. Irons (eds.), pp. 86–132. Duxbury Press, North Scituate, MA.
Chang, M. C., C. R. Austin, J. M. Bedford, B. G. Brackett, R. H. F. Hunter, & R. Yanagimachi. 1977. Capacitation of spermatozoa and fertilization in mammals. In *Frontiers in Reproduction and Fertility Control*, R. O. Greep and M. A. Koblinsky (eds.). pp. 434–451. MIT Press, Cambridge, MA.
Chimbos, P. D. 1978. *Marital Violence: A Study of Interspouse Homicide*. R&E Research Associates, San Francisco.
Cook, J. M. 1975. *In Defense of Homo sapiens*. Farrar, Straus, and Giroux, New York.
Crook, J. H. 1972. Sexual selection, dimorphism, and social organization in the primates. In *Sexual Selection and the Descent of Man*, 1871–1971, B. Campbell (ed.), pp. 231–281. Aldine-Atherton, Chicago.
Daly, M., & M. Wilson. 1978. *Sex, Evolution, and Behavior*. Duxbury Press, North Scituate, MA.
Daly, M., M. Wilson, & S. J. Weghorst. 1982. Male sexual jealousy. *Ethol. Sociobiol.* **3**:11–27.
Daniels, D. 1983. The evolution of concealed ovulation and self deception. *Ethol. Sociobiol.* **4**:69–87.
Davenport, W. H. 1977. Sex in cross-cultural perspective. In *Human Sexuality in Four Perspectives*, F. A. Beach (ed.), pp. 115–163. Johns Hopkins Univ. Press, Baltimore, MD.
Diagram Group, The. 1981. *Sex: A User's Manual*. Perigee Books, New York.
Dickeman, M. 1978. Confidence of paternity mechanisms in the human species. (Circulated and cited unpublished manuscript.)

Dickeman, M. 1979. Female infanticide, reproductive strategies, and social stratification: A preliminary model. In *Evolutionary Biology and Human Social Behavior, An Anthropological Perspective*, N. A. Chagnon and W. G. Irons (eds.), pp. 321–367. Duxbury Press, North Scituate, MA.

Dickeman, M. 1981. Paternal confidence and dowry competition: A biocultural analysis of purdah. In *Natural Selection and Social Behavior*, R. Alexander and D. Tinkle (eds.), pp. 417–438. Chiron Press, New York.

Dirasse, L. 1978. *The Socioeconomic Position of Women in Addis Ababa: The Case of Prostitution*. Ph.D. Dissertation, Boston Univ.

Dorjahn, V. R. 1958. Fertility, polygyny, and their interrelations in Temne socicty. *Am. Anthropol.* **60**:838–860.

Dorjahn, V. R. 1959. The factor of polygyny in African demography. In *Continuity and Change in African Cultures*, W. R. Bascom and M. J. Herskovits (eds.), pp. 87–112. Univ. Chicago Press, Chicago.

Eliasson, R., & N. Posse. 1965. Rubin's test before and after intravaginal application of prostaglandin. *Int. J. Fertil.* **10**:373–377.

Etkin, W. 1954. Social behavior and the evolution of man's mental facilities. *Am. Nat.* **88**:129–134.

Etkin, W. 1963. Social behavior factors in the emergence of man. *Human Biol.* **35**:299–310.

Exner, J. E., J. Wylie, A. Leura, & T. Parrill. 1977. Some psychological characteristics of prostitutes. *J. Personality Assessment* **41**:474–485.

Eyquem, A., M. D'Almeida, & J. Rothman. 1976. Local antispermic immunity in the genital tract. In *Biological and Genical Aspects of Reproduction*, F. J. G. Ebling and I. W. Henderson (eds.), pp. 174–177. Excerpta Medica, Amsterdam.

Fisher, S. 1973. *The Female Orgasm*. Basic Books, New York.

Foote, R. H. 1973. My experience with artificial insemination. In *Fertility and Sterility*. T. Hasegawa, M. Hayashi, F. J. G. Ebling, & I. W. Henderson (eds.), pp. 353–358. Excerpta Medica, Amsterdam.

Ford, C. D., & F. A. Beach. 1951. *Patterns of Sexual Behavior*. Harper & Row, New York.

Fossey, D. 1974. Development of the mountain gorilla (*Gorilla gorilla beringei*) through the first thirty-six months. Paper presented at Berg Wartenstein, symposium no. 62, the behavior of the Great Apes. Wenner-Gren foundation for Anthropological Research. (See Hrdy 1979).

Fossey, D. 1976. *The behavior of the mountain gorilla*. Ph.D. Dissertation, Cambridge Univ.

Fox, C. A., & B. Fox. 1967. Uterine suction during orgasm. *Br. Med. J.* **1**:300–301.

Fox, C. A., & B. Fox. 1969. Blood pressure and respiratory patterns during human coitus. *J. Reprod. Fertil.* **19**:405–415.

Fox, C. A., & B. Fox. 1971. A comparative study of coital physiology, with special reference to the sexual climax. *J. Reprod. Fertil.* **24**:319–336.

Fox, C. A., H. S. Wolff, & J. A. Baker. 1970. Measurement of intra-vaginal and intra-uterine pressures during human coitus by radio-telemetry. *J. Reprod. Fertil.* **22**:243–251.

Franklin, R. R., & C. D. Dukes. 1964*a*. Antispermatozoal antibody and unexpected infertility. *Am. J. Obstet. Gynecol.* **89**:6–9.

Franklin, R. R., & C. D. Dukes. 1964*b*. Further studies on sperm agglutinating antibody and unexplained infertility. *J. Am. Med. Assoc.* **190**:682–683.

Freund, M. 1962. Interrelationships among the characteristics of human semen and factors affecting semen specimen quality. *J. Reprod. Fertil.* **4**:143–159.

Freund, M. 1963. Effect of frequency of emission on semen output and an estimate of daily sperm production in man. *J. Reprod. Fertil.* **6**:269–286.

Galdikas, B. M. F. 1981. Orangutan reproduction in the wild. In *Reproductive Biology of the Great Apes: Comparative and Biomedical Perspectives*. C. E. Graham (ed.), pp. 281–300. Academic Press, New York.

Gaulin, S. J. C., & A. Schlegel. 1980. Paternal confidence and paternal investment: A cross-cultural test of a sociobiological hypothesis. *Ethol. Sociobiol.* **1**:301–309.

Goldberg, V. J., & P. W. Ramwell. 1977. The role of prostaglandins in reproduction. In *Frontiers in Reproduction and Fertility Control*, R. O. Greep and M. A. Koblinsky (eds.), pp. 219–235. MIT Press, Cambridge, MA.

Gould, K. G., & D. E. Martin. 1981. The female ape genital tract and its secretions. In *Reproductive Biology of the Great Apes: Comparative and Biomedical Perspectives*, C. E. Graham (ed.), pp. 105–125. Academic Press, New York.

Gowaty, P. A. 1981. An extension of the Orians-Verner-Willison model to account for mating systems besides polygyny. *Am. Nat.* **118**:851–859.

Grafenberg, E. 1950. The role of the urethra in female orgasm. *Int. J. Sexol.* **3**:145–148.

Graham, C. E., & C. F. Bradley. 1972. Microanatomy of the chimpanzee genital system. In *The Chimpanzee, Vol. 5*, G. C. Bourne (ed.), pp. 77–126. Karger, Basel, Switzerland.

Graham, C. G. 1981. Menstrual cycle of the great apes. In *Reproductive Biology of the Great Apes: Comparative and Biomedical Perspectives*, C. E. Graham (ed.), pp. 1–43. Academic Press, New York.

Green, R. 1980. Variant forms of human sexual behavior. In *Reproduction in Mammals, Book 8, Human Sexuality*, C. R. Austin and R. V. Short (eds.), pp. 68–97. Cambridge Univ. Press, Cambridge.

Halliday, T. 1980. *Sexual Strategy*. Oxford Univ. Press, Oxford.

Hamburg, D. A., & E. McGown (eds.). 1979. *The Great Apes*. Benjamin/Cummings Publ., Menlo Park, CA.

Harcourt, A. H. 1981. Intermale competition and the reproductive behavior of the great apes. In *Reproductive Biology of the Great Apes: Comparative and Biomedical Perspectives*, C. E. Graham (ed.), pp. 301–318. Academic Press, New York.

Harcourt, A. H., P. H. Harvey, S. G. Larson, & R. V. Short. 1981*a*. Testis weight, body weight, and breeding system in primates. *Nature* **293**:55–57.

Harcourt, A. H., K. J. Stewart, & D. Fossey. 1981*b*. Gorilla reproduction in the wild. In *Reproductive Biology of the Great Apes: Comparative and Biomedical Perspectives*, C. E. Graham (ed.), pp. 265–279. Academic Press, New York.

Hawthorn, G. 1970. The Sociology of Fertility. Collier-Macmillan, London.

Heape, W. 1900. The "sexual season" of mammals and the relations of the "pro-oestrum" to menstruation. *Q. J. Microsc. Sci.* **44**:1–70.

Hite, S. 1976. *The Hite Report*. Macmillan, New York.

Hosken, F. P. 1979. *The Hosken Report. Genital and Sexual Mutilation of Females*, 2nd rev. ed. Women's International Network News, Lexington, MA.

Howell, N. 1979. *Demography of the Dobe !Kung*. Academic Press, New York.

Hrdy, S. B. 1977. *The Langurs of Abu*. Harvard Univ. Press, Cambridge, MA.

Hrdy, S. B. 1979. Infanticide among animals. A review, classification, and examination of the implications for the reproductive strategies of females. *Ethol. Sociobiol.* **1**:13–40.

Hrdy, S. B. 1981. *The Woman That Never Evolved*. Harvard Univ. Press, Cambridge, MA.

Hughes, A. L 1982. Confidence of paternity and wife-sharing in polygynous and polyandrous systems. *Ethol. Sociobiol.* **3**:125–124.

Hummel, K. 1979. Das biostatistische Gutachten als forensisches Beweismittel. *Arztl. Lab.* **25**: 131–137.

Hunt, M. M. 1974. *Sexual Behavior in the 1970s*. Playboy Press, Chicago.

Hurault, J. 1961. *The Boni Refugee Blacks of French Guiana*. Institut Francais d'Afrique Noire, Dakar.

Isaac, B. L 1979. Female fertility and marital form among the Mende of rural upper Bambara chiefdom, Sierra Leone. *Ethnology* **19**:297–313.

Isojima, S. 1973. The nature of antibodies against spermatozoa found in women with unexplained sterility. In *Fertility and Sterility*, T. Hasegawa, M. Hayashi, F. J. G. Ebling, & I. W. Henderson (eds.), pp. 105–107. Excerpta Medica, Amsterdam.

Jöchle, W. 1973. Coitus-induced ovulation. *Contraception* **7**:523–565.

Jöchle, W. 1975. Current research in coitus-induced ovulation: A review. *J. Reprod. Fertil. Suppl.* **22**:165–207.

Johnson, M. H., & B. J. Everitt. 1980. *Essential Reproduction*. Blackwell Sci. Publ., Oxford.

Kay, D. J. 1977. Clinical significance of antiodies to antigens of the reproductive tract. In *Immunological Influence on Human Fertility*. B. Boettecher (ed.), pp. 119–123. Academic Press, New York.

Kinsey, A. C., W. B. Pomeroy, & C. E. Martin. 1948. *Sexual Behavior of the Human Male*. W. B. Saunders, New York.

Kinsey, A. C., W. B. Pomeroy, C. E. Martin, & P. H. Gebhard. 1953. *Sexual Behavior of the Human Female*. W. B. Saunders, New York.

Kramer, A. 1932. *Truk*. Friederichsen, De Gruyter, Hamburg.

Kurland, J. A. 1979. Paternity, mother's brother and human sociality. In *Evolutionary Biology and Human Social Behavior,* N. A. Chagnon and W. G. Irons (eds.), pp. 145–180. Duxbury Press, North Scituate, MA.

Levi-Strauss, C. 1969. *The Elementary Structure of Kinship*. Beacon Press, Boston, MA.

Little, K. 1967. *The Mende of Sierra Leone*. Routledge and Kegan Paul Limited, London.

Lobban, C. F. 1972. *Law and Anthropology in the Sudan (An Analysis of Homicide Cases in Sudan).* African Studies Seminar Series No. 13, Sudan Research Unit, Khartoum Univ.

Lovejoy, C. 0.1981. The origin of man. *Science* **211**:341–350.

Malinowski, B. 1929. *The Sexual Life of Savages in North-Western Melanesia*. Routledge, London.

Marshall, D. S. 1971. Sexual behavior on Mangaia. In *Human Sexual Behavior*, D. S. Marshall and R. C. Suggs (eds.), pp. 103–162. Basic Books, New York.

Martin, D. E., & K. G. Gould. 1981. The male ape genital tract and its secretions. In *Reproductive Biology of the Great Apes: Comparative and Biomedical Perspectives*, C. E. Graham (ed.), pp. 127–161. Academic Press, New York.

Masters, W. H., & V. E. Johnson. 1966. *Human Sexual Response*. Little Brown and Co., Boston, MA.

Maynard Smith, J. 1978. *The Evolution of Sex*. Cambridge Univ. Press, Cambridge, MA.

Mendelson, A. R. 1982. From here to paternity. *Barrister* **9**:12–15, 55.

Mitchell, G. 1979. *Behavioral Sex Differences in Nonhuman Primates*. VanNostrand Reinhold, New York.

Mori, I., Y. Maki, Y. Ijuin, A. Tukuda, & M. Takeda. 1973. Psychosomatic aspects of sterility. In *Fertility and Sterility*, T. Hasegawa, M. Hayashi., F. J. G. Ebling, & I. W.Henderson (eds.), pp. 960–961. Excerpta Medica, Amsterdam.

Morris, D. 1967. *The Naked Ape*. McGraw-Hill, New York.

Morris, J. M. 1977. The morning-after pill: A report on postcoital contraception and interception. In *Frontiers in Reproduction and Fertility Control*, R. O. Greep and M. A. Koblinsky (eds.), pp. 203–208. MIT Press, Cambridge, MA.

Murdock, G. P. 1967. *Culture and Society*. Univ. Pittsburgh Press, Pittsburgh, PA.

Musham, H. V. 1956. The fertility of polygamous marriages. *Population Studies* **10**:3–16.

Nadler, R. D., C. E. Graham, D. C. Collins, & O. R. Kling. 1981. Post partum amenorrhea and behavior of apes. In *Reproductive Biology of the Great Apes: Comparative and Biomedical Perspectives*, C. E. Graham (ed.), pp. 69–81. Academic Press, New York.

Neel, J. V., & K. M. Weiss. 1975. The genetic structure of a tribal population, the Yanomama Indians. XIII. Biodemographic studies. *Am. J. Phys Anthropol*. **42**:25–51.

Okamura, Y., M. Kitazia, K. Arakawa, H. Tateyama, M. Nagakawa, T. Goto, A. Kurano, R. Mori, & L Taki. 1973. Psychological aspects of gynecological endocrine diseases. In *Fertility and Sterility*, T. Hasegawa, M. Hayashi, F. J. G. Ebling, & I. W. Henderson (eds.), pp. 965–967. Excerpta Medica, Amsterdam.

Osol, A. (ed.). 1972. Blakiston's *Gould Medical Dictionary, 3rd ed.* McGraw-Hill, New York.

Parker, G. A. 1983. Mate quality and mating decisions In *Mate Choice*, P. Bateson (ed.), pp. 141–166. Cambridge Univ. Press, Cambridge, England.

Parks, G. R. (ed.). 1929. Marco Polo, Travels. *Book League of America*, New York.

Parish, W. E., J. A. Carron-Brown, & C. B. Richards. 1967. The detection of antibodies to spermatozoa and to blood group antigens in cervical mucus. *J. Reprod. Fertil.* **13**:469–483.

Pfeiffer, J. E. 1972. *The Emergence of Man. Harper and Row*, New York.

Pickles, V. R. 1967. Uterine suction during orgasm. *Br. Med. J.* **1**:427.

Pietropinto, A., & J. Simenauer. 1977. *Beyond the Male Myth*. Times Books, New York.

Pilbeam, D. R. 1984. The descent of hominoids and hominids. *Sci. Am.* **150**:84–96.

Porter, D. G., & C. A. Finn. 1977. The biology of the uterus. In *Frontiers in Reproduction and Fertility Control*, R. O. Greep and M. A. Koblinsky (eds.), pp. 146–156. MIT Press, Cambridge, MA.

Portmann, A. 1952. *Animal Forms and Patterns*. Farber and Farber, London.

Rosenblatt, P. C. 1974. Cross-cultural perspective on attraction. In *Foundations of Interpersonal Attraction*, T. L. Huston (ed.), pp. 79–95. Academic Press, New York.

Sanday, P. R. 1981. The socio-cultural context of rape: A cross-cultural study. J. *Social Issues* **37**:5–27.

Sass, S. L. 1977. The defense of multiple access (exceptio plurium concubentium) in paternity suits: A comparative analysis. *Tulane Law Rev*. **51**:468–509.

Schonfeld, W. A. 1943. Primary and secondary sexual characteristics. Study of their development in males from birth through maturity, with biometric study of penis and testes. *Am. J. Dis. Child.* **65**:535–549.

Schwartz, J. H. 1984. The evolutionary relationships of man and orangutans. *Nature* **308**:501–505.

Shanor, K. 1978. *The Shanor Study: The Sexual Sensitivity of the American Male*. Dial Press, New York.

Shearer, L. 1978. Sex sensation. In "Intelligence Report," *Parade Mag*., Sept. 10.

Shields, W. M., & Shields, L. M. 1983. Forcible rape: An evolutionary perspective. *Ethol. Sociobiol.* **4**:155–136.

Short, R. V. 1976. Definition of the problem: The evolution of human reproduction. *Proc. R. Soc. Lond. B* **195**:3–24.

Short. R. V. 1977. Sexual selection and the descent of man. In *Reproduction and Evolution*, J. H. Calaby and C. H. Tyndale-Briscoe (eds.), pp. 3–19. Aust. Acad. Sci., Canberra.

Short. R. V. 1979. Sexual selection and its component parts, somatic and genital selection, as illustrated by man and the great apes. *Adv. Study Behav.* **9**:131–158.

Short. R. V. 1980. The origins of human sexuality. In *Reproduction in Mammals, Book 8, Human Sexuality*, C. R. Austin and R. V. Short (eds.), pp. 1–33. Cambridge Univ. Press, Cambridge.

Short. R. V. 1981. Sexual selection in man and the great apes. In *Reproductive Biology of the Great Apes*, C. E. Graham (ed.), pp. 319–341. Academic Press, New York.

Sorensen, R. C. 1973. *Adolescent Sexuality in Contemporary America*; World Publ. Co., New York.

Sorgo, G. 1973. Das Problem der Superfecundatio im Vaterschaftsgutachten. *Beitr. Gerichtl. Med.* **30**:415–421.

Spielmann, W., & P. Kuhnl. 1980. The efficacy of modem blood group genetics with regard to a case of probable superfecundation. *Haematologia* **134**:75–85.

Stangle, J. J. 1979. *Fertil ity and Conception: An Essential Guide for Childless Couples*. Paddington Press, New York.

Strassmann, B. L 1981. Sexual selection, paternal care, and concealed ovulation in humans. *Ethol. Sociobiol.* **2**:31–40.

Symanski, R. 1981. *The Immoral Landscape: Female Prostitution in Western Societies*. Butterworth, Toronto.

Symons, D. 1979. *The Evolution of Human Sexuality*. Oxford Univ. Press, New York.

Tanner, R. E. S. 1970. *Homicide in Uganda, 1964. Crime in East Africa.* Scandinavian Inst. African Studies, Uppsala.

Terasaki, P. I. 1978. Resolution by HLA testing of 1000 paternity cases not excluded by ABO testing. *J. Fam. Law* **16**:543–557.

Terasaki, P. I., D. Gjertson, D. Bernoco, S. Perdue, M. R. Mickey, & J. Bond. 1978. Twins with two different fathers identified by HLA. *New England J. Med.* **299**:590–592.

Thornhill, R., & J. Alcock. 1983. *The Evolution of Insect Mating Systems*. Harvard Univ. Press, Cambridge, MA.

Thornhill, R., & N. W. Thornhill. 1983. Human rape: An evolutionary analysis. *Ethol. Sociobiol.* **4**:137–173.

Titiev, M. 1972. *The Hopi Indians of Old Oraibi: Change and Continuity*. Univ. Michigan Press, Ann Arbor, MI.

Tortora, G. J., & N. P. Anagnostakos. 1984. *Principles of Anatomy and Physiology*. Harper & Row, New York.

Trivers, R. L 1972. Parental investment and sexual selection. In *Sexual Selection and the Descent of Man, 1871–1971*, B. Campbell (ed.), pp. 136–179. Aldine-Atherton, Chicago.

Trivers, R. L 1974. Parent-offspring conflict. Am. Zool. 14:249–264.

Trivers, R. L, & D. E. Willard. 1973. Natural selection of parental ability to vary the sex ratio of offspring. *Science* **179**:90–92.

Turner, T. L, & S. S. Howards. 1977. Sperm maturation, transport, and capacitation. In *Male Infertility*. A. T. K. Cockett and R. L Urry (eds.), pp. 29–57. Grone and Stratton, New York.

Turnbull, C. M. 1965. *The Mbuti Pygmies: An Ethnographic Survey*. Am. Mus. Nat. Hist., AMNH Board of Trustees, New York.

Tutin, C. E. G. 1975. *Sexual behavior and mating patterns in a community of wild chimpanzees (Pan troglodytes)*. Ph. D. Dissertation, Univ. Edinburgh.

Tutin, C. E. G., & P. R. McGinnis. 1981. Chimpanzee reproduction in the wild. In *Reproductive Biology of the Great Apes: Comparative and Biomedical Perspectives*, C. E. Graham (ed.), pp. 239–264. Academic Press, New York.

Udry, J. R., & N. M. Morris. 1968. Distribution of coitus in the menstrual cycle. *Nature* **220**:593–596.

Uemura, T., & N. Suzuki. 1973. Treatment for psychosomatic sterility. In *Fertility and Sterility*. T. Hasegawa, M. Hayashi, F. J. G. Ebling, & 1. W. Henderson (eds.), pp. 969–970. Excerpta Medica, Amsterdam.

Van den Berghe, P. L, & D. P. Barash. 1977. Inclusive fitness theory and human family structure. *Am. Anthropol.* **79**:809–823.

Washburn, S. L. 1974. *Ape Into Man*. Little, Brown and Co., Boston, MA.

Washburn, S. L, & C. S. Lancaster. 1971. The evolution of hunting. In *Background for Man*. P. Dolhinow and V. M. Sarich (eds.), pp. 386–403. Little, Brown and Co., Boston, MA.

Weatherhead, P. J., & R. J. Robertson. 1979. Offspring quality and the polygyny threshold: "The sexy son hypothesis." *Am. Nat.* **113**:201–208.

Whitlatch, W. G., & R. W. Masters. 1962. Comment: Contribution of blood tests in 734 disputed paternity cases: Acceptance by the law of blood tests as scientific evidence. *Western Reserve Law Rev.* **14**:114–115.

Williams, G. C. 1966. *Adaptation and Natural Selection*. Princeton Univ. Press, Princeton, NJ.

Williams, G. C. 1975. *Sex and Evolution*. Princeton Univ. Press, Princeton, NJ.

Wilson, E. O. 1975. *Sociobiology: The New Synthesis*. Belknap Press, Cambridge.

Winick, C., & P. M. Kinsie. 1972. *The Lively Commerce: Prostitution in the United States*. Quadrangle, Chicago.

Wolfe, L. 1981. *The Cosmo Report*. Arbor House, New York.

Woodland, W. 1903. On the phylogenetic case of the transposition of the testes in mammalia. *Proc. Zool. Soc. Lond.* **1**:319–340.

# 5. 'KAMIKAZE' SPERM IN MAMMALS?

R. Robin Baker[1] and Mark A. Bellis[1]

Current debate over the adaptive significance of the reproductive behaviour of mammals and other animals often derives from characteristics of male ejaculates (Smith 1984). Mammalian ejaculates have two conspicuous characteristics: (1) astronomical numbers of sperm; and (2) a surprisingly high proportion (e.g. up to 40% in normal humans (Wren 1985) and 30–50% in normal lions (Wildt *et al.* 1987)) of apparently deformed sperm. In addition, the ejaculates of many species coagulate and often harden after copulation to form a 'plug' within the female tract (Mann & Lutwak-Mann 1981). Soft plugs, or at least coagulates, are nearly universal among mammals; even hard plugs are common throughout the rodents and similarly situated formations have also been noted in some marsupials, insectivores, bats and primates (Voss 1979).

There have been two major attempts to explain these characteristics. The first, Parker's theory of sperm competition (Parker 1970, 1982), concentrates on the evolution of astronomical numbers. Parker postulates that when ejaculates from different males compete within the reproductive tract of a single female, males with a numerical sperm advantage will be most successful at fertilizing that female's egg(s). Males that produce large numbers of sperm at each ejaculation are the evolutionary result of this competition. Copulatory plugs give the first male to inseminate a female a further advantage, rendering less efficient any copulations by subsequent males (Voss 1979); in most mammals studied, the first male to mate with a female is most likely to fertilize her egg(s) (Huck *et al.* 1985).

Parker's theory of sperm competition has become central to discussions of male courtship behaviour (Smith 1984). However, the alternative theory for ejaculate characteristics (Cohen 1975) is that large numbers of sperm are produced because so many die within the female reproductive tract, suboptimal, such as malformed, sperm being killed and removed by phagocytosis. The existence of such a high proportion of apparently deformed sperm is normally attributed (Cohen 1977) to errors of chiasmata during spermatogenesis; errors that natural selection has failed to eradicate.

---

[1] Department of Environmental Biology, University of Manchester.

Reprinted from *Animal Behaviour*, 36 (3), Baker, R. R. & Bellis, M. A., 'Kamikaze' sperm in mammals? pp. 936–939, Copyright (1988), with permission from Elsevier.

Our studies of the ejaculates of rats suggest to us a different interpretation. We find particularly provocative the observation that the copulatory plug of rats seems to form primarily around a skeleton of intermeshed sperm (Fig. 1). We accept sperm competition as the major evolutionary force behind the characteristics of mammalian ejaculates. However, we suggest that the astronomical sperm numbers are not a simple response to sperm competition. Moreover, and perhaps most importantly, we interpret the different sperm morphologies as being functionally adaptive, not as 'deformities' or 'abnormalities'. In effect we propose that a large number of sperm in an ejaculate are morphologically and physiologically adapted to a 'kamikaze' role: to stay behind (within a plug and/or within sperm aggregations at other strategic sites such as utero-tubal and cervico-uterine junctions) and interfere with the passage of any later sperm ejaculated by a second male, thus protecting the small proportion adapted as 'egg-getters'. These kamikaze sperm may be incapable of fertilization even when directly confronted with an egg, a phenomenon consistent with the observation that a surprisingly small proportion of sperm are physiologically capable of fertilizing an egg, even in controlled laboratory conditions (Cohen & Massey 1984).

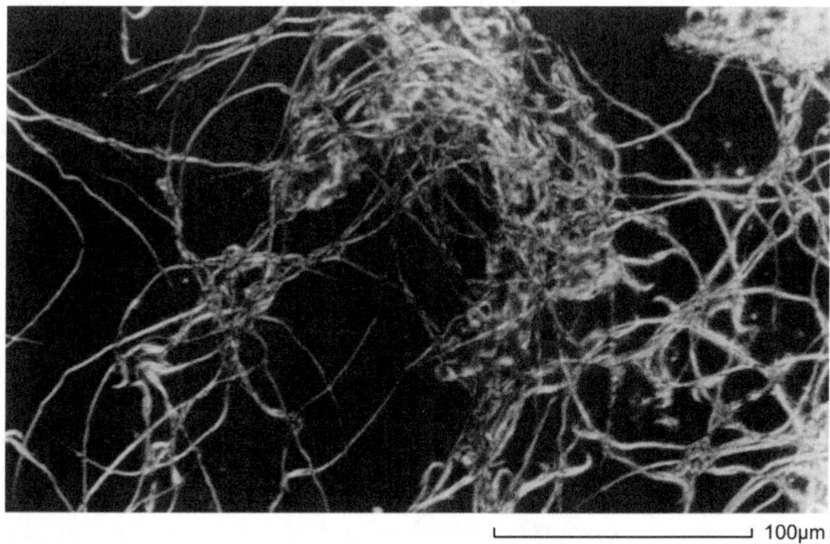


**Figure 1.** Intermeshed sperm digested out from the copulatory plug of the rat by means of human saliva. Part of the coagulated seminal fluids still remains and can be seen around the sperm matrix.

The function of 'kamikaze' sperm may be relatively crude (e.g. to mesh together to form a barrier; sperm such as those with two heads or tails etc., far from being deformed, actually being morphologically adapted for intertwining and barrier formation. Even a proportion of kamikaze sperm as low as 1–5% of an ejaculate of 60 million sperm, each 0.17 mm long, could give a reinforcing matrix up to 510 m long intertwined within a plug 1–2 cm long). Consistent with this hypothesis are casual observations made on other mammals; for example sperm within the plugs of bats are often in some way 'malformed' (Fenton 1984). However, we do not rule out the possibility of more sophisticated functions. For example, while the sperm are still alive, any barrier they form might selectively (perhaps in collaboration with lymphocytes) allow passage to further 'egg-getter' sperm from the original male (following his second or subsequent copulation with the same female) while attempting to bar the passage to sperm inseminated by a different male. For example, in rabbits cervical leucocytosis elicited by the sperm from a previous male may be associated with reduced (though not zero) fertility of a subsequent mating by a different male, though interpretation of the data (Taylor 1982) is open to further debate. We envisage, therefore, two adaptive bases for plugs that consist of seminal fluids coagulated on a sperm matrix rather than simply of seminal fluids alone: (1) the potential for selective passage of sperm; and (2) improved structural strength, much in the way of wire in reinforced concrete.

With a kamikaze system in operation the possibility also exists that the relative proportions of egg-getter and kamikaze sperm may vary with the particular situation in which a male finds himself. For example a male of the European rabbit, *Oryctolagus cuniculus*, produces a vaginal plug only when its status is dominant (Bell & Mitchell 1984). This suggests that males may manipulate the nature of their ejaculate according to their socio-sexual situation. Perhaps in this and other species the proportions of kamikaze and egg-getter sperm in an ejaculate vary depending on whether plugging or fertilization is of major importance. In particular, when a male copulates several times with the same female during a single mating 'session' such a variation could occur between first and subsequent ejaculations.

Our primary interest lies with attempting to understand the characteristics of mammalian ejaculates. Nevertheless we anticipate that the kamikaze sperm hypothesis may well be of universal application to animal ejaculates. For example, insects often produce a number of sperm morphs. In some species, e.g. the stink bug, *Arvelius albopunctatus*, and in flies of the genus *Drosophila*, one of the sperm morphs is unusually large. Sivinski (1984) has suggested that these large morphs may block the passages in the female's reproductive tract preventing access by the sperm of other males. More particularly Silbergleid *et al.* (personal communication, cited in Sivinski 1984) have suggested that apyrene sperm (with only part of the normal chromosomal complement) in Lepidoptera may displace or inactivate the fertile eupyrene sperm from previous matings.

A feature specific to mammalian ejaculation is that it occurs in 'spurts' during a single copulation (Mann & Lutwak-Mann 1981). In humans different spurts are known to have different characteristics (Eliasson & Lindholmer 1973). Our hypothesis would predict that 'egg-getter' sperm would be found in early spurts (and perhaps be deposited higher up the female reproductive tract) while the 'kamikaze' sperm should predominate in later spurts (and perhaps be deposited lower down). It is already known that some time after copulation different proportions of sperm types are found at different positions within the female reproductive tract of laboratory mice (Cohen 1977).

The possibility that mammalian copulations may involve males producing ejaculates containing specific balances of 'kamikaze' and 'egg-getter' sperm could have far-reaching implications for the interpretation of copulatory and other courtship behaviour.

In addition, of course, the kamikaze sperm hypothesis may have implications for the interpretation of infertile ejaculates in humans and other animals.

## ACKNOWLEDGEMENTS

We thank Dr John Kennaugh, Les Lockey, Brian Landamore and Philip Wheater for assistance with this project.

## REFERENCES


Bell, D. J. & Mitchell, S. 1984 Effects of female urine on growth and sexual maturation in male rabbits *J. Reprod. Fert.*, **71**, 155–160.

Cohen, J. 1975. Gamete redundancy: wastage or selection? In *Gametic Competition in Plants and Animals* (Ed by E Mulcahy), pp. 99–112. New York: Elsevier.

Cohen, J. 1977. *Reproduction* London: Butterworths.

Cohen, J. & Massey, B. 1984. *Animal Reproduction, Parents Making Parents*. London: Edward Arnold.

Eliasson, R. & Lindholmer, Chr. 1973 Effects of human seminal plasma on sperm survival and transport. *INSERM.*, **26**, 219–230.

Fenton, M. B 1984 The case of vespertilionid and rhinolophid bats In: *Sperm Competition and the Evolution of Animal Mating Systems* (Ed by R L. Smith), pp 573–587. London: Academic Press.

Huck, U. W., Quinn, R. P. & Lisk, R. D. 1985. Determinants of mating success in the golden hamster (*Mesocricetus auratus*) IV. Sperm competition. *Behav Ecol. Sociobiol.*, **17**, 239–252.

Mann, T. & Lutwak-Mann, C. 1981. *Male Reproductive Function and Semen*. New York: Springer-Verlag.

Parker, G. A. 1970. Sperm competition and its evolutionary consequences in the insects. *Biol. Rev.*, **45**, 525–567.

Parker, G. A. 1982. Why so many tiny sperm? The maintenance of two sexes with internal fertilization. *J. Theor. Biol.*, **96**, 281–294.

Sivinski, J. 1984. Sperm in competition. In: *Sperm Competition and the Evolution of Animal Mating Systems* (Ed. by R. L. Smith), pp. 85–115. London: Academic Press.

Smith, R. L. (Ed.) 1984. Sperm Competition and the Evolution of Animal Mating Systems. London: Academic Press.

Taylor, N. J. 1982. Investigation of sperm-induced cervical leucocytosis by a double mating study in rabbits. *J. Reprod. Fert.*, **66**, 161–168.

Voss, R. 1979. Male accessory glands and the evolution of the copulatory plugs in rodents. *Occ. Papers Mus. Zool., Univ. Michigan*, **No. 689**, 1–27.

Wildt, D. E., Bush, M., Goodrowe, K. L., Packer, C., Pusey, A. E., Brown, J. L., Joslin, P. & O'Brien, S. J. 1987. Reproductive and genetic consequences of founding isolated lion populations. *Nature, Lond.*, **329**, 328–331.

Wren, B. G. 1985. *Handbook of Obstetrics and Gynaecology*. London: Chapman and Hall.

# 6. DEFORMED SPERM ARE PROBABLY NOT ADAPTIVE

Alexander. H. Harcourt[1]

Baker & Bellis (1988) suggest that the high numbers of deformed sperm in ejaculates of many mammalian species are produced to facilitate the formation in the female's reproductive tract of copulatory plugs, objects whose function is to inhibit the passage of the sperm of males who mate subsequently with the same female. I submit that currently available evidence indicates that this hypothesis and its claim for wide applicability are probably wrong.

To begin with a minor quibble, no evidence exists to show that deformed sperm (e.g. ones with two heads, no heads, no tails, two tails) are more likely than normal sperm to facilitate the formation of plugs. Baker & Bellis imply that the presence of deformed sperm in bats' plugs (Fenton 1984) is evidence for their hypothesis. However, instead of being responsible for the plug, the sperm could simply have been trapped in it, perhaps because they were not motile enough to move further up the female's reproductive tract.

Second, the proportion of abnormal sperm in the ejaculate of lions, *Panthera leo*, increases as the level of genetic homozygosity in its population increases (Wildt *et al.* 1987). It would be odd for an adaptive trait to increase in expression due to inbreeding. However, it is feasible that threshold effects could be involved: some abnormal sperm are advantageous: too many might not be.

Third, the male who used seminal fluids, rather than potentially useful sperm to form the plug would be at a major advantage. For example, where the biochemistry of plug production is known, it is formed by a secretion of heat-stable protein from the prostate gland, coagulinogen, which coagulates the ejaculate (Price & Williams-Ashman 1961).

Fourth, an obvious prediction of the 'kamikaze sperm' hypothesis is the existence of a positive association between polyandry and the proportion of deformed sperm in ejaculates. Only polyandrous species (those in which several males mate with a potentially fertile female) benefit from plugs and deformed sperm; species in which only one male mates with a female have no need of a plug and, therefore, by Baker & Bellis'

---

[1] Large Animal Research Group, Department of Zoology, University of Cambridge.

hypothesis, no need of deformed sperm. Yet there seems to be no association, or a negative one, between polyandry and the proportion of deformed sperm in at least one mammalian taxon.

Sperm competition certainly occurs in mammals. For example, in normally polyandrous species, males that mate frequently have a reproductive advantage (Dewsbury 1984); available evidence indicates that polyandrous primates mate more frequently than do monandrous ones (Harcourt 1981); polyandrous primates have a greater proportion of their testes as spermatogenic tissue (Schultz 1938; Harcourt *et al.* 1980); polyandrous primates, deer, and mammals as a whole have larger testes than do monandrous species (Harcourt *et al.* 1980; Clutton-Brock *et al.* 1982; Kenagy & Trombulak 1986); and polyandrous primates inseminate more sperm per ejaculate, because of greater density of sperm or greater volume of ejaculate (Short 1979; Harcourt *et al.* 1980). Thus mammalian reproductive anatomy and physiology provides some of the best evidence for Parker's (1970, 1984) theory of sperm competition.

Nevertheless, while the polyandrous male chimpanzee, *Pan troglodytes*, has larger testes, produces more sperm per ejaculate, has larger ejaculates and possibly mates more frequently than does the monandrous male gorilla, *Gorilla gorilla* (Short 1979), all as predicted by sperm competition theory, the chimpanzee does not appear to have a greater proportion of deformed sperm in the ejaculate than does the gorilla. Thus only 5% of chimpanzee sperm are deformed, compared to 30% of gorilla sperm, from males of proven fertility in captivity (Seuanez 1980).

Fifth and finally, while Baker & Bellis suggest that their 'kamikaze sperm hypothesis may well be of universal application to animal ejaculates', it will, in fact, apply only to a subset of species, and probably a small subset. It will apply only to those species in which a first-male advantage exists, i.e. those in which the first male to mate is more likely to fertilize the ovum than is the second. Baker & Bellis recognize this and claim that, in most mammals, a first-male advantage does exist. However, the very paper that Baker & Bellis use to claim a preponderance of first-male advantage among mammals in fact says 'In mammals the results are variable' (Huck *et al.* 1985). To date, no evidence exists for a bias towards a first-male advantage in mammals (Dewsbury 1984; Huck *et al.* 1985). In addition, in many insects (Parker 1984) and at least two avian species (McKinney *et al.* 1984; Birkhead *et al.* 1988), the last male to mate is at an advantage.

Moreover, the 'kamikaze' hypothesis will work only in those species with first-male advantage in which males can alter the ratio of deformed to normal sperm according to whether they are the first to mate or not. If a male is not the first, he needs to produce as many normal sperm as possible to increase the chances of his sperm penetrating the first male's plug. The evidence is that males cannot alter the ratio in the short term, at least in the cat family (Wildt *et al.* 1987). In the longer term, males might vary the ratio in relation to their dominance status, as Baker & Bellis suggest by implication. However, dominant males of many species by no means always obtain prior access to potentially fertile females (e.g. Smuts 1987). In such cases, and I suspect that they are the majority, the dominant male who produced deformed 'kamikaze' plug sperm instead of normal fertilizing sperm could be at a disadvantage.

In conclusion, I have reservations about the likelihood that the kamikaze sperm hypothesis will "have far-reaching implications for the interpretation of copulatory and other courtship behaviour" and "of infertile ejaculates in humans and other animals"

(Baker & Bellis 1988), interesting as the hypothesis is. Instead, evidence and theory indicate that males should produce as many fertile sperm as energy considerations allow, especially where there is a chance of polyandry (Parker 1982), meaning that abnormal sperm are still best explained by errors in production (Cohen 1973).

## ACKNOWLEDGEMENTS

I thank Kelly J. Stewart and two referees for criticisms that improved the paper.

## REFERENCES

Baker, R. R. & Bellis, M. A. 1988. 'Kamikaze' sperm in mammals? *Anim. Behav.*, **36**, 936–939.

Birkhead, T. R., Pellat, J. & Hunter, F. M. 1988. Extrapair copulation and sperm competition in the zebra finch. *Nature, Lond.*, **334**, 60–62.

Clutton-Brock, T. H., Guinness, F. E. & Albon, S. D. 1982. *Red Deer. Behaviour and Ecology of Two Sexes.* Edinburgh: Edinburgh University Press.

Cohen, J. 1973. Crossovers, sperm redundancy, and their close association. *Heredity*, **31**, 408–413.

Dewsbury, D. A. 1984. Sperm competition in muroid rodents. In: *Sperm Competition and the Evolution of Animal Mating Systems* (Ed. by R.L. Smith), pp. 547–571. New York: Academic Press.

Fenton, M. B. 1984. Sperm competition? The case of vespertilionid and rhinolophid bats. In: *Sperm Competition and the Evolution of Animal Mating Systems* (Ed. by R. L. Smith), pp. 573–587. New York: Academic Press.

Harcourt, A. H. 1981. Intermale competition and the reproductive behavior of the great apes. In: *Reproductive Biology of the Great Apes* (Ed. by C. E. Graham), pp. 301–318. New York: Academic Press.

Harcourt, A. H., Harvey, P. H., Larson, S. G. & Short, R. V. 1980. Testis weight, body weight and breeding system in primates. *Nature, Lond.*, **293**, 55–57.

Huck, U. W., Quinn, R. P. & Lisk, R. D. 1985. Determinants of mating success in the golden hamster (*Mesocricetus auratus*) IV. Sperm competition. *Behav. Ecol. Sociobiol.*, **17**, 239–252.

Kenagy, G. J. & Trombulak, S. C. 1986. Size and function of mammalian testes in relation to body size. *J. Mammal.*, **67**, 1–22.

McKinney, F., Cheng, K. M. & Bruggers, D. J. 1984. Sperm competition in apparently monogamous birds. In: *Sperm Competition and the Evolution of Animal Mating Systems* (Ed. by R. L. Smith), pp. 523–545. New York: Academic Press.

Parker, G. A. 1970. Sperm competition and its evolutionary consequences in the insects. *Biol. Rev.*, **45**, 525–567.

Parker, G. A. 1982. Why are there so many tiny sperm? Sperm competition and the maintenance of two sexes. *J. Theor. Biol.*, **96**, 281–294.

Parker, G. A. 1984. Sperm competition and the evolution of animal mating strategies. In: *Sperm Competition and the Evolution of Animal Mating Systems* (Ed. by R. L. Smith), pp. 2–60. New York: Academic Press.

Price, D. & Williams-Ashman, H. G. 1961. The accessory reproductive glands of mammals. In: *Sex and Internal Secretions. Vol. 1* (Ed. by W. C. Young), pp. 366–448. London: Baillière Tindall & Cox.

Schultz, A. H. 1938. The relative weight of the testes in primates. *Anat. Rec.*, **72**, 387–394.

Seuanez, H. N. 1980. Chromosomes and spermatozoa of the African great apes. *J. Reprod. Fert., Suppl.*, **28**, 91–104.

Short, R. V. 1979. Sexual selection and its component parts, somatic and genital selection, as illustrated by man and the great apes. *Adv. Stud. Behav.*, **9**, 131–158.

Smuts, B. B. 1987. Sexual competition and mate choice. In: *Primate Societies* (Ed. by B. B. Smuts, D. L. Cheney, R. M. Seyfarth, R. W. Wrangham & T. T. Struhsaker), pp. 385–399. Chicago: University of Chicago Press.

Wildt, D. E., Bush, M., Goodrowe, K. L., Packer, C., Pusey, A. E., Brown, J. L., Joslin, P. & O'Brien, S. J. 1987. Reproductive and genetic consequences of founding isolated lion populations. *Nature, Lond.*, **329**, 328–331.

# 7. ELABORATION OF THE KAMIKAZE SPERM HYPOTHESIS: A REPLY TO HARCOURT

R. Robin Baker[1] and Mark A. Bellis[1]

The 'Kamikaze Sperm Hypothesis' (KSH) (Baker & Bellis 1988) proposes that the sperm in animal ejaculates are polymorphic and adapted to a variety of roles. Some are 'egg-getter sperm' adapted to proceed along the female tract and fertilize the egg(s). The remainder are 'kamikaze sperm' adapted to occupy strategic locations in the tract where they attempt to prevent the passage of any sperm inseminated by another male.

Kamikaze sperm may often be morphs in only a physiological or behavioural sense. Sometimes, however, they may be actual structural morphs (e.g. pin heads, long heads). In many mammals, such bizarre sperm consistently occur in large numbers (e.g. humans, 20–40% of sperm/ejaculate). Previously, despite their abundance, these sperm were dismissed as 'abnormalities' which natural selection had failed to eradicate (Cohen 1973). In our view, however, such consistent and abundant structures demand an adaptive explanation such as provided by the KSH. Nevertheless, Harcourt (1989) has argued that the KSH is wrong and that such sperm are still best interpreted as errors in production.

Harcourt follows tradition and assumes that the so-called 'normal' sperm (i.e. the most common) are the egg-getters. We carefully avoided this assumption in our original presentation of KSH (Baker & Bellis 1988) for there is no firm evidence that it is justified for any mammal. On the contrary, some researchers into human *in vitro* fertilization have been forced to conclude that fewer than 0.001% of sperm (i.e. no more than 2000 in an ejaculate of $300 \times 10^6$ sperm) are capable of fertilizing an egg (Lee 1988). Perhaps significantly, this corresponds roughly to the proportion of sperm that are found at or near the site of fertilization, not only in humans but also in a variety of other mammals (Hamner 1973). Evidently egg-getters may be the least, not the most, common sperm in an ejaculate and thus cannot simply be equated with the 'normal' morph. We suspect that, far from being egg-getters, the so-called 'normal' sperm are actually the most important of the kamikaze morphs. Perhaps the less common, more bizarre, structural morphs have more-specific kamikaze roles each at a particular location in the female tract (though not

[1] Department of Environmental Biology, University of Manchester.

Reprinted from *Animal Behaviour*, 37 (5), Baker, R. R. & Bellis, M. A., Elaboration of the kamikaze sperm hypothesis. pp. 865–867, Copyright (1989), with permission from Elsevier.

only in copulatory plugs as assumed by Harcourt). It remains feasible, of course, that egg-getters are visually 'normal' and are unique only in their particular physiology and behaviour. However, the possibility that egg-getters belong to a less common structural type than the 'normal' morph should at least be borne in mind.

At present, there is no visual method for identifying an egg-getter sperm in the ejaculate of any mammal. Consequently, we do not agree with Harcourt that "an obvious prediction of the 'kamikaze sperm' hypothesis is the existence of a positive association between polyandry and proportion of deformed sperm in ejaculates". Rather, we suggest that the KSH predicts the existence of a positive association between level of polyandry and "the number of kamikaze sperm in ejaculates". However, until we understand the different roles of the different sperm morphs (particularly the 'normal' morph), analyses of the type offered by Harcourt are inappropriate as tests of KSH.

Also inappropriate for the same reason are Harcourt's comments concerning the relative proportion of 'normal' and 'deformed' sperm in copulatory plugs. Instead, what is important and consistent with KSH is the observation that morphological diversity of sperm seems to differ at different locations within the female tract (e.g. in humans; Mortimer *et al.* 1982). As far as copulatory plugs are concerned, at least in rats, *Rattus norvegicus,* sperm form a mesh at particular locations within the plug, their precise position varying as a function of socio-sexual situation. Thus, when a male first ejaculates in a female, sperm are equally abundant in the front and back sections of the plug. At the second or subsequent ejaculations, sperm are most abundant in the front of the plug but often also form a neat 'cap' at the back. Such patterns of aggregation are not consistent with Harcourt's suggestion that sperm are simply trapped in the plug but would fit the view that sperm are specifically positioned before the plug begins to harden. The strategic positioning of sperm could make enough difference to a plug's properties (Baker & Bellis 1988) to disadvantage any male who used only seminal fluids for plug formation.

Harcourt's assumption that KSH applies only where there is a first male advantage is incorrect. For example, few behavioural ecologists would deny that copulatory plugs have evolved, at least in part, in response to sperm competition; yet such plugs do not always lead to a first-male advantage (e.g. rats; Dewsbury & Hartung 1980). Evolutionary arms races have produced males of some species that are as adept at removing or penetrating plugs as they are at producing such plugs (Dewsbury 1981). In the same way, evolutionary arms races between ejaculates may well have produced species with kamikaze sperm that are as proficient at clearing access for their egg-getter sperm when ejaculated by a 'second' male as they are at blocking access when ejaculated by a 'first' male. For example, kamikaze sperm could kill 'foreign' sperm either directly (using acrosomal enzymes) or indirectly (by attaching themselves to the sperm then promoting attack by the female's leucocytes). All evolutionary outcomes are possible and the existence of species without a first-male advantage is no more evidence against KSH than it is evidence against sperm competition theory in general.

One of the more interesting predictions of KSH is that males should adjust at each copulation the number of each morph ejaculated into a female according to socio-sexual situation (e.g. risk of the female double-mating). Harcourt noted this prediction but suggested that males cannot alter the ratio of sperm morphs in the short term. However, the evidence for lions, *Panthera leo* (Wildt *et al.* 1987) quoted by Harcourt derives from ejaculates collected by electroejaculation, hardly a technique that allows the male fine control! Indeed, we have now shown for humans that even masturbation is unsuitable as a collection technique for this purpose. However, when ejaculates are collected during copulation, evidence of a short-term response to socio-sexual situation is obtained, both with respect to total sperm (Baker & Bellis 1989) and morph ratios (e.g. the relative proportions of large and pin-headed sperm varies according to the percentage of time male and female spend together between copulations).

Variation in sperm number and morph distribution from copulation to copulation in response to socio-sexual situation is a major prediction of the KSH. The 'average' total and diversity of sperm that is then characteristic of a species would be determined ultimately and proximately by such factors as structure of the female tract, mating system and mean risk of double-mating. KSH naturally assumes that both the 'average' ejaculate and the response curve is under genetic control (cf. Beatty 1970). As usual, inbreeding might be expected to lead to departure from the evolved optimum. Such departure might involve an increase in the proportion of 'non-normal' morphs (as in the study of lions by Wildt *et al.* (1987) cited by Harcourt) or a decrease (as has been shown, for example, in bulls; Hultnas 1959, cited by Beatty 1970). We suggest that such a variable response to inbreeding is more in accordance with an adaptive explanation for sperm polymorphism (such as the KSH) than it is with interpretations based on errors during production.

## ACKNOWLEDGEMENTS

We thank Sandy Harcourt for giving us such prompt access to his paper and Liz Oram for critical comments on various drafts of this manuscript.

## REFERENCES

Baker. R. R. & Bellis, M. A. 1988. 'Kamikaze' sperm in mammals? *Anim. Behav.*, **36**, 936–939.

Baker, R. R. & Bellis, M. A. 1989. Number of sperm in human ejaculates varies in accordance with Sperm Competition Theory. *Anim. Behav.*, **37**, 867–869.

Beatty, R. A. 1970. The genetics of the mammalian gamete. *Biol. Rev.*, **45**, 73–119.

Cohen, J. 1973. Crossovers, sperm redundancy, and their close association. *Heredity*, **31**, 408–413.

Dewsbury, D. A. 1981. On the function of the multiple-intromission, multiple-ejaculation copulatory patterns of rodents. *Bull. Psychonom. Soc.*, **18**, 221–223.

Dewsbury, D. A. & Hartung, T. G. 1980. Copulatory behaviour and differential reproduction of laboratory rats in a two-male, one-female competitive situation. *Anim. Behav.*, **28**, 95–102.

Hamner, C. E. 1973. Physiology of sperm in the female reproductive tract. *Fogarty Inter. Cent. Proc.*, **8**, 203–214.

Harcourt, A. H. 1989. Deformed sperm are probably not adaptive. *Anim. Behav.*, **37**, 863–865.

Lee, S. 1988. Sperm preparation for assisted conception. *Conceive*, **12**, 4–5.

Mortimer, D., Leslie, E. E., Kelly, R. W. & Templeton, A. A. 1982. Morphological selection of human spermatozoa *in vivo* and *in vitro*. *J. Reprod. Fert.*, **64**, 391–399.

Wildt, D. E., Bush, M., Goodrowe, K. L., Packer, C., Pusey, A. E., Brown, J. L., Joslin, P. & O'Brien, S. J. 1987. Reproductive and genetic consequences of founding isolated lion populations. *Nature, Lond.*, **329**, 328–331.

# 8. NUMBER OF SPERM IN HUMAN EJACULATES VARIES IN ACCORDANCE WITH SPERM COMPETITION THEORY

R. Robin Baker[1] & Mark A. Bellis[1]

One of the major predictions of sperm competition theory (Parker 1970, 1982) is that there should be a positive association between number of sperm inseminated and risk of the female double-mating (i.e. mating with a second male while containing fertile sperm from a previous male). A number of across-species studies on mammals support this prediction. For example, polyandrous species of primates have larger testes and may inseminate more sperm per ejaculate than monandrous species (Short 1979; Harcourt *et al.* 1980). However, such studies tell us only that sperm competition may have shaped the general reproductive anatomy and physiology of species during mammalian evolution. They do not show that sperm competition has shaped the variation in copulatory behaviour shown by individual males. Yet the theory would predict that at each copulation the ejaculate should be adjusted according to the risk of the sperm finding themselves in competition with sperm from another male. We have tested this prediction using humans.

Our assumption that modern humans are suitable for testing the predictions of sperm competition theory is based on the following considerations. Double-mating occurs in both overtly promiscuous species and apparently monogamous species (McKinney *et al.* 1984), including humans (Smith 1984). In consequence, sperm competition may have been a selective force in shaping many aspects of human anatomy, physiology and behaviour (Smith 1984). The minimum level of double-mating necessary for sperm competition to generate and maintain selection for competitive ejaculates in humans may be calculated (from formulae in Parker 1982) to be about one copulation in every 7700. In a recent (1987–1989) survey of women in London, Rotherham and Manchester, U.K., more than one in every 1000 copulations was a double-mating (R. R. Baker & M. A. Bellis, unpublished data), at least eight times greater than the threshold level.

---

[1] Department of Environmental Biology, University of Manchester.

Reprinted from *Animal Behaviour*, 37 (5), Baker, R. R. & Bellis, M. A., Number of sperm in human ejaculates varies in accordance with sperm competition theory. pp. 867–869, Copyright (1989), with permission from Elsevier.

Demonstrable levels of cuckoldry and frequent cases of legally disputed paternity (Smith 1984) show that double-matings do lead to reproductive reward for the male whose sperm prevail in such competition, despite modern methods of contraception.

The males of many 'monogamous' species spend much of their time between copulations close to their female partner, thus reducing the risk of that female double-mating. We assume that, for such species, the level of risk is inversely proportional to the percentage of time that the pair are together. Sperm competition theory would predict, therefore, that the males of such pairs should adjust the number of sperm ejaculated according to the percentage of time the pair have been together since their last copulation.

Fifteen male-female human pairs with average sexual activity (one–three copulations per week; Smith 1984) were recruited from staff and postgraduates in the School of Biological Sciences at the University of Manchester. Each pair was asked to collect an ejaculate by condom during copulation and to return the sample along with a completed questionnaire.

One potential problem in using ejaculates collected by condom is that the appropriate psychological and other stimuli may not be present. Except through accident, the ejaculated sperm are not destined for competition: a scenario that could militate against a male responding in accordance with sperm competition theory. Of course, the problem is conservative making it less, not more, likely that a negative correlation between sperm number and the percentage of time together will emerge. Nevertheless, we sought some control by asking each pair also to return a masturbatory ejaculate, again collected in a condom but with no genital contact. Such ejaculations should be even further divorced from the stimuli associated with normal copulation and thus even less likely to fit the predictions of the theory.

An additional advantage of having more than one ejaculate from each pair was that it facilitated a double-blind protocol: (1) subjects could be told only that the aim of the experiment was some form of comparison of copulatory and masturbatory samples; and (2) sperm could be counted by an experimenter unaware of the type of sample being counted.

Subjects were provided with a 'kit' containing instructions and all necessary equipment, including lubricated (non-spermicidal) condoms. After ejaculation, subjects suspended the condom from a cardboard stand for 8 h or more while the ejaculate decoagulated. With scissors, subjects then cut the condom so that the end containing the ejaculate fell into a screw-top jar containing 52 ml of fixative (2% glutaraldehyde in a phosphate buffer, pH 7.2). The jar was shaken to disperse the ejaculate and was then labelled with a cryptic code. The same code was written on a questionnaire and only here was the sample identified as masturbatory or copulatory. Among other information requested for each sample was: time since last ejaculation; time since last copulation; and the percentage time together since last copulation. The completed questionnaire was placed in an envelope which was then sealed.

Pairs were asked to return at least two ejaculates (one copulatory; one masturbatory) collected, as nearly as possible, during their normal sexual activity. Samples were taken from April to October 1988. Between pairs, there was no pattern to the sequence of collection and, in most cases, samples were not consecutive and were interspersed with other ejaculations that were part of the pair's normal activity. Five of the 15 pairs failed to return any sample. The remainder returned a total of 34 (18 copulatory; 16

masturbatory). Median time interval between ejaculation of the sample and the male's previous ejaculation was 56 h (IQR: 18–71) for copulatory samples and 54 h (22–78) for masturbatory samples. Ejaculates were returned for sperm counts to MAB; questionnaires to RRB. Estimates of the number of sperm per ejaculate were based on procedures in the *Human Semen Manual* (Belsey *et al.* 1980) except that we used an 'Improved Neubauer Haemocytometer' and obtained a mean (±SE) from eight (not two) samples per ejaculate.

To avoid problems of independence, we restricted analysis to the 10 copulatory and 10 masturbatory samples first provided by each pair. As predicted, the copulatory samples show a highly significant ($r_s = -0.948$, $P_{one\text{-}tailed} < 0.001$) rank-order correlation between the estimated number of sperm in an ejaculate and the percentage of time a pair have spent together since their last copulation (Fig. 1). The masturbatory samples show no significant rank-order correlation ($r_s = -0.195$, $P_{one\text{-}tailed} = 0.295$). The rank-order correlation coefficients for copulatory and masturbatory samples are significantly different ($z = 3.020$, $P_{one\text{-}tailed} = 0.0013$; $z$-transformation test; Chambers 1958).

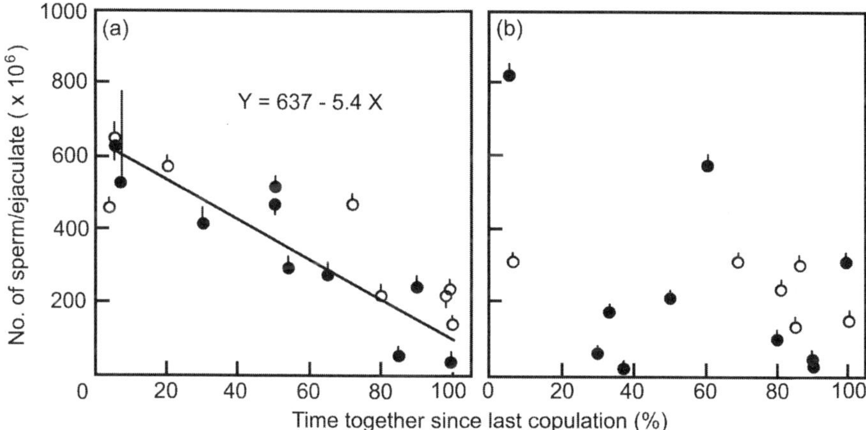


**Figure 1.** Variation in number of sperm in human ejaculates in relation to the percentage of time the male and female have spent together since their last copulation. (a) Copulation; (b) Masturbation. Data points show the estimated number of sperm in a single ejaculate ($\bar{X}$ ± SE based on eight separate counts per ejaculate). ●: first sample of each type provided by a pair; O: subsequent samples.

Regression analysis showed that the percentage of time a pair spent together was a strong (79% of variance explained) and significant ($F_{1,8} = 30.5$, $P_{two\text{-}tailed} < 0.001$) predictor of the number of sperm ejaculated during copulation. In contrast, time since last ejaculation was a weak (32%) and nonsignificant ($F_{1,8} = 3.7$, $P_{two\text{-}tailed} > 0.2$) predictor. Moreover, multiple regression analysis and analysis of residuals failed to find even a secondary influence of time since last ejaculation. Thus, we found no evidence of physiological constraint imposed on number of sperm ejaculated during copulation by rate of sperm production (approximately $300 \times 10^6$ per day; Johnson *et al.* 1980). Of course, this conclusion may not hold at higher levels of sexual activity. Multiple regression analysis of the masturbatory samples showed that the best predictor (70% of variance explained) of the number of sperm ($N \times 10^6$) was time ($Th$) since last copulation ($N = 3.806T + 22.30$; $F_{1,8} = 18.47$, $P_{\text{two-tailed}} < 0.01$).

This study has shown that the number of sperm inseminated into a human female during copulation varies in a way that fits the predictions of sperm competition theory. No part of our data allows us to identify the mechanism of adjustment. Pheromonal effects, short- and/or long-term psychological effects or even variation in the sexual stimuli given by the female are all possible. However, whatever the mechanism, adjustment persists despite, and may even be unaffected by, the wearing of a condom, at least as long as copulation is otherwise normal with full genital penetration. When ejaculation is the result of masturbation, such adjustment is not apparent. Instead, the number of sperm ejaculated relates primarily to the time interval since last copulation.

## ACKNOWLEDGEMENTS

We thank our volunteers, both male and female, for their courage in taking part, Liz Oram for revision of the manuscript and an anonymous referee for statistical advice.

## REFERENCES

Belsey, M. A., Eliasson, R., Gallegos, A. J., Moghissi, K. S., Paulsen, C. A. & Prasad, M. R. N. 1980. *WHO Laboratory Manual for the Examination of Human Semen and Semen/Cervical Mucus Interaction.* Singapore: Press Concern.

Chambers, E. G. 1958. *Statistical Calculation for Beginners*. Cambridge: Cambridge University Press.

Harcourt, A. H., Harvey, P. H., Larson, S. G. & Short, R. V. 1980. Testis weight, body weight and breeding system in primates. *Nature, Lond.*, **293**, 55–57.

Johnson, L., Petty, C. S. & Neaves, W. B. 1980. A comparative study of the daily production and testicular composition in humans and rats. *Biol. Reprod.*, **22**, 1233–1243.

McKinney, F., Cheng, K. M. & Bruggers, D. J. 1984. Sperm competition in apparently monogamous birds. In: *Sperm Competition and the Evolution of Animal Mating Systems* (Ed. by R. L. Smith), pp. 523–540. London: Academic Press.

Parker, G. A. 1970. Sperm competition and its evolutionary consequences in the insects. *Biol. Rev.*, **45**, 525–567.

Parker, G. A. 1982. Why are there so many tiny sperm? Sperm competition and the maintenance of two sexes. *J. Theor. Biol.*, **96**, 281–294.

Short, R. V. 1979. Sexual selection and its component parts, somatic and genital selection, as illustrated by man and the great apes. *Adv. Stud. Behav.*, **9**, 131–158.

Siegel, S. 1956. Nonparametric Statistics for the Behavioural Sciences. London: McGraw-Hill.

Smith, R. L. 1984. Human sperm competition. In: *Sperm Competition and the Evolution of Animal Mating System* (Ed. by R. L. Smith), pp. 601–660. London: Academic Press.

# 9. DO FEMALES PROMOTE SPERM COMPETITION? DATA FOR HUMANS

Mark A. Bellis[1] and R. Robin Baker[1]

By definition, most mating by monogamous species is in-pair copulation (IPC). However, an apparently universal feature of such species (Mock & Fujioka 1990), is that from time to time both sexes engage in extra-pair copulation (EPC). The advantage of infidelity seems straightforward for the male (Trivers 1972) but may be more subtle for the female (e.g. Smith 1984).

A special category of EPC is double-mating (the female mating with a second male while still containing fertile sperm from a previous male). The resulting sperm competition (Parker 1970) is usually analysed from the males' viewpoint but also imposes selection on the female. If sperm competitiveness is heritable, females fertilized by the most competitive sperm will gain through the greater reproductive success of their sons (Fisher 1930). If a female is unable to judge ejaculate quality from a male's appearance (e.g. in sheep; Gibson & Jewell 1982), her best strategy is actively to promote sperm competition (Smith 1984). Costs (e.g. desertion) are minimized and benefits maximized by concentrating double-matings around times of peak fertility (e.g. oestrus in mammals).

We have collected data for humans that allow us to address the function of female infidelity. Our questionnaire was distributed throughout Britain in March/April 1989 by *Company* magazine (Bellis *et al.* 1989). This report is based on those 2708 females (out of 3679 respondents) who: claimed to have a main male sexual partner; indicated whether their last copulation was an IPC or EPC; and provided enough information for us to calculate length of menstrual cycle and day of cycle of last copulation (after conventional standardization to a cycle of 28 days; McCance *et al.* 1937). Results are analysed using non-parametric statistics, particularly the *G*-test (Sokal & Rohlf 1981), and the powerful blocking, frequency-handling and hypothesis-testing facilities of Meddis' (1984) non-parametric analysis of variance (test statistics: *H* for non-specific 'two-tailed' tests; *z* for specific 'one-tailed' tests in which hypotheses are formulated by lambda coefficients).

---

[1] Department of Environmental Biology, University of Manchester.

Reprinted from *Animal Behaviour*, 40 (5), Bellis, M. A. & Baker, R. R. Do females promote sperm competition? pp. 997–999, Copyright (1990), with permission from Elsevier.

Initially, the menstrual cycle was divided into its three major hormonal phases (Hawker 1984): I (days 1–5, menses); II (6–14, proliferative); and III (15–28, secretory). In a standardized cycle, ovulation occurs on day 14 and copulations are fertile on days 9–14 (peak fertility = day 12; Barrett & Marshall 1969). Responses to the questionnaire were non-uniformly distributed with respect to these three phases (I:II:III. Observed = 660 (days 1–5):1135 (6–14):1615 (15–28). Expected = 609: 1096:1705; $G$ = 9.788, $P$ = 0.008). However, there was no significant difference ($H$ = 0.539, $P$ = 0.460) in the mean ± SE response rates in phases I (132 ± 6/day) and II (126 ± 3/day). The significant 10% decrease ($H$ = 6.512, $P$ = 0.010) to 115 ± 8/day in phase III was due entirely to females apparently being 25% less likely to complete questionnaires on days 17–24. None of the major analyses or conclusions of this paper is negated by this pattern of non-uniformity. All conclusions that could be affected are confirmed by tests based on differences (in frequency distributions or correlation coefficients) or on proportions. Such tests are unaffected by non-uniformity of response to the questionnaire.

IPCs show peak occurrence (mean ± SE of number reported/day out of a total of 2546) in phase III (I:II:III = 80 ± 3:89 ± 4:96 ± 5). This pattern matches that for married women in North Carolina (Udry & Morris 1968) and is significantly non-uniform (lambda coefficients 1–2–3; $z$ = 1.876, $P$ = 0.039). The distribution of IPCs (I:II:III = 399:797:1350) differs significantly ($G$ = 6.687, $P$ = 0.035) from EPCs (28:63:71). As anticipated, EPCs show a pre-ovulatory peak (I:II:III = 5.6 ± 0.5/day: 7.0 ± 1.2:5.1 ± 0.6) that is also significantly non-uniform (lambda coefficients 2–3–1; $z$ = 1.706, $P$ = 0.044). Most importantly, as a proportion of total copulations in each phase, this pre-ovulatory peak of EPCs (I:II:III = 6.6%:7.3%:5.0%) is highly significant (lambda coefficients 2–3–1; $z$ = 2.671, $P$ = 0.004).

Barrett & Marshall (1969) have measured the probability of conception for copulations on different days of the menstrual cycle (Fig. 1). Their probabilities may be used both as independent variables (in correlation analysis) and lambda coefficients (in analysis of variance). To be conservative, we avoided undue favourable bias from zero probabilities of conception by restricting analysis to days 6–20. Pairing these days as in Fig. 1, EPCs ($r_S$ = 0.911, $N$ = 8, $P$ = 0.001, one-tailed) but not IPCs ($r_S$ = –0.417, $P$ = 0.848, one-tailed) show a highly significant positive correlation with probability of conception. These correlation coefficients are very different ($z$ = 3.127, $P$ = 0.001, two-tailed; $z$-transformation test; Sokal & Rohlf 1981).

The fertile life of human sperm, once inseminated, is still uncertain, but is at least 5 days (Barrett & Marshall 1969; Ferin *et al.* 1973) and may be longer (Smith 1984). Of the 162 women who claimed their last copulation was an EPC, 50 indicated the event was within 5 days of the last copulation with their main partner. We considered these to be double-matings. If EPCs on days 6–20 are separated into double-matings and non-double-matings (Fig. 1), double-matings ($z$ = 2.519, $P$ = 0.006; Meddis' test) but not non-double-matings ($z$ = 1.562, $P$ = 0.059) show a significant association with probability of conception.

None of these trends is changed by contraceptive practice. The distribution (phases I:II:III) of 2154 'protected' IPCs (336:684:1134) did not differ ($G$ = 1.071, $P$ = 0.591) from 392 'unprotected' IPCs (63: 113:216), Similarly, the distribution of 122 'protected' EPCs (18:49:55) did not differ ($G$ = 1.369, $P$ = 0.509) from 40 'unprotected' EPCs (10:14:16), Double-matings and non-double-mating EPCs showed similar levels of contraception (78 and 74% respectively; $G = 0.432$, $P = 0.518$), but EPCs as a whole were less likely to involve contraception than IPCs. Over the entire cycle, 24.7% of 162 EPCs were 'unprotected' compared with only 15.4% of 2546 IPCs ($G = 10.016$, $P = 0.002$), Even during the fertile period (days 6–15), 26.0% of 77 EPCs were 'unprotected' compared with only 14.0% of 895 IPCs ($G = 8.223$, $P = 0.005$).

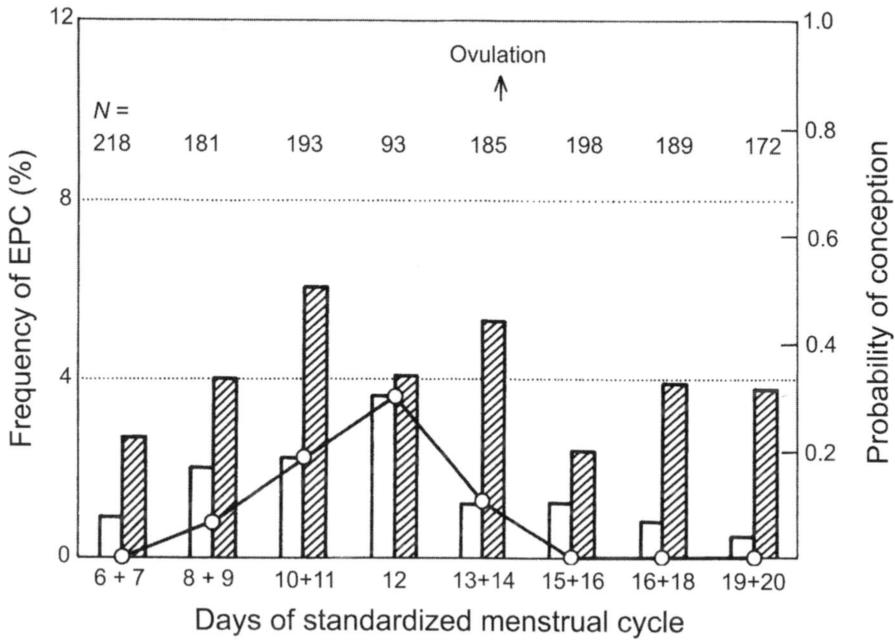


**Figure 1.** Frequency distribution of extra-pair copulations (EPCs) by female humans in relation to probability of conception on days 6–20 of a standardized 28–day menstrual cycle. EPCs are divided into double-matings (□ i.e. within 5 days of last in-pair copulation, IPC) and non-double-matings (▨ i.e. more than 5 days after last IPC). Frequency is expressed as a percentage of total copulations (*N*) reported on each day (or day-pair). The line shows the probability of conception on each day (from Barrett & Marshall 1969, based on a study of 1898 menstrual cycles by 241 women).

The sample of females on which our analyses are based is non-random. This should not invalidate our analysis of trends in the data but could affect our measures of average rates. As far as level of EPC is concerned, however, we suspect that our sample is not unrepresentative. EPCs constitute 13.8% of 145 'unprotected' copulations during the fertile period. Most EPCs will have been preceded by IPCs but few IPCs will have been preceded by EPCs. On average, EPCs that generate sperm competition should have a 50% chance of fertilization by the extra-pair male. Other EPCs should have a higher chance. Our study thus predicts a level of paternal discrepancy (i.e. offspring sired by males other than their putative fathers) of between 6.9 and 13.8%. Blood group studies in

Britain indicate levels of paternal discrepancy of from 5.7% ($N$ = 2578; Southern England; Edwards 1957) up to 20–30% (sample size unknown; 'Liverpool Flats' study; McLaren cited in Cohen 1977) and 30% ($N$ = approximately 250; Southern England; Philipp 1973). There is no indication here that the level of EPC reported in our nationwide study is either unrepresentative or high. Moreover, at 110 ± 23/year, the mean rate of IPCs in our study is in the middle of the range (1–3/week) reported for humans by other studies (Smith 1984).

One interpretation of our data is that the pattern of EPCs is male-driven, extra-pair males having been selected to seek out oestrous females. However, for this view to be tenable, our data require acceptance of two unlikely phenomena: (1) extra-pair males prefer oestrous females who contain sperm from IPCs; and (2) in-pair males have been selected for peak copulation when the female is least fertile. We suggest our data are more consistent with a female-driven phenomenon. Anecdotal observations (Mason 1966) imply similar behaviour by the females of another monogamous primate,

*Callicebus moloch.* Although EPCs may well result in greater genetic variety among the female's offspring, 'better' genes, and/or additional paternal care (see Smith 1984), the primary consequence of the behaviour revealed by our analysis appears to be the promotion of sperm competition.

## ACKNOWLEDGEMENTS

We thank Dr J. Cohen and Dr K. Richardson for help with literature.

## REFERENCES

Barrett, J. C. & Marshall, J. 1969. The risk of conception on different days of the menstrual cycle. *Popul. Stud.*, **23**, 455–461.

Bellis, M. A., Baker, R. R., Hudson, G., Oram, E. & Cook, V. 1989. *Company,* **April**, 90–92.

Cohen, J. 1977. *Reproduction.* London: Butterworths.

Edwards, J. H. 1957. A critical examination of the reputed primary influence of ABO phenotype on fertility and sex ratio. *Br. J. Prev. Soc. Med.* **11**, 79–89.

Ferin, J., Thomas, K. & Johansson, E. D. B. 1973. Ovulation detection. In: *Human Reproduction: Conception and Contraception* (Ed. by E. S. E. Hafez & T. N. Evans), pp. 260–283. New York: Harper & Row.

Fisher, R. A. 1930. *The Genetical Theory of Natural Selection.* Oxford: Clarendon Press.

Gibson, R. M. & Jewell, P. A. 1982. Semen quality, female choice and multiple mating in domestic sheep: a test of Trivers' sexual competence hypothesis. *Behaviour,* **80**, 9–31.

Hawker, R. W. 1984. *Notebook of Medical Physiology: Endocrinology.* 2nd edn. London: Churchill Livingstone.

McCance, R. A., Luff, M. C. & Widdowson, E. E. 1937. Physical and emotional periodicity in women. *J. Hygiene,* **37**, 571–611.

Mason, W. 1966. Social organisation of the South American monkey, *Callicebus moloch:* a preliminary report. *Tulane Stud. Zool.,* **13**, 23–28.

Meddis, R. 1984. Statistics Using Ranks: a Unified Approach. Oxford: Blackwell.

Mock, D. W. & Fujioka, M. 1990. Monogamy and long-term pair bonding in vertebrates. *Trends Ecol. Evol.,* **5**, 39–43.

Parker, G. A. 1970. Sperm competition and its evolutionary consequences in the insects. *Biol. Rev.,* **45**, 525–567.

Philipp, E. E. 1973. Discussion: moral, social and ethical issues. In: *Law and Ethics of A.I.D. and Embryo Transfer. Ciba Foundation Symposium. Vol.* 17 (Ed. by G. E. W. Wolstenholme & D. W. Fitzsimons), pp. 63–66. London: Associated Scientific.

Smith, R. L. 1984. Human sperm competition. In: *Sperm Competition and the Evolution of Animal Mating Systems* (Ed. by R. L. Smith), pp. 601–660. London: Academic Press.

Sokal, R. R. & Rohlf, F. J. 1981. *Biometry.* 2nd edn. New York: W. H. Freeman.

Trivers, R. L. 1972. Parental investment and sexual selection. In: *Sexual Selection and the Descent of Man* (Ed. by B. Campbell), pp. 136–179. London: Heinemann.

Udry, J. R. & Morris, N. M. 1968. Distribution of coitus in the menstrual cycle. *Nature, Lond.,* **220**, 593–596.

Siegel, S. 1956. Nonparametric Statistics for the Behavioural Sciences. London: McGraw-Hill.

Smith, R. L. 1984. Human sperm competition. In: *Sperm Competition and the Evolution of Animal Mating System* (Ed. by R. L. Smith), pp. 601–660. London: Academic Press.

# 10. HUMAN SPERM COMPETITION: EJACULATE ADJUSTMENT BY MALES AND THE FUNCTION OF MASTURBATION

R. Robin Baker[1] and Mark A. Bellis[1]

## ABSTRACT

Sperm competition theory argues that the number of sperm inseminated into a female by a male is a trade-off between two opposing pressures. On the one hand, the risk that sperm may find themselves in competition with the sperm from another male favours the male inseminating more sperm. On the other hand, ejaculates are costly to produce and males are favoured who economize over the number of sperm inseminated. This paper analyses: (1) sperm numbers and other ejaculation data for 35 human couples; and (2) data relating to the most recent copulation reported by 3587 women. Three sets of predictions based on sperm competition theory are tested. These are that the number of sperm inseminated should be a function of: (1) risk of sperm competition; (2) female reproductive value; and (3) optimum partitioning of ejaculates between successive in-pair copulations. During in-pair copulation the male used successive inseminations to 'top-up' the population of sperm in his female partner. In accordance with sperm competition theory: (1) individual males inseminated more sperm when the pair had spent a smaller proportion of their time together and hence risk of sperm competition was greater; and (2) larger females were inseminated with more sperm than smaller females. In apparent contradiction to sperm competition theory, the number of sperm inseminated did not vary according to female orgasm pattern or the probability of conception. This apparent failure of the theory may instead be due to the male's lack of necessary information. Paradoxically, male mammals seem to waste huge numbers of sperm through spontaneous emission and self-masturbation. Such shedding of sperm could be adaptive if it led to more competitive and/or more fertile inseminates at the next copulation. The data showed that a recent male masturbation reduced the number of sperm inseminated at the

---

[1] Department of Environmental Biology, University of Manchester.

Reprinted from *Animal Behaviour*, 46 (5), Baker, R. R. & Bellis, M. A., Human sperm competition: Ejaculate adjustment by males and the function of masturbation. pp. 861–885, Copyright (1993), with permission from Elsevier.

next copulation but not the number retained by the female. It is concluded that masturbation is a male strategy to increase sperm fitness without increasing sperm numbers in the female tract. The possibility that, in the absence of sperm competition, the probability of fertilization decreases if too many sperm are inseminated is discussed. This latter factor may be more important than ejaculate cost in favouring male restraint over the number of sperm inseminated.

---

It is conventional, for species in which males and females form long-term pair bonds with some form of mate guarding, to distinguish between in-pair copulations and extra-pair copulations (i.e. with an individual other than a long-term partner). A particular category of extra-pair copulation is 'double-mating' which occurs when a female mates with a second male while still containing fertile sperm from one or more previous males. The result is 'sperm competition' (Parker 1970*a*) as the sperm from different males compete to fertilize the female's egg(s).

There is some debate over whether sperm competition takes the form of a lottery (Parker 1982, 1984) or warfare (Sivinski 1980; Silberglied *et al.* 1984; Baker & Bellis 1988, 1989*b*; Harcourt 1991). On either mechanism, however, it is predicted that sperm competition will favour males who inseminate larger ejaculates (i.e. more sperm/ejaculate; Parker 1990). Although individual males differ in the general competitiveness of their sperm, the more sperm inseminated from an individual male, the greater his chances of success in sperm competition (Martin *et al.* 1974).

Current sperm competition theory states that the number of sperm a male inseminates into a female is a trade-off between two opposing pressures. On the one hand, the risk of sperm competition favours an increase in the number of sperm inseminated (Parker 1982); on the other, as sperm are inseminated in 'packages' ( = ejaculates), non-trivial costs involved in ejaculate production (Dewsbury 1982) require restraint over the allocation of sperm to individual ejaculates. The predicted result is that males will inseminate a female with the number of sperm that is the optimum trade-off between the risk of sperm competition and the need for economy.

This theory has generated at least three predictions: (1) the greater the risk of sperm competition, the more sperm should be inseminated (Parker 1982); (2) the greater the reproductive value of a female, the more a male should invest in sperm competition and hence the more sperm should be inseminated (Dewsbury 1982); and (3) under some circumstances males do better in sperm competition to partition sperm between a succession of in-pair copulations rather than inseminate all available sperm in a single in-pair copulation (Parker 1984).

The first of these predictions (that males should inseminate more sperm when the risk of sperm competition is greater) has now been tested at three levels: interspecific (butterflies: Svard & Wiklund 1989; birds: Møller 1991; primates: Harvey & Harcourt 1984; and ungulates: Ginsberg & Rubenstein 1990); intraspecific (beetles, *Tenebrio molitor*: Gage & Baker 1991; flies, *Ceratitis capitata*: Gage 1992; rats, *Rattus norvegicus*: Bellis *et al.* 1990; and humans: Baker & Bellis 1989*a*); and intra-male (beetles, Gage & Baker 1991). So far, all tests are consistent with the predictions of sperm competition theory: males inseminate more sperm when the risk of sperm competition is higher.

In contrast to the number and variety of tests of this first prediction, there has been minimal attempt to test the second (that males should invest more in sperm competition when the female's reproductive value is higher). One initial problem is that the two predictions may sometimes be difficult to separate. For example, females of greater reproductive value may be inseminated by more males and thus offer greater risk of sperm competition. Even so, there are some circumstantial data consistent with the prediction, at least for insects. Thus, male dung flies, *Scatophaga stercoraria*, copulate for shorter periods with females containing fewer eggs (Parker 1970b). Male mormon crickets, *Anabrus simplex*, are less likely to transfer a spermatophore to lighter females containing fewer eggs (Gwynne 1981). In both cases, it seems likely that, on average, females containing fewer eggs are inseminated with fewer sperm.

Female mammals may differ in their reproductive value in many ways. In this paper we consider four: (1) body size; (2) sperm retention; (3) stage of menstrual cycle; and (4) use of oral contraception.

In many animals heavier females contain more eggs (Halliday 1983). Insofar as dizygotic twinning rates are higher in heavier human females (MacGillivray & Campbell 1978) the same may, on average, be true for humans. In addition, heavier human females have reduced risk of miscarriage during pregnancy, faster growth of the fetus and heavier birth weight (Lechtig & Klein 1981; Bongaarts & Potter 1983; Garn & LaVelle 1983; Gray 1983). In some societies, heavier women have higher offspring survival and, at anyone time, more living children (Hill & Kaplan 1990). A male preference for larger females is reported to exist, both in humans (Ford & Beach 1952; Mulder 1990) and other animals (Halliday 1983) and male humans are reported to compete more for, and to show greater defence of, more fertile and fecund women (Flinn 1988). Sperm competition theory would predict, therefore, that males inseminate larger females with more sperm.

The function of the female orgasm in humans is discussed in detail in a separate paper (Baker & Bellis 1993). There we show a significant association between sperm retention and the occurrence and timing of the female orgasm relative to the timing of male ejaculation. It follows that, at any given copulation, the timing of female orgasm is likely to have some association with the reproductive value of the female to the male. We might expect males to vary the number of sperm inseminated accordingly.

Both stage of menstrual cycle and use and efficacy of contraception influence the probability of conception and hence, at any given copulation, the reproductive value of the female to the male.

Conventionally, studies of the menstrual cycle of human females use conversion to a standardized 28-day cycle (McCance *et al.* 1937). This standardized cycle may be divided into three major hormonal phases (Hawker 1984): I (days 1–5, menses); II (6–14, proliferative); and III (15–28, secretory). In this standardized cycle, ovulation occurs on day 14 and copulations are fertile on days 9–14 (peak fertility day 12; Barrett & Marshall 1969). The rank-order of the three phases for fertility is II > I > III (i.e. phase II is the most fertile).

As far as contraception is concerned, the probability of conception (expressed as % pregnant women in 100 fertile women-years; i.e. 100 women for 1 year or 10 women for 10 years) is 5–30% when couples use a condom and < 1% when the female uses an oral contraceptive (combined pill; Johnson & Everitt 1988). When couples use both a condom and an oral contraceptive, the probability of conception should be correspondingly lower than when either method of contraceptive is used on its own. We might expect males to adjust the number of sperm inseminated in response to these altered levels of probability of conception.

The third prediction made by sperm competition theory is that males should partition sperm numbers strategically between successive in-pair copulations during the course of their partner's reproductive cycle. The most detailed consideration of the optimum strategy was by Parker (1984) though other authors (e.g. Ginsberg & Rubenstein 1990) have at least made assumptions as to how partitioning would influence the course of sperm competition. As yet, however, there has been no empirical study of the allocation of sperm to successive in-pair copulations which might allow the theoretical models to be evaluated.

In this paper we present data on the number of sperm inseminated by humans that allow us to test all three types of prediction derived from sperm competition theory. Thus, we evaluate predictions concerning: intra-male variation; variation in female reproductive value; and the partitioning of sperm between successive in-pair copulations.

In contrast to the wide range of evidence that increased risk of sperm competition favours larger ejaculates, there is as yet no unequivocal evidence that the cost of producing ejaculates is the major restraining factor in determining the number of sperm inseminated per ejaculate. On the contrary, behaviour exists, so far unconsidered, that is intuitively inconsistent with ejaculate cost being a major factor favouring male restraint, at least in mammals. Thus, male mammals appear to be extremely wasteful of sperm. Not only do they void large numbers of sperm in their urine (Mann & Lutwak-Mann 1981) but they also shed vast numbers during spontaneous (in humans, usually nocturnal) emissions and through the more directed behaviour of self-masturbation.

Such shedding of sperm is a characteristic of all male mammals studied (e.g. rhesus monkeys, *Macaca mulatta*: Carpenter 1942; cats, *Felis catus*: Rosenblatt & Schneirla 1962; deer, *Cervus elaphus*: Darling 1963; cattle, *Bos taurus*: Hafez *et al.* 1969*a*; horses, *Equus caballus*: Hafez *et al.* 1969*b*; and rats: Beach 1975) and is a virtually universal facet of the sexual behaviour of male humans (Kinsey *et al.* 1948). Such spontaneous and/or deliberate shedding of sperm is by no means restricted to males denied access to females, either in humans or other mammals. Thus, stags with harems frequently shed whole ejaculates following antler rubbing (Darling 1963) and dominant male rhesus monkeys masturbate to ejaculation even with full access to oestrous females (Carpenter 1942).

This apparent paradox could be resolved if masturbation and spontaneous emissions produced a future ejaculate with a fitness enhanced beyond the level necessary to offset the numeric and energetic costs of the lost sperm. Previous authors have suggested that sperm have a limited 'shelf-life' and, if not used within a certain storage time, become suboptimal (e.g. Smith 1984). Masturbation is thus suggested to be a mechanism for shedding suboptimal sperm and reducing the mean age of sperm inseminated at the next copulation. The advantage to the male could be that younger sperm are more acceptable to the female and/or are better able to reach a secure position in the female tract. Moreover, once retained in the female tract, younger sperm could be more fertile in the absence of sperm competition and/or more competitive in the presence of sperm competition. Finally, if younger sperm live longer in the female tract, any enhanced fertility and competitiveness would also last longer. However, no evidence exists that either masturbation or spontaneous emission improves the fitness of a future ejaculate. The only support for the 'sperm age' hypothesis is the observation that after collection rabbit, *Oryctolagus cuniculus*, sperm aged outside of the male rapidly become poorer at competing for eggs against sperm from freshly collected ejaculates (Roche *et al.* 1968).

The data presented in this paper allow the first empirical test of the 'sperm age' hypothesis. The result of our analysis is the emergence of another potential constraint on number of sperm inseminated that we suggest may be more important than the cost of ejaculate production.

## METHODS

Our data come from four different investigations: (1) counts of sperm in whole ejaculates collected by condom during either copulation or male masturbation; (2) counts of sperm in 'flow backs' (the mixture of seminal fluids, sperm, female secretions and female tissue that flows back out of the vagina after copulation); (3) subjective estimates of flowback volume; and (4) a U.K. nationwide survey of female sexual behaviour.

In our nationwide survey, 3679 females each answered 57 questions on their sexual behaviour, including information relating to their most recent copulation. The copulations for which we obtained data included 2744 in-pair copulations and 126 extra-pair copulations (of which 76 were double-matings).

In our ejaculate study, 35 male-female human pairs (involving 33 males and 33 females; two male and two female contributors changed partners during the study) provided material or estimates relating to 323 in-pair copulations and 67 male masturbations. Our 66 volunteers were recruited through staff, postgraduates and undergraduates in the School of Biological Sciences at the University of Manchester. Pair codes and the number of different types of usable samples and/or estimates they each provided are listed in Table I.

### Whole Ejaculates

Whole ejaculates were collected in condoms during copulation or masturbation. Subjects were provided with a 'kit' containing instructions and all necessary equipment, including a mixture of lubricated (non-spermicidal) and non-lubricated condoms. Ejaculate collection and fixation and the counting of sperm followed the double-blind protocol used previously (Baker & Bellis 1989*a*), based on the World Health Organization Human Semen Manual (Belsey *et al.* 1987).

In addition to information relating to each sample, volunteers were asked their age, weight and height. Males also measured the length (L) and width (W) of their left testis in centimetres (to one decimal place), using callipers. Volume (V $cm^3$) was then calculated using the formula for a spheroid: $(\pi/6) \times L \times W^2$.

### Flowbacks

The detailed instructions given to volunteers for the collection of flowbacks are published in full in Baker & Bellis (in press). The flowback emerges 5–120 min after copulation as a relatively discrete event over a period of 1–2 min in the form of three to eight white globules (for further details see Baker & Bellis 1993). With practice, females can recognize the sensation of the beginning of flowback and can collect the material by squatting over a 250 ml glass beaker. Once the flowback is nearly ready to emerge, it can be hastened by, for example, coughing. If the flowback has not yet emerged, it invariably does so at the female's first urination after copulation and on these occasions the flowback is often ejected with some force (cf Ginsberg & Rubenstein 1990 for zebra *Equus* spp). Again with practice, the female can collect the flowback during urination with minimal, or no, contamination from urine. Volunteers were asked to record which method of collection was used.

After flowback collection, the protocol followed was the same as for whole ejaculates.

Estimates of flowback volume are not used in this paper, but we do use the associated information on the in-pair copulations recorded (Table I).

**Nationwide Survey of Sexual Behaviour**

Our survey was based on a questionnaire developed between 1987 and 1989 in three pilot studies (two self-selected; one by interview), involving 250 females (Baker & Bellis 1989*a*). The final version (seeking 57 answers) was distributed throughout Britain in March/April 1989 by *Company* magazine (Bellis *et al.* 1989). Female readership of the relevant issue was estimated by the publishers to be 439 000. Our 3679 replies (excluding seven overtly spoilt) therefore represent 0.84% of the potential respondents who were themselves roughly 5% of the U.K. population of females of reproductive age. Ninety-two respondents claimed to be virgins leaving a sample of 3587 sexually experienced females aged between 13 and 72 years (mode = 21 years). Other major characteristics of our sample have been published elsewhere (Baker *et al.* 1989; Bellis & Baker 1990).

**Statistics: Calculation of Probabilities**

Throughout, medians are used instead of means. Variation about the median is expressed in terms of inter-quartile range. In figures and tables, as appropriate, the medians presented are the medians of medians (i.e. the median value is calculated for each male or female, then the median is calculated for the group). The calculation of probabilities, however, uses the techniques described below.

At one point, we used the *z*-transformation test (Sokal & Rohlf 1981) to compare correlation coefficients. Otherwise, to avoid making any assumptions concerning the normality of our data, we calculated probability values using only non-parametric statistical tests.

We consider (following Siegel 1956 and Meddis 1984) that the use of parametric statistics requires justification on each application. This justification should involve demonstration that the data are significantly similar to a normal distribution, not merely that they do not significantly depart from a normal distribution. Not only could we not justify the use of parametric statistics, in some parts such statistics were clearly unjustified due to skewed, bi- or multi-modality etc., unsuitable for transformation. We

**Table I.** Codes and number of different types of samples provided by couples collecting whole ejaculates and/or flowbacks.

| | | | No. of samples | | | |
|---|---|---|---|---|---|---|
| | | | | | Flowbacks | |
| Couple | Male | Female | Masturbation ejaculates | In-pair copulation ejaculates | Samples | Estimates |
| A | A | A | 6 | 4 | 9 | 0 |
| B | B | B | 17 | 27 | 93 | 8 |
| C | C | C | 1 | 2 | 0 | 0 |
| D | D | D | 1 | 1 | 0 | 0 |
| E | E | E | 1 | 1 | 0 | 0 |
| F | F | F | 1 | 1 | 0 | 0 |
| G | G | G | 2 | 1 | 1 | 0 |
| H | H | H | 1 | 2 | 1 | 0 |
| I | I | I | 1 | 1 | 0 | 0 |
| J | J | J | 1 | 1 | 0 | 0 |
| K | K | K | 1 | 1 | 0 | 0 |
| L | L | L | 2 | 3 | 0 | 0 |
| M | M | M | 1 | 1 | 0 | 0 |
| N | N | N | 2 | 22 | 2 | 0 |
| O | O | O | 0 | 1 | 0 | 0 |
| P | P | P | 1 | 1 | 0 | 0 |
| Q | Q | Q | 1 | 1 | 1 | 0 |
| R | R | R | 0 | 1 | 8 | 0 |
| S | S | S | 2 | 1 | 6 | 0 |
| T | T | T | 5 | 3 | 0 | 0 |
| U | U | U | 1 | 3 | 0 | 0 |
| V | V | V | 0 | 2 | 0 | 0 |
| W | W | I | 0 | 0 | 3 | 0 |
| X | X | X | 8 | 0 | 0 | 0 |
| Y | Y | Y | 2 | 1 | 0 | 0 |
| Z | Z | Z | 0 | 0 | 2 | 26 |
| AA | A | W | 0 | 0 | 1 | 0 |
| AB | O | C | 8 | 2 | 0 | 0 |
| AC | AC | AC | 0 | 0 | 0 | 19 |
| AD | AD | AD | 0 | 0 | 0 | 6 |
| AE | AE | AE | 0 | 0 | 0 | 14 |
| AF | AF | AF | 0 | 0 | 0 | 6 |
| AG | AG | AG | 0 | 0 | 0 | 16 |
| AH | AH | AH | 0 | 0 | 0 | 4 |
| AI | AI | AI | 0 | 0 | 0 | 16 |
| Totals (pairs:ejaculates/flowbacks) | | | 22:67 | 24:84 | 11:127 | 9:115 |

particularly wished to avoid any reliance on some unspecified 'robustness' of parametric tests. For all of these reasons we considered a distribution-free test to be essential. The only such test with the power and elegance to cope with the complexity of our data and analysis is the Meddis rank sum test (Meddis 1984).

**Table II.** Six equations used to calculate residuals or to predict number of sperm ejaculated (for derivations, see Results)

| Equation |
|---|
| (1) NSM = 12 + 4.63 × HEJ |
| (2) NSC = 452 – 2.92 × PCT |
| (3) NSC = 357 + 1.94 × HIPC – 3.40 × PCT |
| (4) NSC = 357 – 3.40 × PCT + 1.94 × HIPC + ((2.41 × HMAS – 228) × MC) |
| (5) NSC = 1.94 × HIPC – 3.40 × PCT + ((2.41 × HMAS – 228) × MC) – 1008 + 23.37 × FW |
| (6) NSC = PVequ5 – (((PVequ5 – mid)/(limC – mid))2 × (limC – limO)) |

NSM, Number of sperm ejaculated during masturbation in millions; NSC, number of sperm inseminated during in-pair copulation in millions; HEJ, hours since last ejaculation; HIPC, hours since last in-pair copulation up to 192 h (192 h is the longest time interval for which we have data and up to which analysis supports a linear relationship); PCT, % time a pair have spent together since their last in-pair copulation; HMAS, hours (up to 72) since last masturbation; MC, 0 for IPC–IPC ejaculates and 1 for MAS–IPC ejaculates; FW, female weight (kg); PVequ5, the predicted value for NSC from equation (5); mid, mid-point of number of sperm in observed ejaculates ($350 \times 10^6$); limC, the minimum or maximum calculable limit to NSC from equation (5); limO, the minimum or maximum observed number of sperm in in-pair copulation ejaculates. When: PVequ5 > mid, maximum limits should be used (see Table XII); PVequ5 > mid, minimum values should be used. When female weight is not available, equation (4) may be used to calculate PVequ5 instead of equation (5).

## Statistics: Residuals and Predictive Equations

When the data allowed, we used Meddis' blocking technique to provide statistical control for one variable while analysing for an independent influence of another. Most often, however, we used regression or multiple regression to calculate values for the dependent variable with respect to one or more independent variables, then calculated residuals. These residuals were then analysed further for the influence of some other independent variable not included in the original regression. This procedure removed the influence of those independent variables included in the regression. However, even when we have used the parametric procedures of regression and multiple regression as intermediate steps in data handling, we still base probability statements only on non-parametric tests, for the reasons given above. It seems unlikely that any autocorrelation in the residuals would cause a problem for the Meddis procedures though we know of no formal analysis of this possibility.

We used five different equations to generate residuals. Derivation of the equations is described under Results and the equations are collected together in Table II. To avoid giving undue weight to samples from the more active couples, equations were calculated from only the first sample in a category provided by each couple. Similarly, when necessary to avoid pseudoreplication of data, we followed our previous procedure (Baker & Bellis 1989*a*) and made between-couple comparisons also using only the first sample from each couple. Finally, we used regression and residual analysis to develop an equation to predict the number of sperm inseminated by males during in-pair copulation in different socio-sexual situations (Table II; equation 6).

## RESULTS

### Risk of Sperm Competition and Number of Sperm Inseminated

As previously (Bellis & Baker 1990), we take the conservative view that human sperm remain competitive for only 5 days (ca 7–9 days; Smith 1984). For humans, therefore, a double-mating is a copulation with one male within 5 days of copulation with a different male. Claimed total number of male sexual partners in life (so far) per female and the proportion of women who claimed ever to have double-mated (data from our nationwide survey) increased in relation to sexual experience (i.e. number of lifetime copulations; Table III).

**Table III.** Lifetime number of male sexual partners/female and number of females who have ever double-mated in relation to sexual experience

| | | Total number of male sexual partners | | | |
|---|---|---|---|---|---|
| Lifetime number of copulations | *N* | Median (inter-quartile range) | > 1 partner (%) | > 50 partners (%) | Number of females having double-mated at least once (%) |
| ≤ 50 | 481 | 2 (1–4) | 64.5 | – | 17.5 |
| 51–200 | 796 | 4 (2–7) | 80.6 | 0.5 | 36.3 |
| 201–500 | 896 | 5 (3–10) | 86.9 | 0.8 | 49.4 |
| 501–1000 | 631 | 7 (3–13) | 92.7 | 1.9 | 61.9 |
| > 1000 | 585 | 8 (4–20) | 93.9 | 5.6 | 71.8 |

Double-mating = copulation with one male within 5 days of copulation with a different male.

The probability for a female of (1) having more than one current male partner, (2) the last copulation being an extra-pair copulation, and (3) the last copulation being a double-mating were all significant negative functions of the average proportion of time (including sleeping time) the pair spent together (Table IV). The functions remained significant even when restricted to those 407 females who claimed that their last copulation was unprotected by contraception (Table IV). We conclude that there is a negative association between the proportion of time a pair spends together and the probability that the female will engage in extra-pair copulation (including double-mating).

**Table IV.** Influence of % time with male partner on incidence of extra-pair copulation and double-mating by human females

| | Time spent with main male partner (%) | | | | | | | | | | Meddis' specific test | |
|---|---|---|---|---|---|---|---|---|---|---|---|---|
| | 100–91 | 90–81 | 80–71 | 70–61 | 60–51 | 50–41 | 40–31 | 30–21 | 20–11 | 10–0 | *z* | *P* |
| Lambda | 1 | 2 | 3 | 4 | 5 | 6 | 7 | 8 | 9 | 10 | | |
| **Total sample** | | | | | | | | | | | | |
| *N* | 26 | 52 | 189 | 228 | 381 | 361 | 269 | 534 | 454 | 314 | | |
| > 1 male partner (%) | 3.8 | 3.8 | 5.8 | 8.3 | 66.7 | 5.8 | 5.2 | 7.1 | 7.9 | 13.7 | 3.04 | 0.001 |
| Last copulation an extra-pair copulation (%) | 0.0 | 0.0 | 3.2 | 4.0 | 1.8 | 4.2 | 2.0 | 5.1 | 5.6 | 9.8 | 4.69 | < 0.001 |
| Double-mating (%) | 0.0 | 0.0 | 1.1 | 1.3 | 1.3 | 2.8 | 1.2 | 1.5 | 1.8 | 4.0 | 2.17 | 0.015 |
| **Females not using contraception** | | | | | | | | | | | | |
| *N* | 4 | 13 | 28 | 28 | 55 | 65 | 26 | 67 | 69 | 52 | | |
| > 1 male partner (%) | 0.0 | 7.7 | 7.1 | 3.6 | 5.5 | 6.2 | 11.5 | 7.5 | 10.2 | 19.2 | 2.36 | 0.009 |
| Last copulation an extra-pair copulation (%) | 0.0 | 0.0 | 3.6 | 7.1 | 1.8 | 3.1 | 3.8 | 9.1 | 3.3 | 20.5 | 2.85 | 0.002 |
| Double-mating (%) | 0.0 | 0.0 | 0.0 | 0.0 | 1.8 | 1.6 | 4.0 | 0.0 | 1.7 | 9.1 | 2.2 | 0.017 |

Twenty-five couples contributed 84 ejaculates collected in condoms during in-pair copulation (Table I). Of these ejaculates; 50 (from 15 males) were preceded by an ejaculate also produced during in-pair copulation (henceforth IPC–IPC ejaculates). Median time interval between in-pair copulations for these 50 ejaculates was 48 h (inter-quartile range = 25–70 h). Five males contributed more than one IPC–IPC ejaculate ($N =$ 40). The clear tendency (Table V) for individual males to inseminate fewer sperm as % time with partner increased is highly significant ($z = 3.535$, $P < 0.001$). In part, this tendency is due to an equally strong tendency for inter-copulation intervals to be shorter when couples spent more time together ($z = 3.830$, $P < 0.001$). However, the former is not an artefact of the latter. When the data are blocked to remove the influence of time since last copulation, the association between % time together and number of sperm inseminated remains highly significant ($z = 2.438$, $P = 0.007$).

We conclude that for any given time interval since last in-pair copulation, individual males inseminate more sperm per ejaculate if they have spent less time with their partner and hence, on average, the risk of sperm competition is higher.

**Table V.** Variation in number of sperm inseminated during in-pair copulation (IPC) by five males in relation to % time with partner (IPC–IPC ejaculates only)

| | No. of sperm (millions) inseminated | | | | | | | | | |
|---|---|---|---|---|---|---|---|---|---|---|
| | Pair A | | Pair B | | Pair N | | Pair T | | Pair V | |
| Time together since last IPC (%) | *N* | Median | *N* | Median | *N* | Median | *N* | Median | *N* | Median |
| 1–20 | 1 | 507 | | | 1 | 296 | | | | |
| 21–40 | | | | | 6 | 196 (147–319) | | | | |
| 41–60 | | | 1 | 495 | 6 | 210 (194–231 ) | | | | |
| 61–80 | 1 | 220 | 8 | 477 (268–530) | 2 | 143 (128–159) | 1 | 109 | 1 | 341 |
| 81–100 | | | 5 | 206 (143–219) | 4 | 67 (48–82) | 2 | 183 (161–205) | 1 | 32 |

Inter-quartile range is shown in parentheses. IPC–IPC ejaculate: in-pair copulation ejaculate with no inter-IPC masturbation.

## Time Since Last In-pair Copulation and Number of Sperm Inseminated

On average, human pairs engage in in-pair copulation at median intervals of about every 3 days (72 h; Britain: U.K. Family Planning Research Network 1988; U.S.A.: Kinsey *et al.* 1953). Our own nationwide survey produced a slightly longer median interval of 90 h (inter-quartile range = 51–184 h; $N = 2835$) whereas the 34 couples in our in-pair copulation and flow back studies showed a slightly shorter median interval of 62 h (interquartile range = 43–119 h).

Table VI shows the first IPC–IPC ejaculate produced by each of 15 males in relation to % time with partner and inter-copulation interval. Even for this small subset of our data, when the influence of inter-copulation interval is removed by blocking, % time together is still significantly negatively associated with the number of sperm ejaculated during in-pair copulation ($z = -1.854$, $P = 0.032$). The converse Meddis analysis (blocking to remove the influence of % time together) shows that time since last in-pair copulation is also a significant influence on the number of sperm ejaculated ($z = 1.976$, $P = 0.024$). We conclude that % time together and inter-copulation interval have significant but independent influences on the number of sperm ejaculated during in-pair copulation.

Next, we calculated the least squares regression line for the 15 IPC–IPC ejaculates in Table VI relating number of sperm inseminated during in-pair copulation to % time together (Table II; equation 2). We assumed a straight-line relationship (see Baker & Bellis 1989*a*). We then applied equation (2) to the complete set of 50 IPC–IPC ejaculates and calculated residuals for each ejaculate. This procedure removes the confounding influence of % time together.

Figure 1 shows the relationship between the residuals from equation (2) and time (h) since last in-pair copulation. Five males produced more than one IPC–IPC ejaculate (total = 40 ejaculates). We divided these ejaculates according to time since last in-pair copulation ( ≤ 24h; 25–48; … 169–192h) and block by male (five blocks) to test for within-male variation. Time since last in-pair copulation has a significant positive association with the number of sperm ejaculated during in-pair copulation ($z$ = 2.232, $P$ = 0.013). $z$ is maximized by rank order lambda coefficients which specify that, within the range of inter-copulation intervals in our data (1–192 h), males continued to increase the number of sperm inseminated the longer the time since the last in-pair copulation. We cannot determine from our samples whether the number inseminated would level off after in-pair copulation intervals greater than 8 days (192 h).

**Table VI.** First IPC–IPC ejaculate contributed by each of 15 males in relation to % time together and inter-copulation interval

| Male | Time since last IPC (h) | Time with partner since last IPC (%) | No. of sperm inseminated (× $10^6$) | No. of sperm inseminated/h since last IPC ( × $10^6$) |
|---|---|---|---|---|
| A | 60 | 20 | 570 | 9.5 |
| B | 149 | 98 | 219 | 1.5 |
| D | 70 | 50 | 485 | 6.9 |
| F | 168 | 50 | 516 | 3.1 |
| K | 48 | 20 | 448 | 9.3 |
| L | 32 | 100 | 60 | 1.9 |
| M | 48 | 2 | 282 | 5.9 |
| N | 56 | 37 | 455 | 8.1 |
| P | 31 | 30 | 76 | 2.5 |
| Q | 38 | 45 | 228 | 6.0 |
| R | 48 | 75 | 76 | 1.6 |
| T | 54 | 75 | 109 | 2.0 |
| U | 32 | 60 | 295 | 9.2 |
| V | 48 | 80 | 341 | 7.1 |
| Y | 44 | 75 | 225 | 5.1 |

(IPC = in-pair copulation. IPC–IPC ejaculate = in-pair copulation ejaculate with no inter-IPC masturbation.

The multiple regression that best describes the number of sperm inseminated in terms of time since last in-pair copulation and % time together (first in-pair. copulation per couple; Table VI) is equation (3) (Table II). Although equation (3) is the best to express the association between sperm number, % time together and time since last in-pair copulation, some later discussions are facilitated by considering the number of sperm inseminated during in-pair copulation for each hour since the last in-pair copulation. This rate is not fixed but is significantly negatively correlated with % time together ($r_S$ = –0.587, $P$ = 0.011, one-tailed; data from Table VI). When % time together is ≤ 25%, the rate of insemination is 9.3 (inter-quartile range = 5.9–9.5) million sperm/h since last in-pair copulation. Percentage times together of 25–75% and ≥ 75% are associated with insemination rates of 6.5 (inter-quartile range = 3.1–8.1) and 2.0 (1.6–5.1) million sperm/h, respectively (from data in Table VI).

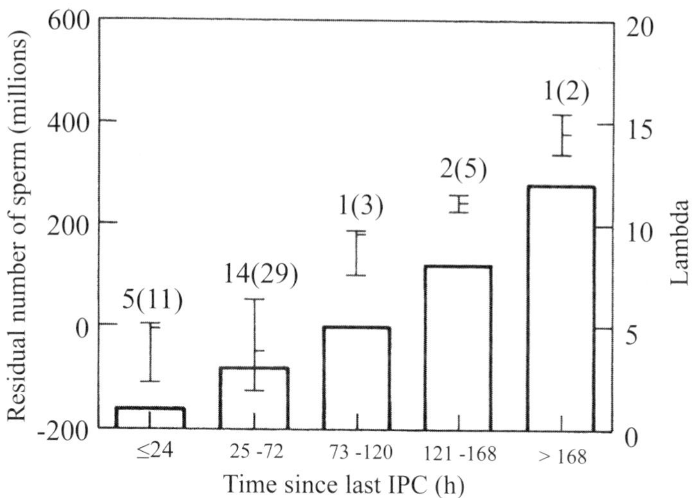


**Figure 1.** Variation in number of sperm ejaculated during in-pair copulation with time since last in-pair copulation. Bars show the median and inter-quartile range residual number of sperm after application of equation (2). Histograms show lambda coefficients which maximize *z* in a Meddis' specific test after blocking by couple. Numbers above bars: number of pairs (number of in-pair copulations).

## Reproductive Value of Female and Variation in Number of Sperm Inseminated

We applied equation (4) to our 84 in-pair copulation (50 IPC–IPC; 34 MAS–IPC; see section on masturbation) ejaculates and calculated residuals. These residuals were then used as the dependent variables for tests of the prediction that the number of sperm inseminated should vary according to the reproductive value of the female.

Seven couples returned 30 in-pair copulation ejaculates complete with information on the occurrence and timing of female orgasm during the copulation sequence. We could find no evidence that the numbers of sperm (residuals) inseminated varied in relation to occurrence and timing of orgasm. There was no significant heterogeneity across six categories (no orgasm; orgasm during foreplay; during copulation but before ejaculation; during ejaculation; during copulation but after ejaculation; after copulation) whether we analysed the first sample in each category by each couple ($H_5 = 3.662$, $P = 0.602$) or blocked by couple to test for within-couple variation ($H_5 = 5.575$, $P = 0.350$). Exploratory post-hoc analyses comparing different orgasm categories with each other (e.g. orgasm versus no orgasm; orgasm during foreplay versus no orgasm during foreplay; etc.) and using different dependent variables (raw data; residuals from equation 5; etc.) also failed to find any significant differences in number of sperm inseminated. We conclude that the number of sperm inseminated does not vary in association with variation in the occurrence and timing of female orgasm.

Similarly, we could find no evidence of variation in the number of sperm inseminated at different phases (I, II, III; see Introduction) of the menstrual cycle. We restricted analysis to those five pairs (52 ejaculates) who provided multiple samples while the female was not using oral contraceptives and who was thus on an unaltered hormonal cycle. Blocking by male to test for within-male response, neither analysis of residuals ($z = -0.630$, $P = 0.736$) nor of actual sperm numbers ($z = 0.191$, $P = 0.425$) shows a significant positive association with the relative fertilities of the different phases.

Of the 23 first ejaculates provided by each male, 14 were collected at a time when the female partner was also taking an oral contraceptive. There was no significant difference in the number of sperm (residuals) inseminated by males whose female partner was taking an oral contraceptive and those whose partner was not ($H_1 = 1.433$, $P = 0.230$). The trend was for females to be inseminated with more sperm when they were taking oral contraceptives. Three of the males produced ejaculates ($N = 34$) before and after their female partner changed from taking to not taking an oral contraceptive or vice versa. Blocking by male to test for within-male response, there is no significant difference in number of sperm (residuals) inseminated ($H_1 = 0.305$, $P = 0.588$). The trend was for females to be inseminated with more sperm when they were not taking oral contraceptives.

In contrast to the failure of predictions that the number of sperm inseminated should vary according to female orgasm pattern and/or probability of conception, the prediction that the number inseminated should be a function of female body size is strongly supported. There is significant heterogeneity between the 24 pairs who contributed in-pair copulation samples in the number of sperm inseminated during in-pair copulation ($H_{23} = 38.50$, $P = 0.022$; Table VII). Using the residuals from equation (4) as the dependent variable, the different couples as samples, and stature measurements (Table VIII) as lambda coefficients, we tested the prediction that there would be a positive association between female body size and between-pair variation in number of sperm inseminated (Fig. 2). A very significant association was found (Table IX), female weight showing a slightly greater association with sperm number than height. Comparison of pairs with the influence of female weight removed (residuals from equation 5; Table II), shows that there is no longer a significant heterogeneity in the number of sperm inseminated during in-pair copulation ($H_{23} = 29.354$, $P = 0.173$).

## The Dynamics of Male Self-masturbation

### *Pattern of timing of masturbation*

Figure 3 shows the frequency distribution of in-pair copulations preceded by male masturbation in relation to inter-copulation interval. When the inter-copulation interval was less than 72 h (3 days), the incidence of masturbation by males was low (< 5%). There was then a sharp rise to around 50% once time since last in-pair copulation exceeded 96 h (4 days). All 18 of the in-pair copulations for which the inter-copulation interval was more than 240 h (10 days) were preceded by the shedding of sperm through masturbation. When masturbation did occur, there was rarely a gap of more than 72 h (3 days) between in-pair copulation and the last masturbation (Fig. 4). On over 50% of occasions, this time interval was less than 48 h (2 days).

We conclude that when time since last copulation exceeds 72 h, males become increasingly likely to shed sperm through masturbation before they next copulate. On the majority of occasions, their last masturbation is within 48 h of their next in-pair copulation.

**Table VII.** Median observed and residual number of sperm ejaculated during in-pair copulation by 24 pairs

| | | No. of sperm ($\times 10^6$) | | | |
|---|---|---|---|---|---|
| | | Observed | | Residuals (from equation 4) | |
| Pair | N | Median | (Inter-quartile range) | Median | (Inter-quartile range) |
| A | 4 | 514 | (220–570) | 80 | (–59 to 92) |
| B | 27 | 393 | (244–507) | 81 | (1 to 133) |
| C | 2 | 305 | (301–308) | –174 | (–222 to –126) |
| D | 1 | 485 | | 162 | |
| E | 1 | 422 | | –127 | |
| F | 1 | 516 | | 3 | |
| G | 1 | 244 | | 293 | |
| H | 2 | 65 | (46–83) | –140 | (–531 to 251) |
| I | 1 | 525 | | –95 | |
| J | 1 | 55 | | 44 | |
| K | 1 | 448 | | 66 | |
| L | 3 | 39 | (2–60) | –233 | (–407 to –233) |
| M | 1 | 282 | | –161 | |
| N | 22 | 175 | (128–231) | –56 | (–150 to 15) |
| O | 1 | 213 | | –179 | |
| P | 1 | 76 | | –239 | |
| Q | 1 | 228 | | –50 | |
| R | 1 | 76 | | –119 | |
| S | 1 | 87 | | –129 | |
| T | 3 | 161 | (109–205) | 63 | (–98 to 63) |
| U | 3 | 283 | (193–295) | 80 | (24 to 80) |
| V | 2 | 182 | (32–341) | 86 | (9 to 163) |
| Y | 1 | 225 | | 38 | |
| AB | 2 | 242 | (173–310) | –109 | (–148 to –70) |

The number of sperm ejaculated during masturbation shows a highly significant correlation with time since the male's last ejaculation, both between males (analysis of first sample from each male: $r_S = 0.558$, $N = 22$, $P = 0.003$, one-tailed) and within males ($z = 2.853$, $N = 10$ males, 55 ejaculates, $P = 0.002$). $z$ is maximized by lambda coefficients which specify that the number of sperm ejaculated continued to increase linearly with time since last ejaculation, at least up to 9 days (216 h) after last ejaculation. The best fit to the masturbation data in regression analysis (using only the first masturbation ejaculate produced by each of our 22 males) is given by equation (1) (Table II).

*Number of sperm ejaculated during masturbation*

We could find no association between the percentage of time a male had spent with his partner since their last in-pair copulation and the number of sperm ejaculated during either: (1) self-masturbation in the absence of a female ($r_S = -0.193$, $N = 20$, $P = 0.415$, two-tailed); or (2) masturbation (by self, partner or both) in the presence of a female ($r_S = 0.190$, $N = 8$, $P = 0.762$, two-tailed; data from Table X; cf Baker & Bellis 1989*a*). There is no significant difference between the two correlation coefficients ($z = 0.762$, $P = 0.448$, two-tailed, *z*-transformation test). Six males contributed masturbation ejaculates under both circumstances (female present, $N = 17$; female absent, $N = 24$). Blocking by individual male to test for within-male variation, there is no significant difference in the number of sperm ejaculated according to whether the female partner was present or absent ($H_1 = 1.088$, $P = 0.297$). The non-significant trend is for more to be ejaculated when the female was present. Thus, for the remainder of this analysis, no distinction is made between whether the female partner was present or absent.

**Table VIII.** Age and stature details for males and females of all pairs who contributed in-pair copulation and masturbation ejaculates.

| | Female | | | Male | | | |
|---|---|---|---|---|---|---|---|
| Pair | Age (years) | Height (cm) | Weight (kg) | Age (years) | Height (cm) | Weight (cm) | Testes (cm$^3$) |
| A | 25 | 170 | 64 | 25 | 188 | 95 | 28 |
| B | 24 | 175 | 58 | 44 | 180 | 79 | 20 |
| C | 22 | 180 | 57 | 20 | 185 | 76 | – |
| D | 27 | 183 | 66 | 30 | 183 | 67 | – |
| E | 26 | 163 | 54 | 25 | 183 | 67 | 14 |
| F | 30 | 166 | 59 | 32 | 175 | 73 | 18 |
| G | 20 | 173 | 63 | 31 | 180 | 75 | 31 |
| H | 21 | 164 | 57 | 24 | 175 | 70 | 15 |
| I | 20 | 155 | 56 | 18 | 178 | 75 | 23 |
| J | 21 | 163 | 54 | 28 | 173 | 68 | – |
| K | 31 | 170 | 57 | 31 | 180 | 76 | – |
| L | 27 | 152 | 52 | 31 | 174 | 67 | 7 |
| M | 19 | 163 | 54 | 19 | 163 | 57 | 12 |
| N | 19 | 163 | 54 | 22 | 185 | 73 | 17 |
| O | 22 | 165 | 54 | 20 | 180 | 70 | 12 |
| P | 21 | 170 | 53 | 21 | 180 | 67 | – |
| Q | 21 | 170 | 54 | 21 | 172 | 71 | – |
| R | 21 | 166 | 56 | 21 | 166 | 60 | – |
| S | 26 | 168 | 58 | 26 | 174 | 86 | 29 |
| T | 22 | 171 | 56 | 22 | 187 | 64 | 11 |
| U | 19 | 157 | 52 | 23 | 183 | 68 | 15 |
| V | – | – | – | 21 | – | – | – |
| X | – | – | – | 20 | – | – | – |
| Y | – | – | – | 20 | – | – | – |
| AB | 20 | 180 | 57 | 20 | 180 | 70 | 12 |

Measurements refer to time of contribution or first sample.

We applied equation (1) to our full set of 67 masturbation ejaculates, then calculated residuals. Seven males produced ejaculates in both IPC–MAS (masturbation preceded by in-pair copulation) and MAS–MAS (masturbation preceded by a further masturbation) categories. Blocking by male to test for within-male variation, there is no significant difference between the residual number of sperm in IPC–MAS and MAS–MAS ejaculates ($H_1$ = 1.364, $N$ = 7 males, 49 ejaculates, $P$ = 0.241). The non-significant trend is for males to ejaculate more when masturbations were preceded by a masturbation than when preceded by an in-pair copulation.

We conclude that the primary factor associated with the number of sperm ejaculated by a given male during masturbation is the length of time since the male's last ejaculation. The circumstances of the last and current ejaculation have no significant influence.

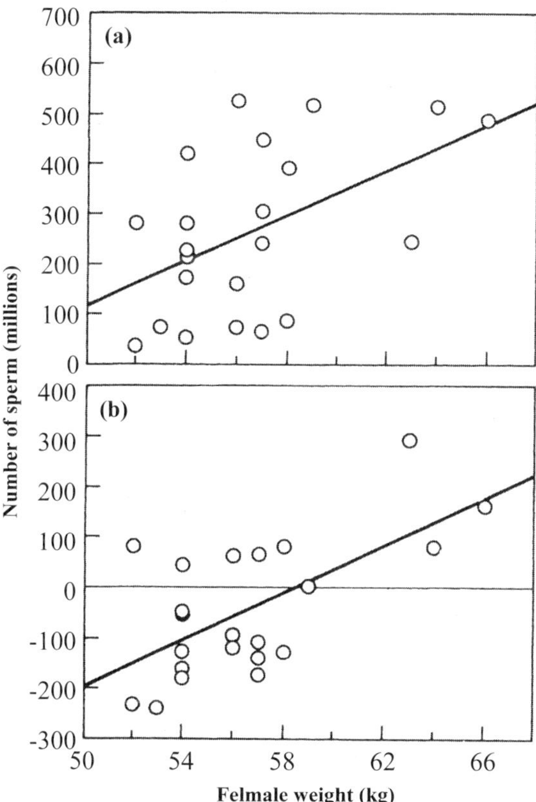


**Figure 2.** Variation in number of sperm ejaculated during in-pair copulation with weight of female. (a) Actual number of sperm: $r_S$ = 0.484, $P$ = 0.012; (b) residual number of sperm after application of equation (4): $r_S$ = 0.504, $P$ = 0.009.

There is a just significant heterogeneity between males in the number of sperm ejaculated during masturbation when time since last ejaculation is controlled ($H_{21}$ = 33.127, $P$ = 0.045; residuals from equation 1; Table XI). However, when tested against measurements of male stature (Table VIII), these differences were not a significant function of male height ($z = 0.916$, $P = 0.180$), weight ($z = 1.024$, $P = 0.153$) or volume of testes ($z = -0.066$, $P = 0.526$). Only male age approached significance ($z = -1.880$, $P = 0.060$, two-tailed), the tendency being for older males to ejaculate fewer sperm during masturbation. We conclude that although males may differ in the number of sperm ejaculated during masturbation, the differences are not associated with any of the factors we have so far investigated.

**Table IX.** Analysis of number of sperm inseminated during in-pair copulation by 22 pairs in relation to male and female stature

| | *N* | | Meddis' test | |
|---|---|---|---|---|
| | Pairs | Ejaculates | *z* | *P* |
| Female weight | 22 | 81 | 3.902 | < 0.001 |
| Male weight | 22 | 81 | 2.824 | 0.003 |
| Female height | 22 | 81 | 3.038 | 0.001 |
| Male height | 22 | 81 | 0.897 | 0.185 |
| Testis volume | 15 | 73 | 3.181 | 0.001 |

No. of sperm inseminated = residual from equation (4). Stature measurements from Table VIII.

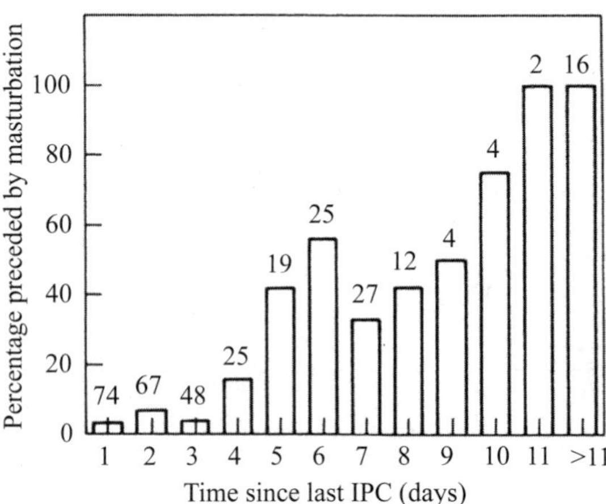


**Figure 3.** Frequency distribution of in-pair copulations (IPCs) preceded by male masturbation as a function of time since last in-pair copulation, based on 323 in-pair copulations recorded by 34 pairs. Numbers above bars: total number of in-pair copulations in time interval.

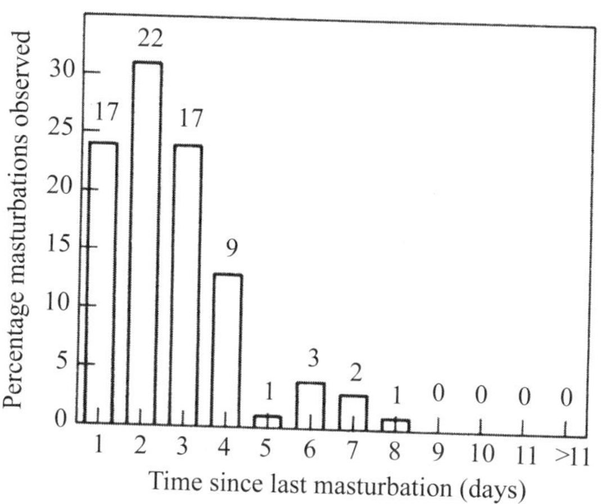


**Figure 4.** Frequency distribution of time intervals from male's last masturbation (MAS) to next in-pair copulation (IPC), based on 72 MAS–IPC events recorded by 34 pairs. Numbers above bars: total number or MAS–IPC events in time interval.

## *Masturbation and sperm at next in-pair copulation*

During our study, 15 males produced 50 IPC–IPC ejaculates and 13 males produced 34 MAS–IPC ejaculates (MAS–IPC = in-pair copulations preceded by a masturbation; no cases of nocturnal emission were reported). Equation (3) (Table II explains 73% of observed variation in sperm number in the first IPC–IPC ejaculates produced by each male and 76% of variation in all 50 IPC–IPC ejaculates. It is reasonable, therefore, to apply equation (3) to the 34 MAS–IPC ejaculates to predict the number of sperm that would have been inseminated had the male not masturbated in the inter-copulation interval (assuming that equation 3 is as applicable on the 34 MAS–IPC occasions as on the 50 IPC–IPC occasions). The difference between expected and observed numbers of sperm is then a measure of the influence of masturbation on the number of sperm inseminated during in-pair copulation.

When blocked by male to test for within-male variation there is a significant decrease in the difference between observed and expected number of sperm with increase in time since last masturbation ($z = 2.274$, $N = 13$ males, 34 MAS–IPC ejaculates, $P = 0.011$). $z$ is maximized by lambda coefficients that specify that masturbation had no further influence on the difference between observed and expected number of sperm inseminated at the next in-pair copulation after 72 h. When masturbation preceded in-pair copulation by less than 72 h, the observed number of sperm in in-pair copulation inseminates was significantly lower than expected ($z = 3.291$, $P = 0.001$). After 72h, however, there was no significant decrease ($z = -2.087$, $P = 0.982$). Allocating a value of 72 to times since last masturbation > 72 h, then performing regression analysis on the residuals (from equation 3) for the first MAS–IPC provided by each male allows calculation of a reduction factor. This reduction factor may then be used to modify equation (3) such that it becomes applicable to all in-pair copulation ejaculates whether or not they are IPC–IPC or MAS–IPC ejaculates. This modification of equation (3) gives equation (4) (Table II) which suggests that masturbation between in-pair copulations is associated with a reduction in the number of sperm the male inseminates into his female partner at the next in-pair copulation by 228 million sperm minus 2.41 million sperm for every hour since last masturbation (up to 72 h).

**Table X.** First masturbation ejaculate contributed in each of two circumstances (female partner present or absent) by 22 males.

| | Female absent | | | Female present | | |
|---|---|---|---|---|---|---|
| | Time since last ejaculation (h) | Time with partner since IPC (%) | No. sperm ejaculated ($\times 10^6$) | Time since last ejaculation (h) | Time with partner since IPC (%) | No. sperm ejaculated ($\times 10^6$) |
| A | 72 | 5 | 825 | | | |
| B | 103 | 68 | 415 | 53 | 81 | 235 |
| C | 81 | 33 | 168 | | | |
| D | 55 | 50 | 213 | | | |
| E | 24 | 30 | 61 | | | |
| F | 62 | 60 | 582 | | | |
| G | 24 | 90 | 20 | | | |
| H | 1 | 90 | 37 | | | |
| I | 9 | 37 | 12 | | | |
| J | 50 | 80 | 103 | | | |
| K | | | | 24 | 75 | 205 |
| L | 20 | 5 | 150 | | | |
| M | 36 | 1 | 175 | | | |
| N | 48 | 8 | 52 | 70 | 30 | 87 |
| O | 15 | 20 | 112 | 14 | 11 | 155 |
| P | 23 | 30 | 110 | | | |
| Q | 34 | 62 | 228 | | | |
| S | 60 | 50 | 140 | | | |
| T | 36 | 20 | 102 | 48 | 70 | 343 |
| U | | | | 17 | 65 | 112 |
| X | 48 | 1 | 213 | 27 | 1 | 261 |
| Y | 36 | 100 | 119 | 40 | 90 | 211 |

IPC = in-pair copulation

### *Male masturbation and female sperm retention*

In the rabbit, the number of sperm remaining in a female after flowback is a meaningful index of the number of sperm that attain functional positions within the female tract (Morton & Glover 1974*a*,*b*). In the following analysis, we assume that, on average, the same is true for humans. Full details of our flowback studies are given in our companion paper (Baker & Bellis 1993).

**Table XI.** Median observed and residual number of sperm ejaculated during masturbation by 22 males

| | | No. of sperm ( × $10^6$) | | | |
|---|---|---|---|---|---|
| | | Observed | | Residuals (from equation 1) | |
| Male | *N* | Median | (Inter-quartile range) | Median | (Inter-quartile range) |
| A | 6 | 195 | (156–254) | 117 | (70 to 147) |
| B | 17 | 323 | (282–415) | –74 | (–267 to –22) |
| C | 1 | 168 | | –219 | |
| D | 1 | 213 | | –54 | |
| E | 1 | 61 | | –62 | |
| F | 1 | 582 | | 283 | |
| G | 2 | 54 | (20–87) | –60 | (–103 to –18) |
| H | 1 | 37 | | 20 | |
| I | 1 | 12 | | –42 | |
| J | 1 | 103 | | –141 | |
| K | 1 | 205 | | 82 | |
| L | 2 | 116 | (82–150) | 21 | (–4 to 45) |
| M | 1 | 175 | | –4 | |
| N | 2 | 70 | (52–87) | –216 | (–249 to –182) |
| O | 8 | 97 | (64–125) | –10 | (–31 to 20) |
| P | 1 | 110 | | –8 | |
| Q | 1 | 228 | | 59 | |
| S | 2 | 118 | (96–140) | –218 | (–286 to –150) |
| T | 6 | 236 | (102–335) | –60 | (–77 to 68) |
| U | 1 | 112 | | 21 | |
| X | 8 | 149 | (83–185) | 13 | (–23 to 50) |
| Y | 2 | 165 | (119–211) | –23 | (–60 to 14) |

**Table XII.** Range of parameter values for which equation (6) was calculated

| Parameter | Measure | Minimum value | Maximum value | Imposed maximum |
|---|---|---|---|---|
| Time together between IPCs | % | 1 | 100 | 100 |
| Time since last IPC | h | 1 | 2500 | 192 |
| Time since last masturbation | h | 1 | 186 | 72 |
| Female weight | kg | 52 | 66 | |
| Observed no. sperm | ( × $10^6$) | 2 | 692 | 692 |
| Calculable no. sperm | ( × $10^6$) | –359 | 919 | 919 |

**IPC** = in-pair copulation. Mid-point of observed ejaculates = 350 × $10^6$ sperm. When ejaculate parameter exceeds the value of the imposed maximum, the imposed maximum value should be used. When female weight is outside the range of 52–66 kg, equation (6) may no longer be valid.

Combined with equation (4) for those couples for whom female weight was unavailable, equation (5) explains 72% of the variation in the first in-pair copulation ejaculate contributed by each of our 24 couples and 56% of the total sample of 84 in-pair copulation ejaculates. We modified equation (5) to a form (equation 6; Table II) that always gives biologically realistic (i.e. never negative) predictions of the number of sperm inseminated during any given in-pair copulation. Naturally, equation (6) should be used only within the ranges and limits encountered in our study. These ranges and limits are listed in Table XII. Extrapolation beyond these limits should be done only with caution.

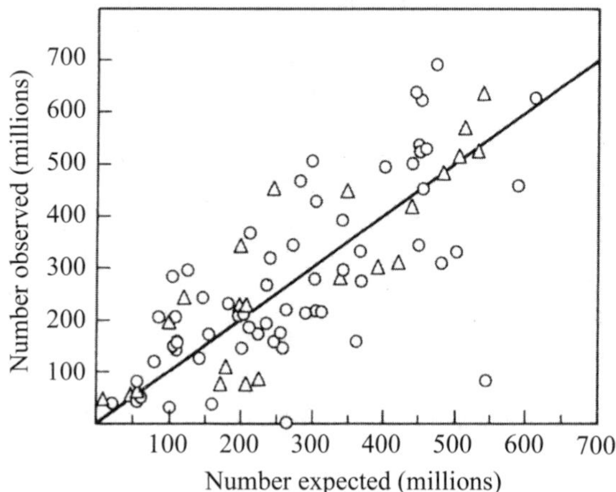


**Figure 5.** Number of sperm observed in 84 in-pair copulation ejaculates compared with the number predicted by equation 6. △: First sample contributed by each of 23 males; O: subsequent samples. Diagonal line shows 1:1 relationship.

Figure 5 shows the relationship between predicted (from equation 6) and observed numbers of sperm in in-pair copulation ejaculates. This equation accounts for 76% of the observed variation in the first in-pair copulation ejaculate contributed by each of the 23 males in our study, 51% of all subsequent samples ( $N = 61$) and 58% of the total data set ($N = 84$). We assume that equation (6) (while not necessarily the best) is sufficiently accurate and robust to be used as a tool to predict the number of sperm inseminated during in-pair copulation for pairs involved in our study. Whether it is robust for other groups of subjects remains to be tested.

We used equation (6) to estimate the number of sperm that should have been inseminated into the female during each of the 121 non-pregnancy in-pair copulations from 11 couples for which we received flow back samples. Subtraction of the observed number of sperm ejected from the predicted number inseminated provides a measure of the number of sperm retained on each occasion.

Analysing just the first flowback from each couple, there is a significant positive correlation between the number of sperm predicted to be inseminated (equation 6) and both the number of sperm observed in the flowback ($r_S = 0.655$, $P = 0.026$, one-tailed) and the number calculated to be retained by the female ($r_S = 0.791$, $P = 0.004$, one-tailed). Using all 121 flowbacks and blocking by couple, we again find a significant increase in number of sperm retained with number sperm inseminated ($z = 6.666$, $P < 0.001$), $z$ being maximized by the lambda coefficients shown in Fig. 6. Over the whole range in our data, the more sperm the male inseminates, the more are retained.

We removed the influence of the number of sperm inseminated by the male in order to test for the relative retention of sperm by the female using the method described in detail in our companion paper (Baker & Bellis 1993). Briefly, we calculated the least squares regression of number of sperm retained on number of sperm inseminated, then calculated the residuals from that line for each in-pair copulation. Positive residuals indicate that the female retains an above average number of sperm; negative residuals that she retains a below average number. Actual regression lines and equations are given in Baker & Bellis (1993) along with a discussion of the merits of this measure of the female role in sperm retention compared with the alternative measure of % retention.

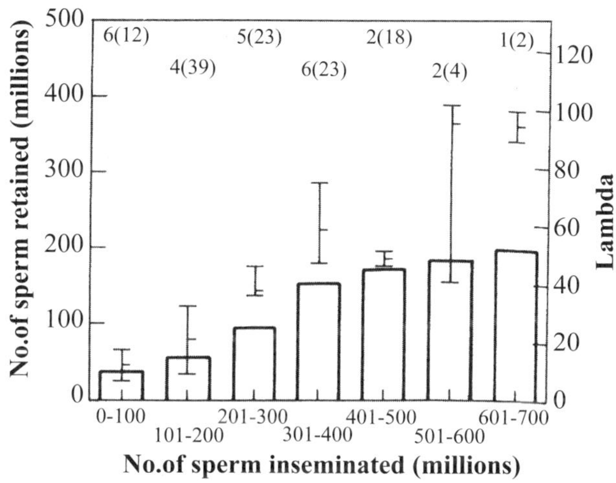


**Figure 6.** Influence of ejaculate size on number of sperm retained by the female during in-pair copulation. Number of sperm inseminated calculated from equation (6). Bars show median and inter-quartile range number of sperm retained. Histograms show lambda coefficients which maximize $z$ in a Meddis' specific test after blocking by couple. Numbers above bars: number of females (number of flowbacks).

Table XIII shows the number of sperm ejected and retained after in-pair copulation by females according to time since their partner's last ejaculation and whether that last ejaculation was the preceding in-pair copulation (IPC–IPC inseminates) or a masturbation (MAS–IPC inseminates). In Table XIV, the results are tested against the predictions of the sperm age hypothesis (see Introduction).

Whether the male's last ejaculation was an in-pair copulation or a masturbation, the number of sperm in the flowback increased significantly with increase in time since last ejaculation. The number of sperm we calculated to be retained by the female also increased with time since the male's last ejaculation if that ejaculation was during in-pair copulation but not if the male's last ejaculation was masturbatory. The number of sperm retained did not change significantly with time since last masturbation though the tendency was for more sperm to be retained the more recent the masturbation.

**Table XIII.** Median number of sperm retained and ejected by females following in-pair copulation as a function of whether the male's previous ejaculation was a copulation or masturbation

| | Previous ejaculation | | | | | | | |
|---|---|---|---|---|---|---|---|---|
| | In-pair copulation | | | | Masturbation | | | |
| | *N* | No. sperm ejected ($\times 10^6$) | No. sperm retained ($\times 10^6$) | Residual sperm retained (residual) | *N* | No. sperm ejected ($\times 10^6$) | No. sperm retained ($\times 10^6$) | Residual sperm retained (residual) |
| ≤ 24 | 7 (30) | 21 | 64 | 61 | 3 (6) | 121 | 120 | 55 |
| | | (12–77) | (34–133) | (13 to 75) | | (98–132) | (45–214) | (–20 to 69) |
| 25–72 | 6 (26) | 121 | 120 | 55 | 3 (13) | 121 | 120 | 55 |
| | | (85–130) | (42–180) | (–2 to 30) | | (141–252) | (154–260) | (35 to 14) |
| > 72 | 5 (39) | 121 | 120 | 55 | 1 (7) | 121 | 120 | 55 |
| | | (44–194) | (151–285) | (–84 to 46) | | (220–320) | (44–92) | (–162 to –50) |

Inter-quartile range is shown in parentheses. Format for sample sizes (*N*): number of females (number of flowbacks).

**Table XIV.** Influence of time since last ejaculation on number of sperm retained and ejected by females following in-pair copulation: results of Meddis' specific test on data in Table XIII

| | Previous ejaculation | | | | | |
|---|---|---|---|---|---|---|
| | In-pair copulation ($N = 97$) | | | Masturbation ($N = 26$) | | |
| | No. sperm ejected | No. sperm retained | Residual sperm retained | No. sperm ejected | No. sperm retained | Residual sperm retained |
| *z* | 5.689 | 5.485 | –1.888 | 3.429 | –0.400 | –1.805 |
| *P* | < 0.001 | < 0.001 | 0.029 | < 0.001 | 0.344 | 0.035 |

Positive *z*-values indicate that number increases with time since last ejaculation; negative *z*-values that number decreases. *P*-values for residual sperm retention are one-tailed against the specific hypothesis that a recent ejaculation improves sperm retention by reducing sperm age.

When we compare the number of sperm retained by the female following IPC–IPC and MAS–IPC inseminations, no differences are significant ($H_1$ = 0.201, $P$ =0.659; blocked by order). This remains true when blocked by time since last in-pair copulation ($H_1$ = 2.436, $P$ = 0.114). On average, therefore, males suffer no disadvantage from masturbation in terms of the number of sperm retained by the female at their next in-pair copulation, despite inseminating fewer (on average, by 228 million minus 2.4 million for every hour since last masturbation; equation 4). In fact, over all in-pair copulations, the non-significant trend is for more sperm to be retained when the male masturbates between copulations ($z$ = 0.448, $P$ = 0.632, two-tailed).

There is good support in Table XIV for the sperm age hypothesis. Assuming that the more recently the male has ejaculated, the younger the average age of sperm in the next ejaculate, there is a decline in residual sperm retention with age of sperm in both IPC–IPC and MAS–IPC inseminates. Over all inseminates, the greater relative retention of younger sperm is highly significant ($z$ = –3.039, $N$ = 11 females, 121 in-pair copulations, $P$ = 0.001; blocked by female).

## DISCUSSION

### Partitioning of Sperm between Successive In-pair Copulations

There has been no previous study for humans of the relationship between time since last in-pair copulation and the number of sperm inseminated during in-pair copulation. However, there are four models of the possible relationship.

#### *Fixed inseminate model*

This assumes that males inseminate a relatively fixed number of sperm per in-pair copulation and are relatively unconstrained by rate of sperm maturation. On this model, the number of sperm inseminated at each in-pair copulation is relatively constant and the total number inseminated during a fixed interval is a function of the number of in-pair copulations during that interval. With some qualification, this is essentially the model assumed by Ginsberg & Rubenstein (1990) for zebras (*Equus* spp.).

Our analysis has demonstrated a significant increase in the number of sperm inseminated with increase in time since last in-pair copulation. As such, the data are clearly inconsistent with the ‘fixed ejaculate’ model. Nevertheless, the form of the relationship (equation 3; Table II) means that the total number of sperm inseminated through in-pair copulation during a given time interval is slightly greater if there are more in-pair copulations during that interval. Thus, if we assume male and female spend 100% of their time together, we calculate that a male inseminates 1372 million sperm during a complete (28 day) menstrual cycle of his female partner if the pair have four in-pair copulations compared with 1440 million sperm if they have eight in-pair copulations. Doubling the in-pair copulation rate thus increases the total number of sperm inseminated by 5%. If male zebra show the same pattern as humans, our data provide some support for Ginsberg & Rubenstein’s (1990) assumption that the number of sperm inseminated is a function of in-pair copulation rate. However, our data do not support their assumption that in-pair copulation rate is more important than any other adjustment the male may be making. Thus, for humans, relative to four in-pair copulations per menstrual cycle with 100% time together, halving the % time together raises the total number of sperm inseminated per menstrual cycle by 50% compared with the 5% increase due to doubling the in-pair copulation rate.

*Physiological constraint model*

This model assumes that, at each in-pair copulation, males inseminate all of the stored sperm mature enough to be ejaculated. On this model, number of sperm inseminated at each in-pair copulation will be a function of time since last ejaculation and the rate at which sperm mature. Total number of sperm inseminated during a fixed interval, such as one oestrous cycle or other fertile phase, will be the number of sperm matured during that interval.

Our data are not consistent with this model. The relationship illustrated in Fig. 1 suggests an increase in number of sperm inseminated of only about 57 million sperm per day since the last in-pair copulation. Yet adult human males probably manufacture nearly 300 million sperm per day (Johnson *et al.* 1980). Moreover, the number of sperm inseminated during in-pair copulation for each hour since the last in-pair copulation is not fixed but is significantly negatively correlated with % time together. Even the greatest rate of insemination (9.3 million sperm/h since last in-pair copulation when couples spend less than 25% of their time together) is lower than the estimated rate of sperm production by humans of 12.5 million/h (Johnson *et al.* 1980). It thus seems likely that observed insemination rates are in some way strategic and not simply due to physiological constraint.

*Parker's partitioning model*

Parker (1984) assumed that a given number of sperm are available for insemination during one fertile phase of the female partner. He then identified circumstances in which the optimum strategy for the male would be to use these sperm during in-pair copulation in a series of smaller inseminations spread through the female's fertile phase rather than in a single large insemination at the beginning of that phase. His model predicts that in species, such as humans, with cryptic ovulation and a sperm life that is short relative to the inter-ovulation interval, males should show multiple insemination. On Parker's model, the total number of sperm inseminated during one fertile phase is relatively constant and independent of the in-pair copulation rate. Number of sperm per insemination is given by the total number of sperm available divided by the number of in-pair copulations during the phase. On average, therefore, there must be a positive association between inter-IPC interval and number of sperm inseminated, as we have shown.

*The 'topping-up' model*

On our own 'topping-up' model, the total number of sperm inseminated during a given time interval is not fixed. Instead, males attempt to maintain an optimum-sized population of sperm in their partner's tract as a defence against sperm competition. Optimum size of sperm population will be a function of the risk of sperm competition. Successive in-pair copulations thus become 'toppings-up'. The number of sperm inseminated is that necessary either simply to replace sperm lost through death and phagocytosis since last insemination and/or, if risk of sperm competition has changed, to adjust the total size of the sperm population. On this model, number of sperm inseminated during in-pair copulation is a function of time since last in-pair copulation and risk of sperm competition. As in Parker's model, the total number of sperm inseminated during a fixed interval will be relatively independent of the in-pair copulation rate (as long as the in-pair copulation rate is greater than zero) but, rather than be fixed, will be a function of the risk of sperm competition.

The data are consistent with this 'topping-up' model. Thus, total number of sperm inseminated during a fixed time interval increase relatively little with an increase in the number of in-pair copulations during that interval. Instead, males appear strategically to adjust the number of sperm with which they inseminate their partner according to time since last insemination. The number of sperm ejaculated during any given insemination increases with time since last in-pair copulation for at least the 8 days (192 h) that some active sperm are known to remain in the female tract (Austin 1975). Any in-pair copulation less than 192 h since the previous in-pair copulation is effectively, therefore, a 'topping-up' to some particular level, rather than a complete insemination in its own right. To what level the male tops-up the female depends on the risk of sperm competition. From equation (3) (Table II), if we assume an 8–day (192 h) inter-IPC interval (to allow all sperm from previous inseminations to die), males inseminate 389 million sperm/ejaculate when % time with the female is 100% compared with 712 million/ejaculate when % time with the female is only 5%.

One possible implication of our data is that the concepts of extra-pair and in-pair copulation may be irrelevant to male strategies of sperm adjustment. Perhaps, when a male inseminates a female, his strategic concern is not whether she is or is not his partner. Rather, the number of sperm inseminated may instead be determined simply by how long it is since he last copulated with that particular female (if ever) and the extent to which he has associated with her within the past 8 days (the active life of sperm she may already contain).

Our data on top-up rates at different levels of association between male and female allow us to make one further calculation. The rate of sperm production in humans (12.5 million/h; Johnson *et al.* 1980) should allow a male to inseminate: (1) six females in a group such that he could associate with them all simultaneously and continuously (because of the 'topping-up phenomenon' this would be independent of inter-copulation interval with each female); (2) two spatially separate females between whom he equally divided his time; or (3) one female with whom he spent 70% of his time while, during the other 30%, seeking occasional extra-pair copulations with other females with whom he spends relatively little time. All of these patterns are found in the anthropological literature (see Smith 1984).

## Sperm Number and Female Reproductive Value

Our demonstration that individual male humans vary the number of sperm they inseminate according to the proportion of time they have spent with the female (and hence the risk of sperm competition) is the first demonstration of intra-male variation for a mammal. As such, it adds to the now considerable body of evidence that, in accordance with sperm competition theory, males inseminate more sperm when the risk of sperm competition is higher. In contrast, the prediction that males should inseminate more sperm when copulating with females of higher reproductive value has met with much more mixed success.

As predicted, heavier females are inseminated with more sperm (Table IX). Even this apparent success, however, has the caveat that the association between sperm number and female body size may be an artefact due to the influence of some third factor, such as male stature. The data suggest male stature may not be important as Table IX shows that female body size, particularly weight, has a stronger association with sperm number than any corresponding measure of the males. From our current data set, however, we are statistically unable totally to rule out a separate influence of male stature, particularly testis volume (Table IX).

The best test, of course, would be to investigate whether any given male inseminates more sperm into larger females. However, during our study, only one male (male O) changed partners, moving from a larger female (female C; height 180 cm; weight 57 kg) to a smaller (female O; 165 cm; 54 kg). Data are minimal (Table VII; pairs O and AB) though the trend is at least in the direction expected. Median number of sperm inseminated decreased (absolute numbers from 242 million to 213 million; residuals from –109 million to –179 million; Table VII).

Detailed consideration of the interaction between male and female stature must await further study and more data. Procedurally, however, we could find no statistical justification for adding any stature parameter to our predictive equation (equation 5; Table II) other than female weight.

In contrast to the positive association between sperm number and female weight, no association emerged between the number of sperm inseminated and the other measures of female reproductive value we considered. One possible explanation for this apparent failure of sperm competition-theory is that whereas males can reliably judge female body size they cannot obtain reliable information concerning female orgasm and probability of conception.

As far as female orgasm is concerned, there is a major difference in sperm retention between a copulation involving no female orgasm and one with a female orgasm after male ejaculation (Baker & Bellis 1993). Yet unless, on ejaculation without a female orgasm, males could predict whether the female will or will not experience orgasm before flow back, there would seem to be no opportunity for the male to adjust sperm number appropriately. Similarly, the difference in sperm retention between an orgasm during foreplay and an orgasm within a minute of ejaculation is also major (Baker & Bellis 1993). However, in this context, the time available for the male to adjust sperm number is very short (a few minutes or even seconds). Add to these difficulties the possibility of the female 'faking' an orgasm and the opportunity of the male to make appropriate strategic adjustments in sperm number in response to the occurrence and timing of female orgasm seems minimal.

Similarly, there is some doubt over whether males can actually perceive when a female is fertile. The hypothesis that females produce pheromones that lead to synchronization of menstruation and which could be used by males to detect phase of cycle has been discounted (Little *et al.* 1989). Most authors now assume that ovulation by humans really is cryptic (e.g. Small 1989) and that such crypsis is a strategy which allows the female more control over the timing of copulations with different males than might be possible if a male partner could detect ovulation. We have shown previously that whereas in-pair copulations peak at the least fertile phase (phase III) of the menstrual cycle, extra-pair copulations, particularly double-matings, peak during the most fertile phase (phase II; Bellis & Baker 1990). Now it seems that in-pair males also fail to adjust the number of sperm inseminated at different phases of their partner's menstrual cycle. However, rather than represent a failure of sperm competition theory, these data could perhaps provide further indication that menstrual fluctuation in fertility of female humans really is cryptic to their male partner (Small 1989; Bellis & Baker 1990).

Finally, given the recent origin of modern contraceptive techniques, it is perhaps not surprising that males may not have the psychophysiological repertoire to vary the number of sperm inseminated according to the use of oral contraceptives by their female partner. Even if male humans had such a repertoire, adjustment of sperm number may not be advantageous. Even if the female has taken oral contraceptives up to the moment of copulation, the male cannot be certain that she will continue to do so for the next 5–8 days while the sperm he inseminates are competitive. Thus, males can never assume that the sperm being inseminated have no chance of entering competition or no chance of fertilization.

Perhaps, therefore, our failure to find any association between number of sperm inseminated and either orgasm pattern, stage of menstrual cycle, or use of oral contraceptives should not be considered a failure of sperm competition theory. Rather it could reflect the extent to which males cannot obtain the information necessary for optimum adjustment of sperm number.

## Masturbation, Costly Ejaculates and Sperm Number

Our analysis of the dynamics and consequences of masturbation suggests that the behaviour is a functional strategy. The function, however, appears to be more to increase the 'fitness' of sperm retained by the female at the next in-pair copulation than to increase the number retained.

As time since last copulation increases beyond 72 h, males become increasingly likely to masturbate (Fig. 3) and are most likely to do so less than 48 h before their next in-pair copulation (Fig. 4). One implication is that they could be anticipating the next in-pair copulation. In any case, the result is a significant improvement in the residual number of sperm retained by the female at the next in-pair copulation (Table XIV). Thus, masturbation appears to remove from the next ejaculate sperm that are either less acceptable to the female and/or less able to attain a 'secure' position in the female tract. Either way, our analysis of flowbacks implies that the sperm removed by masturbation are in some way suboptimal and largely destined to be ejected by the female. Table XIV strongly implies that the primary, though not necessarily the only, way in which these sperm may be suboptimal is that they are older.

Suppose, at least for descriptive convenience, that sperm move along the male's production line on a 'conveyor belt' system and that ejaculation uses the oldest section of sperm. After development and maturation in the seminiferous tubules, vas deferens and early part of the epididymis (Mann & Lutwak-Mann 1981); sperm pass the point at which they are eligible for ejaculation at the rate of 12.5 million/h (Johnson *et al.* 1980). They then accumulate in the wider, distal end of the vas deferens. Some of these waiting sperm may be removed by phagocytosis (Mann & Lutwak-Mann 1981). At ejaculation, a segment of the waiting sperm is moved from the vas deferens and mixed with seminal fluid (Mann & Lutwak-Mann 1981).

The rate at which sperm become suboptimal through waiting to be ejaculated at the storage end of the conveyor belt is unknown. It should be less than 12.5 million/h otherwise there could be no build-up of usable sperm waiting for ejaculation. According to equation (1) (Table II), an extra 4.63 (say 5) million sperm are shed during masturbation for every hour since the male's last ejaculation. It is not unreasonable to suppose that this approximates to the rate at which unused sperm become suboptimal while waiting to be ejaculated (though, to this number, should be added the unknown number that are phagocytosed and voided in the urine).

We have shown that when a male inseminates a female he does not ejaculate all of the sperm available but instead inseminates a number that is a function of the risk of sperm competition inside that female. Thus, some sperm which could have been used are conserved and, we presume, move up on the conveyor belt to be next in line for ejaculation. The important question is why these sperm should be conserved instead of being inseminated at the current copulation. Essentially, interpretation rests on whether restraint generates: (1) a future advantage through use of the conserved sperm; and/or (2) an advantage or disadvantage at the current copulation.

*Is there an advantage in saving sperm for a future copulation?*

The sperm conserved by a male through restraint at the current copulation have four possible destinies. They may be: (1) inseminated into the same female in a future copulation; (2) inseminated into a different female in a future copulation; (3) shed in a future masturbation or nocturnal emission; or (4) phagocytosed or voided in the male's urine.

Clearly, if the conserved sperm are eventually shed or voided their conservation was of no future advantage. Indeed, any cost attached to postcopulation storage and then shedding renders their conservation a disadvantage. Phagocytosis may recoup some of this cost. However, only if the sperm are eventually inseminated into a female is there any real chance that their conservation could have been an advantage. In part, therefore, the value of conservation depends on the probability that the conserved sperm will be inseminated into a female rather than be shed or voided. In part, also, it depends on the value of the sperm for fertilization in a future copulation relative to what would have been their value in the present copulation.

We consider first the possible value of conserving sperm for a future copulation with the same female. For each hour's delay before the next copulation, at least 5 million of the conserved sperm will become suboptimal and, even if inseminated into the female at the pair's next copulation, will probably be ejected in the flowback. At least these sperm, therefore, were of no future value. Our data show that males top-up their female by between 2 million and 9 million sperm (depending on the proportion of time the pair are together) for each hour between copulations. Yet new sperm are becoming available at the rate of 12.5 million/h, more than enough to top-up the female at the next copulation, no matter what proportion of time the pair have been together. It seems unlikely, therefore, that males conserve sperm at one copulation for use at some future copulation with the same female.

This conclusion could be complicated by the possibility that, if the female subsequently engages in extra-pair copulation, the male (as for birds: Birkhead & Møller 1992) may need to show rapid 'unscheduled' in-pair copulation and inseminate large numbers of sperm to counteract the certainty of sperm competition. However, unlike birds which show a clear second male advantage (Birkhead & Møller 1992), mammals may show first male (house mice, *Mus musculus*), last male (prairie voles, *Microtus ochrogaster*) and often no order effects (rats and swine, *Sus scrofa*; see Dewsbury 1984), in which case, the male has little to gain from conserving sperm for insemination after the female has engaged in extra-pair copulation. This is especially so if any sperm conserved have, through greater age, a high probability of being ejected in the flowback.

The only possible use for conserved sperm, therefore, is to increase the chances of securing fertilization of a different female. To be beneficial, the conserved sperm must increase the chances and/or value of fertilization of the second female by more than they would have increased the chances and/or value of fertilization of the first female. A major factor will be the relative probability of sperm competition in the two females. However, even when the conserved sperm would be more beneficial in the second female than the first, this greater benefit has to outweigh three other factors: (1) the cost of storage; (2) the conserved sperm would be younger when inseminated into the first female; and (3) the conserved sperm must be inseminated into the second female before they become suboptimal, are shed, or are inseminated into the first female. As an initial approximation, they have at most about 3 days (median inter-copulation interval and time to increase in probability of masturbation; Fig. 3).

Without knowing the relative costs and benefits of these different parameters, no decision can be reached. Until it can, we have at least to entertain the possibility that conserved sperm are of no future advantage to the male and that the advantage of restraint may derive, not from economy and conservation, but from some direct benefit at the current insemination.

### *Is restraint an advantage or disadvantage at the current copulation?*

There are several possibilities why males could use restraint to gain maximum advantage from the current copulation, irrespective of what happens in the future to any sperm conserved.

Our analysis of ejaculate adjustment has identified two main elements to male strategy. First, successive in-pair copulations are 'toppings-up' to some maximum level that is lower when risk of sperm competition is also lower. Second, masturbation seems to be a strategy to increase the fitness (perhaps longevity, competitiveness and/or fertility) of the sperm retained by the female at the next in-pair copulation without increasing the number retained. The overall impression is that there is some disadvantage to a male in placing too large a population of sperm in his partner's tract.

No data exist for any mammal on the relationship between large numbers of sperm in the female tract and the probability of fertilization. However, a known correlate of fertility impairment in humans is polyzoospermy (ejaculation of too many sperm; Wolf *et al.* 1984). At present, clinical diagnosis places the lower level of polyzoospermy at $250 \times 10^6$ sperm/ml of ejaculate (when inter-ejaculation interval is about 72 h and ejaculates are collected during masturbation). With an average ejaculate volume of about 3 ml, this density is equivalent to an ejaculate containing about 750 million sperm. Why polyzoospermy should lead to impaired fertility is unknown. Apart from their concentration of sperm, polyzoospermic ejaculates have clinically normal parameters and, as long as concentration around the egg is controlled, their sperm fertilize eggs in vitro.

Our own data show that the more sperm a male inseminates, the more are retained (Fig. 6). It is also known that, in rabbits, the more sperm retained, the more are found at all positions throughout the female tract (Morton & Glover 1974*a*,*b*). The implication is that, on average, the more sperm are inseminated the more arrive at all positions in the female tract including, presumably, around the egg.

In vitro studies show clearly that there is an optimum sperm:egg ratio in the vicinity of the egg, above which the probability of obtaining a viable zygote declines. Thus, in laboratory mice the probability of fertilization peaks (78%) at sperm:egg ratios of about 16000:1 (Tsunoda & Chang 1975). Increasing the sperm:egg ratio about ten-fold decreases the probability of fertilization by more than 50%. The optimum sperm:egg ratio for humans in vitro is around 50000:1 (see Lee 1988).

There are at least two potential disadvantages of a high sperm:egg ratio. First, above a certain ratio, enzymes released by sperm have an increasing chance of killing the egg (Adams 1969). Second, the presence of too many sperm around the egg could lead to polyspermic fertilization and a non-viable zygote. In rats and rabbits, higher concentrations of sperm at the site of fertilization are associated with larger numbers of sperm in eggs (Braden & Austin 1954). In humans, the incidence of polyspermy in vitro is directly related to the concentration of sperm (Simpson *et al.* 1982; Wolf *et al.* 1984). In vivo, 10% of all first trimester abortions are polyploid (Simpson *et al.* 1982). A retrospective study of 1499 spontaneously aborted fetuses showed that 39% had abnormal karyotypes of which 25% were polyploid (Boue *et al.* 1975; see also Wolf *et al.* 1984). In other mammals, level of polyspermy in vivo is around 2% (Austin 1961).

### *Conclusion*

We suggest that in the presence of sperm competition, a male's chances of fertilizing a female are increased by increasing sperm number but, in the absence of sperm competition, his chances are increased by decreasing sperm number to some optimum level. We suggest further that this latter factor (the optimum sperm number in the female tract for fertilization in the absence of sperm competition) could be a major constraint to insemination of too many sperm at any given copulation. On this model, the primary trade-off in determining ejaculate size is between the probabilities that the male will fertilize the female if the inseminate does or does not encounter sperm competition. Optimum number of sperm is thus determined for each inseminate simply by the risk of sperm competition.

At present, we cannot determine the relative importance of this factor and any constraint on sperm numbers due to the cost of producing ejaculates (Dewsbury 1982). Inevitably, ejaculate cost must have been a factor in the evolution of species specific rates of sperm production. It is possible, however, that at least for mammals ejaculate cost could be less important than the factors discussed here in influencing restraint over the number of sperm inseminated on any given occasion.

## ACKNOWLEDGMENTS

This study has required the voluntary cooperation, and no small amount of courage, from many people. We are grateful to them all. Valuable statistical advice was offered by Ray Meddis, Laurence Cook, Mike Hounsome and Kevin McConway. The condoms used were kindly donated by Durex. We thank: Emma Creighton, Dominic Shaw and Viki Cook for help with our pilot surveys; Gill Hudson and *Company* magazine for publishing our final questionnaire on sexual behaviour; and Jo Bell, Kath Griffin and Phil Wheater for help with the mammoth task of transferring the data onto computer. Jo Moffitt organized the recruitment and collection of the subjective estimates of flowback volume. Professor Tim Halliday and Professor Tim Birkhead made comments on an early draft which improved the manuscript considerably. We are particularly grateful to Liz Oram for drawing our attention to the potential value of studying flowbacks and for developing the techniques of flowback collection. Many aspects of the work were funded by the Science and Engineering Research Council.

## REFERENCES

Adams, C. E. 1969. Intraperitoneal insemination in the rabbit. *J. Reprod. Fert.*, **18**, 333–339.

Austin, C. R. 1961. *The Mammalian Egg*. Springfield, Illinois: Charles C. Thomas.

Austin, C. R. 1975. Sperm fertility, viability and persistence in the female tract. *J. Reprod. Fert., Suppl.* **22**, 75–89.

Baker, R. R. & Bellis, M. A. 1988. 'Kamikaze' sperm in mammals? *Anim. Behav.*, **36**, 937–980.

Baker, R. R. & Bellis, M. A. 1989a. Number of sperm in human ejaculates varies in accordance with sperm competition theory. *Anim. Behav.*, **37**, 867–869.

Baker, R. R. & Bellis, M. A. 1989b. Elaboration of the kamikaze sperm hypothesis: a reply to Harcourt. *Anim. Behav.*, **37**, 865–867.

Baker, R. R. & Bellis, M. A. 1993. Human sperm competition: ejaculate manipulation by females and a function for the female orgasm. *Anim. Behav.*, **46**, 887–909.

Baker, R. R. & Bellis, M. A. In press. *Human Sperm Competition: Copulation, Masturbation and Infidelity*. London: Chapman & Hall.

Baker, R. R., Bellis, M. A., Hudson, G., Oram, E. & Cook, V. 1989. *Company*, September, 60–62.

Barrett, J. C. & Marshall, J. 1969. The risk of conception on different days of the menstrual cycle. *Pop. Studies*, **23**, 455–461.

Beach, F. A. 1975. Variables affecting 'Spontaneous' seminal emissions in rats. *Physiol. Behav.*, **15**, 91–95.

Bellis, M. A. & Baker, R. R. 1990. Do females promote sperm competition: data for humans. *Anim. Behav.*, **40**, 997–999.

Bellis, M. A., Baker, R. R. & Gage, M. J. G. 1990. Variation in rat ejaculates is consistent with the kamikaze sperm hypothesis. *J. Mammal.*, **71**, 479–480.

Bellis, M. A., Baker, R. R., Hudson, G., Oram, E. & Cook, V. 1989. *Company*, April, 90–92.

Belsey, M. A., Eliasson, R. Gallegos, A. J., Moghissi, K. S., Paulsen, C. A. & Prasad, M. R. N. 1987. *WHO Laboratory Manual for Examination of Human Semen and Semen-Cervical Mucus Interaction.* 2nd edn. Cambridge: Cambridge University Press.

Birkhead, T. R. & Møller, A. P. 1992. *Sperm Competition in Birds*. London: Academic Press.

Bongaarts, J. & Potter, R. G. 1983. *Fertility, Biology and Behaviour: An Analysis of the Proximate Determinants*. New York: Academic Press.

Boue, J., Boue, A. & Lazar, P. 1975. Retrospective and prospective epidemiological studies of 1500 karyotyped spontaneous human abortions. *Teratology*, **12**, 11–25.

Braden, A. W. H. & Austin, C. R. 1954. The number of sperms about the eggs in mammals and its significance for normal fertilization. *Austral. J. Biol. Sci.*, **7**, 543–551.

Carpenter, C. R. 1942. Sexual behaviour of free ranging rhesus monkeys (*Macaca mulatta*). II. Periodicity of oestrous, homosexual, autoerotic and non-conformist behaviour. *J. Comp. Physiol.*, **33**, 143–162.

Darling, F. F. 1963. *A Herd of Deer*. London: Oxford University Press.

Dewsbury, D. A. 1982. Ejaculate cost and male choice. *Am. Nat.*, **119**, 601–610.

Dewsbury, D. A. 1984. Sperm competition in muroid rodents. In: *Sperm Competition and the Evolution of Animal Mating Systems* (Ed. by R. L. Smith), pp. 547–571. London: Academic Press.

Flinn, M. V. 1988. Mate guarding in a Caribbean village. *Ethol. Sociobiol.*, **9**, 1–28.

Ford, C. S. & Beach, F. A. 1952. *Patterns of Sexual Behaviour*. London: Eyre & Spottiswoode.

Gage, M. J. G. 1992. Risk of sperm competition directly affects ejaculate size in the Mediterranean fruit fly. *Anim. Behav.*, **42**, 1036–1037.

Gage, M. J. G. & Baker, R. R. 1991. Ejaculate size varies with socio-sexual situation in an insect. *Ecol. Entomol.*, **16**, 331–337.

Garn, S. M. & LaVelle, M. 1983. Reproductive histories of low weight girls and women. *Am. J. Clin. Nutr.*, **37**, 862–866.

Ginsberg, J. R. & Rubenstein, D. I. 1990. Sperm competition and variation in zebra mating behaviour. *Behav. Ecol. Sociobiol.*, **26**, 427–434.

Gray, R. H. 1983. The impact of health and nutrition on natural fertility. In: D*eterminants of Fertility in Developing Countries, Vol. I. Supplement and Demand for Children* (Ed. by R. A. Bulatao & R. D. Lee), pp. 139–162. New York: Academic Press.

Gwynne, D. T. 1981. Sexual difference theory: mormon crickets show role reversal in mate choice. *Science*, **213**, 779–780.

Hafez, E. S. E., Schein, M. W. & Ewbank, R. 1969*a*. The behaviour of cattle. In: *The Behaviour of Domestic Animals*, 2nd edn (Ed. by E. S. E. Hafez), pp. 235–295. Baltimore: Williams and Wilkins.

Hafez, E. S. E., Williams, M. & Wierzbowski, S. 1969*b*. *The behaviour of horses. In: The Behaviour of Domestic Animals*. 2nd edn (Ed. by E. S. E. Hafez), pp. 380–416. Baltimore: Williams and Wilkins.

Halliday, T. R. 1983. The study of mate choice. In: *Mate Choice* (Ed. by P. Bateson), pp. 3–32. Cambridge: Cambridge University Press.

Harcourt, A. H. 1991. Sperm competition and the evolution of non fertilizing sperm in mammals. *Evolution,* **45**, 314–328.

Harvey, P. H. & Harcourt, A. H. 1984. Sperm competition, testis size, and breeding systems in primates. In: *Sperm Competition and the Evolution of Animal Mating Systems* (Ed. by R. L. Smith), pp. 589–600. London: Academic Press.

Hawker, R. W. 1984. *Notebook on Medical Physiology: Endocrinology*. London: Churchill Livingstone.

Hill, K. & Kaplan, H. 1990. Tradeoffs in male and female reproductive strategies among the Ache: part 2. In: *Human Reproductive Behaviour: A Darwinian Perspective* (Ed. by L. Betzig, M. B. Mulder & P. Turke), pp. 291–306. Cambridge: Cambridge University Press.

Johnson, L., Petty, C. S. & Neaves, W. B. 1980. A comparative study of daily sperm production and testicular composition in humans and rats. *Biol. Reprod.*, **22**, 1233–1243.

Johnson, M. B. & Everitt, B. J. 1988. *Essential Reproduction.* 3rd edn. London: Blackwell Scientific Publications.

Kinsey, A. C., Pomeroy, W. B., & Martin, C. E. 1948. *Sexual Behavior in the Human Male.* Philadelphia: W. B. Saunders.

Kinsey, A. C., Pomeroy, W. B., Martin, C. E. & Gebhard, P. H. 1953. *Sexual Behavior in the Human Female.* Philadelphia: W. B. Saunders.

Lechtig, A. & Klein, R. E. 1981. Pre-natal nutrition and birth weight: is there a causal association? In: *Maternal Nutrition in Pregnancy: Eating for Two?* (Ed. by J. Dobbing), pp. 131–174. New York: Academic Press.

Lee, S. 1988. Sperm preparation for assisted conception. *Conceive*, **12**, 4–6.

Little, B. B., Guzick, D. S., Malina, R. M. & Rocha Ferreira, M. D. 1989). Environmental influences cause menstrual synchrony, not pheromones. *Am. J. Human Biol.*, **1**, 53–57.

McCance, R. A., Luff, M. C. & Widdowson, E. E. 1937. Physical and emotional periodicity in women. *J. Hygiene*, **37**, 571–611.

MacGillivray, 1. & Campbell, D. M. 1978. The physical characteristics and adaptations of women with twin pregnancies. In: *Twin Research: Clinical Studies* (Ed. by W. E. Nance), pp. 81–86. New York: Alan R. Liss.

Mann, T. & Lutwak-Mann, C. 1981. *Male Reproductive Function and Semen*. New York: Springer-Verlag.

Martin, P. A., Reimers, T. J., Lodge, J. R. & Dzuik, P. J. 1974. The effect of ratios and numbers of spermatozoa from two males on proportions of offspring. *J. Reprod. Fert.*, **39**, 251–258.

Meddis, R. 1984. *Statistics Using Ranks: A Unified Approach*. Oxford: Blackwell.

Møller, A. P. 1991. Sperm competition, sperm depletion, paternal care and relative testis size in birds. *Am. Nat.*, **137**, 882–906.

Morton, D. B. & Glover, T. D. 1974a. Sperm transport in the female rabbit: the role of the cervix. *J. Reprod. Fert.*, **38**, 131–138.

Morton, D. B. & Glover, T. D. 1974b. Sperm transport in the female rabbit: the effects of inseminate volume and sperm density. *J. Reprod. Fert.*, **38**, 139–146.

Mulder, M. B. 1990. Kipsigis bridewealth payments. In: *Human Reproductive Behaviour: A Darwinian Perspective* (Ed. by L. Betzig, M. B. Mulder & P. Turke), pp. 65–82. Cambridge: Cambridge University Press.

Parker, G. A. 1970a. Sperm competition and its evolutionary consequences in the insects. *Biol. Rev.*, **45**, 525–567.

Parker, G. A. 1970b. Sperm competition and its evolutionary effect on copula duration in the fly *Scatophaga stercoraria*. *J. Insect Physiol.*, **16**, 1301–1328.

Parker, G. A. 1982. Why are there so many tiny sperm? Sperm competition and the maintenance of two sexes. *J. Theor. Biol.*, **96**, 281–294.

Parker, G. A. 1984. Sperm competition and the evolution of animal mating strategies. In: *Sperm Competition and the Evolution of Animal Mating Systems* (Ed. by R. L. Smith), pp. 1–60. London: Academic Press.

Parker, G. A. 1990. Sperm competition: games, raffles and roles. *Proc. R. Soc. Lond. B.*, **242**, 120–126.

Roche, J. F., Dzuik, P. J. & Lodge, J. R. 1968. Competition between fresh and aged spermatozoa in fertilizing rabbit eggs. *J. Reprod. Fert.*, **16**, 155–157.

Rosenblatt, J. S. & Schneirla, T. C. 1962. The behaviour of cats. In: *The Behaviour of Domestic Animals.* 1st edn (Ed. by E. S. E. Bafez), pp. 453–488. Baltimore: Williams and Wilkins.

Siegel, S. 1956. *Non-parametric Statistics for the Behavioural Sciences*. London: McGraw-Hill.

Silberglied, R. E., Shepherd, J. G. & Lou Dickinson, J. 1984. Enuchs: the role of apyrene sperm in Lepidoptera? *Am. Nat.*, **123**, 255–265.

Simpson, J. L., Golbus, M. S., Martin, A. O. & Sarto, G. E. 1982. *Genetics in Obstetrics and Gynaecology.* New York: Grune & Stratton.

Sivinski, J. 1980. Sexual selection and insect sperm. *Fla Entomol.*, **63**, 99–111.

Small, M. F. 1989. Aberrant sperm and the evolution of human mating patterns. *Anim. Behav.*, **38**, 544–546.

Smith, R. L. 1984. Human sperm competition. In: *Sperm Competition and the Evolution of Animal Mating Systems* (Ed. by R. L. Smith), pp. 601–660. London: Academic Press.

Sokal, R. R. & Rohlf, F. J. 1981. *Biometry*. 2nd edn. New York: W. H. Freeman.

Svard, L. & Wiklund, C. 1989. Mass production rate of ejaculates in relation to monandry-polyandry in butterflies. *Behav. Ecol. Sociobiol.*, **24**, 395–402.

Tsunoda, Y. & Chang, M. C. 1975. Penetration of mouse eggs in vitro: optimal sperm concentration and minimal number of spermatozoa. *J. Reprod. Fert.*, **44**, 139–142.

U. K. Family Planning Research Network. 1988. Patterns of sexual behaviour among sexually experienced women attending family planning clinics in England, Scotland and Wales. *Br. J. Fam. Plann.*, **14**, 74–82.

Wolf, D. P., Byrd, W., Dandekar, P. & Quigley, M. M. 1984. Sperm concentration and the fertilization of human eggs in vitro. *Biol. Reprod.*, **31**, 837–848.

# 11. HUMAN SPERM COMPETITION: EJACULATE MANIPULATION BY FEMALES AND A FUNCTION FOR THE FEMALE ORGASM

R. Robin Baker[1] and Mark A. Bellis[1]

## ABSTRACT

Behavioural ecologists view monogamy as a subtle mixture of conflict and cooperation between the sexes. In part, conflict and cooperation is cryptic, taking place within the female's reproductive tract. In this paper the cryptic interaction for humans was analysed using data from both a nationwide survey and counts of sperm inseminated into, and ejected by, females. On average, 35% of sperm were ejected by the female within 30 min of insemination. The occurrence and timing of female orgasm in relation to copulation and male ejaculation influenced the number of sperm retained at both the current and next copulation. Orgasms that climaxed at any time between 1 min before the male ejaculated up to 45 min after led to a high level of sperm retention. Lack of climax or a climax more than 1 min before the male ejaculated led to a low level of sperm retention. Sperm from one copulation appeared to hinder the retention of sperm at the next copulation for up to 8 days. The efficiency of the block declined with time after copulation but was fixed at its current level by an inter-copulatory orgasm which thus reduced sperm retention at the next copulation. Inter-copulatory orgasms are either spontaneous ( = nocturnal) or induced by self-masturbation or stimulation by a partner. It is argued that orgasms generate a blow-suck mechanism that takes the contents of the upper vagina into the cervix. These contents include sperm and seminal fluid if present; acidic vaginal fluids if not. Inter-copulatory orgasms will therefore lower the pH of the cervical mucus and either kill or reduce the mobility of any sperm that attempt to penetrate from reservoirs in the cervical crypts. Intercopulatory orgasms may also serve an antibiotic function. Copulatory and inter-copulatory orgasms endow females with considerable flexibility in their manipulation of inseminates. The data suggest that, in

[1] Department of Environmental Biology, University of Manchester.

Reprinted from *Animal Behaviour*, 46 (5), Baker, R. R. & Bellis, M. A., Human sperm competition: Ejaculate manipulation by females and a function for the female orgasm. pp. 887–909, Copyright (1993), with permission from Elsevier.

purely monandrous situations, females reduced the number of sperm retained, perhaps as a strategy to enhance conception. During periods of infidelity, however, females changed their orgasm pattern. The changes would have been cryptic to the male partners and would numerically have favoured the sperm from the extra-pair male, presumably raising his chances of success in sperm competition with the female's partner.

---

By definition, most mating by monogamous species is in-pair copulation. However, an apparently universal feature of such species (Mock & Fujioka 1990) is that from time to time both sexes engage in extra-pair copulation. A special category of extra-pair copulation is double-mating (the female mating with a second male while still containing fertile sperm from one or more previous males). The result is 'sperm competition' (Parker 1970) as the sperm from different males compete to fertilize the female's egg(s).

Models of sperm competition (e.g. Parker 1990) tend to view the female tract as a passive receptacle in which males play out their sperm competition games. Females have the potential, however, to influence the outcome of the contest in several different ways. The sequence and frequency with which the female mates with different males and the time interval between in-pair copulations and extra-pair copulations often have a major influence on the outcome of sperm competition (Birkhead & Hunter 1990). More directly, females eject sperm (e.g. birds: Howarth 1971; Davies 1983; mammals: Sumption 1961; Morton & Glover 1974; Tilbrook & Pearce 1986; Ginsberg & Huck 1989; Ginsberg & Rubenstein 1990). On average, about 80% of the sperm inseminated into the rabbit, *Oryctolagus cuniculus*, are ejected in the flowback (Overstreet 1983).

Female humans eject up to 3 ml of seminal fluid from their vagina after copulation (Baker & Bellis 1993). This 'flowback' emerges from the vagina as a discrete series of three to eight white globules and consists of a mixture of sperm, seminal fluid and female tissue and secretions. Flowback either occurs while the female is still horizontal after copulation, when she next begins to walk, or, perhaps most often and most forcefully, when she next urinates. No attempt has previously been made to determine the proportion of sperm ejected and retained by human females following normal copulation. Nor has any attempt been made to investigate variation in this proportion in relation to socio-sexual situation or in relation to female physiological events, such as the female orgasm.

Female orgasms occur in four main situations (Fisher 1973): (1) spontaneously during sleep (equivalent to the nocturnal emissions of males; see Wells 1986); (2) through direct self-stimulation (e.g. of the clitoris) in the absence of a male; (3) through stimulation (either by self and/or by a male or another female) in the presence of a partner but without copulation; and (4) through self, manual or penile stimulation as part of a copulation episode (orgasm occurring during foreplay, postplay, or copulation itself). For convenience, we refer to types (1) – (3) as 'non-copulatory' orgasms, and to type (4) as 'copulatory'. No study of humans, or other mammals, has yet attempted to quantify the relative occurrence of these four types of female orgasm.

Currently, there are two favoured hypotheses concerning the function of copulatory orgasms in females: (1) the 'poleaxe' hypothesis (Morris 1967); and (2) the 'upsuck' hypothesis (Fox *et al.* 1970). The poleaxe hypothesis proposes that, as humans are bipedal, it is important for the female to lie down after copulation in order to reduce sperm loss. The orgasm thus functions to induce fatigue and sleep (see Levin 1981). The upsuck hypothesis proposes that the orgasm functions to suck up sperm during copulation (Fox *et al.* 1970). In an elaboration of the upsuck hypothesis, Singer (1973) proposed that orgasms could be typed by whether they involved contractions of the uterus as well as the vagina. Uterine orgasms were suggested to facilitate conception whereas non-uterine orgasms did not. Singer further proposed that by a systematic shift from one type of orgasm to another a female could, consciously or subconsciously, influence the probability of conception. The occurrence of female orgasms in quadrupeds (Evans 1933; Ford & Beach 1952; Hartman 1957) is inconsistent with the poleaxe hypothesis. Nevertheless, in this paper, we provide the first direct test of both of these hypotheses.

Non-copulatory orgasms are usually considered either to be substitutes for copulatory orgasms or to represent mechanisms by which a female can train her body to orgasm during copulation (Lopiccolo & Lobitz 1972). Neither hypothesis, however, can explain why masturbation is more frequent when copulation is more frequent (Baker *et al.* 1989). Moreover, the 'substitute' hypothesis cannot explain why copulatory and masturbatory orgasms show different patterns in the female menstrual cycle (Baker *et al.* 1989).

Our own hypothesis (Baker *et al.* 1989) is much more firmly based in behavioural ecology. It is that the timing of orgasm, both during and between copulations, is the key feature of the mammalian female's armoury in male:female conflict and cooperation within the female tract. We propose that nocturnal, masturbatory and copulatory orgasms are the primary mechanisms by which the female influences the ability of sperm in the next and/or current ejaculate to remain in, and travel through, her reproductive tract. Thus, we predict that by altering the occurrence, sequence and timing of the different types of orgasm, the female can influence both the probability of conception in monandrous situations and the outcome of sperm competition in polyandrous situations. We also expect that as for the female's timing of copulation during the menstrual cycle (Bellis & Baker 1990), much of this influence will be cryptic to the male partner(s). This paper presents a test of this hypothesis and an exploration of the dynamics of male:female conflict and cooperation over sperm.

## METHODS

### Data

Our data came from four different sources: (1) counts of sperm in whole ejaculates collected by condom during copulation; (2) counts of sperm in 'flowbacks'; (3) subjective estimates of flowback volume; and (4) a U.K. nationwide survey of female sexual behaviour (Bellis *et al.* 1989; Bellis & Baker 1990). Full details of these different sources are presented in Baker & Bellis (1993) and here we give only a brief resumé, describing in detail only those methodological details that were not relevant to our previous study.

In our nationwide survey, 3679 females each answered 57 questions on their sexual behaviour, including information relating to their last copulation and to their last non-copulatory orgasm. This survey provided information on 2745 in-pair copulations and 126 extra-pair copulations. In our ejaculate study, 34 male–female human pairs (involving 32 males and 32 females) provided material or estimates relating to 323 in-pair copulations.

The methods for the collection, fixation, processing and counting of sperm followed the double-blind protocol used previously (Baker & Bellis 1989, 1993) based on the procedures described in the WHO Human Semen Manual (Belsey *et al.* 1987). All standard errors for sperm counts are within the range 5–25% and are not used further in this paper. A complete list of the number of whole ejaculates and flowbacks donated by each couple, plus the female's weight are given in tables in Baker & Bellis (1993).

Seven pairs, while willing to record details of in-pair copulations, preferred not to collect flowbacks directly but volunteered instead to estimate the volume of flowback subjectively. In addition, two pairs who collected flowbacks also made some subjective estimates of volume (Table I in Baker & Bellis 1993). All nine pairs recorded flowback volume as either 'normal' (for that female) ( = 2), heavier than normal ( = 3), lighter than normal ( = 1), or none ( = 0). Crudely, these estimates approximate to the volume of the flowback in ml. None of the subjects knew the results of the investigation of flowback samples that was being carried out simultaneously.

### Statistics: Calculation of Probabilities

Throughout, to be consistent with our non-parametric statistics, medians are used instead of means. In Figures and Tables, the medians presented are the medians of medians (i.e. we first calculate the median for each contributing female, then calculate the median of the medians). Variation about the median is expressed in terms of inter-quartile range. On occasions when only one female contributed to a particular category, rather than give no indication of variability; we present the inter-quartile range for that female's data.

On two occasions we used the $z$-difference test (modified for two samples from Fugle *et al.* 1984) to calculate whether the difference between two $z$-values is greater than expected by chance. Otherwise, to avoid making any assumptions concerning the normality of our data, we calculated probability values using only non-parametric statistical tests. In particular, we used Meddis' (1984) rank-sum test. The reasons and justification for our choice of test are given in detail in Baker & Bellis (1993). All analyses in this paper are 'blocked' by female unless stated otherwise. Results thus indicate whether within-female variation is also consistent between females. All probabilities associated with Meddis' non-specific tests (statistic $= H$) are two-tailed. Probabilities associated with Meddis' specific test (statistic $= z$) are one-tailed unless stated otherwise.

### Statistics: Residuals and a Predictive Equation

As in our previous analyses (Baker & Bellis 1993), we used regression analysis to calculate values for the dependent variable with respect to one or more independent variables, and then calculated residuals. These residuals may then be analysed further for the influence of some other independent variable not included in the original regression. We also used regression and residual analysis to develop an equation to predict the number of sperm retained by females during in-pair copulations under different socio-sexual situations. The equations used in this paper are gathered together in Table I.

## Male and Female Contributions to Sperm Retention

The number of sperm retained by a female from any given in-pair copulation is essentially the product of two influences. Insofar as the male inseminates a finite number of sperm, he places an upper limit on the number that may possibly be retained. The number actually retained then depends on what happens in the female reproductive tract. The behaviour of the sperm and/or the female's response to insemination may both influence the number of sperm actually retained.

The raw data on which this paper is based are counts of the number of sperm ejected in the flowback. Primary biological interest, however, lies not with the number of sperm ejected but with the number retained. Unfortunately, variation in the number of sperm ejected has no certain relationship with the number retained.

**Table I.** Six equations used to calculate residuals or to predict number of sperm retained (for derivations, see Results)

| Equations |
|---|
| (1) $NINS = PV - (((PV - mid)/(limC - mid))^2 \times (limC - limO))$ |
| (2) $NSR = -0.9 + (0.65 \times NINS)$ |
| (3) $NSR = 33.4 + (0.65 \times NINS) - (0.71 \times HEJ)$ |
| (4) $NSR = -33.7 + (0.65 \times NINS) - (0.71 \times HEJ) + (31.2 \times HLOR)$ |
| (5) $NSR = -59.7 + (0.65 \times NINS) - (0.71 \times HEJ) + (31.2 \times HLOR) + (0.6 \times HFEV)$ |
| (6) $NSR = (-34.0 + 0.65 \times NINS) - (0.71 \times HEJ) + (31.2 \times HLOR) + ((1 - ICM) \times (1 - HLOR) \times (HFEV \times 0.60 - 34)) - (ICM \times 2)) \times (1 - PREG)$ |

NINS, Number of sperm inseminated in millions; PV, $1.94 \times HIPC - 3.40 \times PCT + ((2.41 \times HMAS - 228) \times MC) - 1008 + 23.37 \times FW$; HIPC, hours since last in-pair copulation up to 192 h (192 h is the longest time interval for which we have data and up to which analysis supports a linear relationship); PCT, the % time a pair have spent together since their last in-pair copulation; HMAS, hours (up to 72) since last masturbation; MC, 0 for IPC– IPC ejaculates and 1 for MAS–IPC ejaculates; FW, female weight (kg); mid, mid-point of number of sperm in observed ejaculates ($350 \times 10^6$); limC, the minimum or maximum calculable limit to PV; limO, the minimum or maximum observed number of sperm in in-pair copulation (IPC) ejaculates; NSR, number of sperm (in millions) retained by the female after flowback; HEJ, hours since last ejaculation (range 1–72 h; values >72 should be given a value of 72); HLOR, 1 when the orgasm regime at $C_2$ favours a high level of sperm retention and 0 when a low level; HFEV, hours to first event after $C_1$ (i.e. to first inter-copulatory orgasm or, in the absence of an inter-copulatory orgasm, to $C_2$; maximum value = 192); ICM, 1 if the female begins to menstruate between $C_1$ and $C_2$ and 0 when she does not; PREG, 1 if the female is pregnant, 0 if she is not. For derivation and use of equation (1), see Baker & Bellis (1993).

As an illustration, suppose that on two separate occasions the number of sperm in the flowback (in millions) was 100 and 200. If the number of sperm inseminated on these two occasions had been 150 and 300, respectively, the number of sperm retained (50 and 100) would have been positively associated with observed variation in the flowback. On the other hand, if the numbers inseminated had been 200 and 250, the number retained (100 and 50) would have been negatively associated with number in the flowback. Finally, if the numbers inseminated had been 150 and 250, the number retained would have remained constant despite variation in the number in the flowback.

It follows that, in order to draw biologically meaningful conclusions from flowbacks, it is essential that some estimate is made of the number of sperm retained. To do this, it is necessary to make some estimate of the number of sperm inseminated into the female during each of the in-pair copulations for which number of sperm in the flowback is known. The number of sperm inseminated is a measure of the male contribution to the number of sperm retained.

We have shown in our companion paper (Baker & Bellis 1993) that equation (1) (Table I) explains 58% of the observed variance in number of sperm ejaculated during 84 in-pair copulations by 24 couples (see Fig. 5 in Baker & Bellis 1993). In our study of flowbacks, we used equation (1) to calculate the number of sperm inseminated into the female at each in-pair copulation. We then estimated the number of sperm retained by the female by subtracting the number of sperm observed in the flowback from the number calculated to have been inseminated. In the remainder of this paper, to avoid unnecessarily long phrases, we refer simply to 'the number of sperm inseminated' and 'the number of sperm retained' even though both of these measures are the result of calculation rather than direct sampling.

There is a significant positive relationship between the number of sperm inseminated and the number of sperm in the flowback on the one hand and the number of sperm retained on the other ($N = 11$; first sample per couple; Table II). Blocking by couple to test for within-couple response, both relationships are still highly significant (versus number in flowbacks: $z = 6.062$, $N = 11$ females, 127 in-pair copulations, $P < 0.001$, Fig. 1; versus number retained $z = 6.500$, $P < 0.001$). Both between and within couples, therefore, the more sperm that are inseminated the more are ejected and the more retained.

**Table II.** Relationships between the number of sperm inseminated into a female during in-pair copulation and number of sperm ejected, number of sperm retained, and volume of flowback (analysis of only the first sample from each couple, $N = 11$ for sperm number, $N = 9$ for flowback volume)

| | Sperm number | | Flowback volume |
|---|---|---|---|
| | Ejected | Retained | Ejected |
| **Correlation** | | | |
| $r_s$ | 0.655 | 0.791 | – 0.113 |
| $P_1$ one-tailed | 0.026 | 0.004 | 0.560 |
| **Regression** | | | |
| Intercept | 0.90 | – 0.90 | |
| Slope | 0.35 | 0.65 | |
| Percentage explained | 24 | 58 | |

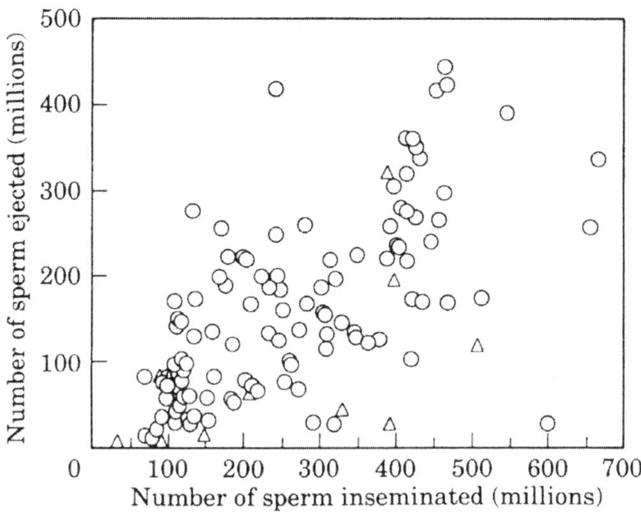


**Figure 1.** Relationship between number of sperm inseminated and number ejected in the flowback. △: First sample from each of 11 females; ○: subsequent samples.

We exploited this positive relationship to remove the influence of the male on the number of sperm retained. Using only the first sample produced by each couple (to avoid pseudoreplication from the more prolific couples), we calculated the least squares regression line for numbers of sperm ejected and retained against numbers inseminated (Table II). We then applied these regression lines to each of the 127 in-pair copulations in our total data set and calculated residuals. Variation in these residuals is thus independent of the variation in the number of sperm inseminated by the males and the residuals are, in effect, a measure of variation in the female's contribution to sperm retention.

Use of residuals to measure the female's influence on sperm retention has a number of advantages over the alternative which would be to use percentage sperm retention. Residuals are more robust, being less prone to wild fluctuations when the denominator is numerically small. Even so, at the end of our analyses, we checked the conclusions reached using residuals against percentage retention as reassurance that our major conclusions are not dependent on the parameter analysed.

The regression equations for sperm ejection and retention (Table II) are mathematically dependent (intercepts sum to zero; slopes sum to one). There is no value, therefore, in analysing residuals from both. Thus, we begin this paper by using residuals from equation (2) (Table I) as a measure of the female's contribution to the fate of the inseminated sperm. Discussion of this contribution is thus phrased in terms of the female's contribution to sperm retention rather than sperm ejection.

## Correcting for the Age of Sperm

We have shown (Baker & Bellis 1993) that younger sperm are retained in the female tract in relatively, but not absolutely, greater numbers than older sperm (assuming that hours since last ejaculation is a measure of the age of the sperm inseminated; $z = 3.039$, $P = 0.002$, two-tailed). $z$ is maximized (at 3.091) by a model that assumes there is no further influence of sperm age once time since last ejaculation exceeds 72 h.

**Table III.** Number of sperm (millions) inseminated, ejected and retained by 11 pairs (127 in-pair copulation flowbacks).

| | | No. inseminated | | No. ejected | | No. retained | | Percentage retained | |
|---|---|---|---|---|---|---|---|---|---|
| Pair | N | Median | (Inter-quartile range) | Median | (Inter-quartile range) | Median | (Inter-quartile range) | Median | (Inter-quartile range) |
| A | 9 | 508 | (211–546) | 173 | (73–257) | 206 | (138–339) | 61 | (46–66) |
| B | 93 | 242 | (121–388) | 158 | (83–224) | 67 | (26–136) | 35 | (14–50) |
| G | 1 | 398 | | 194 | | 204 | | 51 | |
| H | 1 | 90 | | 7 | | 83 | | 92 | |
| N | 2 | 343 | (293–393) | 28 | (27–28) | 315 | (265–366) | 92 | (90–93) |
| Q | 1 | 147 | | 14 | | 133 | | 90 | |
| R | 8 | 185 | (117–206) | 60 | (32–67) | 124 | (66–134) | 71 | (49–75) |
| S | 6 | 214 | (101–309) | 81 | (27–115) | 101 | (16–194) | 66 | (16–76) |
| W | 3 | 80 | (72–80) | 13 | (11–44) | 69 | (59–286) | 86 | (82–87) |
| Z | 2 | 81 | (69–92) | 84 | (83–85) | –3 | (–13 to 8) | –6 | (–19 to 8) |
| AA | 1 | 311 | | 130 | | 181 | | 58 | |
| Overall | | 214 | (90–343) | 81 | (14–158) | 124 | (69–204) | 66 | (51–90) |

Negative values for number of sperm retained (e.g. pair Z) occur when, owing to sampling error, the number of sperm in the flowback exceeds the number calculated to have been inseminated.

It is a moot point whether the age of sperm is a male or female influence on sperm retention. Younger sperm may be more acceptable to the female (female influence) or they may be better able to attain a position in the female tract that makes them more resistant to ejection (male influence; see Discussion in Baker & Bellis 1993). Either way, before analysing contributions to sperm retention that are clearly female, such as timing of orgasm, it is necessary to correct the residuals from equation (2) to make them independent of sperm age. For this, we used equation (3) (Table I). Residuals from equation (3) are thus independent of both the number of sperm inseminated and sperm age.

### Analysis Check

In unravelling the influence of female orgasm on the number of sperm retained, we made ever-increasing use of residuals from increasingly elaborate equations as different influential factors were identified and controlled (Table I). Such an approach is essential if artefacts are not to appear in our analysis due to cross-correlation of independent variables. However, our approach brings its own dangers in that mathematical artefacts may appear that generate their own spurious, non-biological relationships. The further the dependent variable departs from the raw data, the greater the danger becomes. As a final check on our conclusions, therefore, we return to the raw data and consider the extent to which the apparent influence of female orgasm is evident in the variation of sperm numbers directly measured in the flowbacks.

## RESULTS

### Human Flowbacks: General Features

All 127 of the flowbacks collected contained sperm and 94% (103/109) of the in-pair copulations monitored subjectively for flowback volume were followed by noticeable flowbacks (i.e. value > 0).

Sperm numbers in the 127 flowbacks we examined ranged from $7 \times 10^6$ to $443 \times 10^6$. Number of sperm inseminated, ejected and retained by each couple (Table III) suggest that humans have a median retention of about 65%.

There was a significant association between the number of sperm inseminated and both the number of sperm in the collected flowbacks ($z = 6.062$, $N = 11$ females, 127 flowbacks, $P < 0.001$) and the subjectively estimated volume of flowback ($z = 1.922$, $N =$ 9 females, 109 flowbacks, $P = 0.027$).

Median time to emergence of the flowback after male ejaculation was 30 min (inter-quartile range = 15–44; calculating the median for each of the 17 females who recorded time to flow back, then calculating the median for the group) with a range of 5–120 min. The number of sperm ejected in the flowback was not a linear (rank-order) function of the time-interval from ejaculation to flowback collection ($z = 1.608$, $N = 11$ females, 127 flowbacks, $P = 0.108$, two-tailed), the non-significant trend being for more sperm to be ejected the longer the time to ejection. In contrast, the volume of flowback was a significant negative function of time to flowback collection ($z = -2.844$, $N = 8$ females, 103 flowbacks, $P = 0.004$, two-tailed) suggesting that there is either a gradual loss or resorption of fluid without sperm before the main flowback emerges. Finally, time to flowback was not a function of the number of sperm inseminated ($z = 0.936$, $N = 18$ females, 230 in-pair copulations, $P = 0.350$, two-tailed), the non-significant trend being for the time from ejaculation to flowback to be longer when fewer sperm were inseminated.

We calculate that 12% (15/127) of all of the in-pair copulations for which we have flowback samples were followed by virtually 100% ejection of sperm (i.e. <1% retained). Females thus appear capable of total or near-total ejection of in-pair copulation ejaculates.

## Oral Contraceptives, Pregnancy and Flowbacks

Most of the flowbacks collected (11 females, 112 flowbacks) and all of those estimated for volume (nine females, 109 flowbacks) were from females who were taking oral contraceptives. However, three of the females volunteered to collect flowbacks ($N$ = 9) when, for different reasons, they were not using oral contraceptives. Of these, two became pregnant. One of the two females who became pregnant then volunteered to collect six further flowbacks. As this female had monitored her cycle using the basal temperature method (Parsons & Sommers 1978), time of conception could be estimated to within ± 24 h. Ultrasonic scan of the fetus after 14 weeks provided strong support for the estimate. Two pregnancy flowbacks were collected in the first week after conception, four 35–42 days after conception. These six pregnancy flowbacks could be matched with seven flowbacks collected by the same female while not taking oral contraceptives (four pre-conception, three post-partum).

This albeit limited data set shows a significant reduction in sperm retention during pregnancy ($z$ = 2.143, $N$ = 1 female, 13 in-pair copulations, $P$ = 0.032). The estimated median number (millions) of sperm actually retained were: pre-conception post-partum (68; inter-quartile range = 11–139), 1 week post-conception (116; 75–157) and 5–6 weeks post-conception (–42; –178 to +3). In a Meddis' specific test of residuals, $z$ is maximized (at 2.239) by a model that assumes retention during the first week post-conception is intermediate between retention levels pre-conception and 5–6 weeks post-conception.

Insofar as the median retention is negative, these data suggest a virtually 100% ejection of sperm even as early as 5–6 weeks after conception. This assumes, of course, that males do not inseminate more sperm in in-pair copulations during their partner's pregnancy. Unfortunately, no condom samples were collected during the female's pregnancy so we cannot test this assumption with our current data.

Females taking oral contraceptives showed no significant variation in retention or ejection during the 'menstrual' cycle, the tendency being to retain more sperm on days 6–15 (day 1 = first day of bleeding) than at other times ($z$ = 1.091, $N$ = 11 females, 110 in-pair copulations, $P$ = 0.276, two-tailed). The three females not taking oral contraceptives, however, showed a significant decrease in retention during their most fertile phase (days 6–15; $z$ = 2.443, $P$ = 0.016, two-tailed). Finally, when taking oral contraceptives, the same three females tended to retain more sperm than when they were not taking oral contraceptives, though the difference was not significant ($z$ = 0.911, $P$ = 0.362).

## Are Male Strategies Generally Successful?

Our previous studies of whole ejaculates (Baker & Bellis 1989, 1993) identified three male strategies involving adjustment of the number of sperm inseminated during in-pair copulation. First, the number of sperm inseminated is a positive function of time since last in-pair copulation. Adjustment of sperm numbers fits best a model based on the male 'topping-up' his partner to a particular level, inseminating only enough to make up the number of sperm likely to have died since the last insemination. Second, males top-up larger females with more sperm. Third, males top-up females with more sperm when the male spends a lower proportion of his time with his partner and thus, on average, the risk of sperm competition is higher.

On any given occasion, it appears quite possible that females could override any attempt by the male to place more sperm in her tract in response to a particular socio-sexual situation. On average, however, we should expect male strategies to be successful otherwise they would not have been promoted by natural selection. This appears to be the case (Table IV). Thus, on average, males did succeed in placing more sperm in the tract of larger females, when % time together was lower, and when time since last in-pair copulation was greater. On average, when males inseminated more sperm, more were ejected in the flowback but more were also retained. The volume of flowback showed the same response as number of sperm in the flowback. As a function of time since last in-pair copulation, $z$ is maximized (at 2.326) by the model that volume of flowback increased only for the first 72 h after the last in-pair copulation, thereafter remaining constant.

**Table IV.** Male and female influence on sperm ejection and retention in relation to male strategies concerning % time with partner, inter-IPC interval and female weight (Meddis' specific test)

| | Percentage time together | | Inter-IPC interval | | Female weight | |
|---|---|---|---|---|---|---|
| | $z$ | $P$ | $z$ | $P$ | $z$ | $P$ |
| **Male influence** | | | | | | |
| Number inseminated | –7.109 | < 0.001 | 8.637 | < 0.001 | 3.876 | < 0.001 |
| **Female influence** | | | | | | |
| Residual no. retained | –0.519 | 0.604 | –1.429 | 0.154 | –1.218 | 0.224 |
| **Result** | | | | | | |
| Number retained | –5.704 | < 0.001 | 6.327 | < 0.001 | 2.861 | 0.004 |
| Number ejected | –3.549 | < 0.001 | 5.723 | < 0.001 | 3.241 | 0.002 |
| Volume of flowback | –1.749 | 0.040 | 1.930 | 0.027 | No data | |

Meddis' test: blocked by pair. Positive $z$-values indicate that an increase in time together, inter-IPC interval, or female weight are associated with an increase in number of sperm or volume of flow back; negative $z$-values indicate the converse. $P$-values are one-tailed against the hypothesis that the more sperm a male inseminates, the more will be ejected, the more will be retained and the greater will be the volume of flowback (except for analysis of female influence for which $P$-values are two-tailed). Analysis is based on 121 flowback samples from 11 females and 109 estimates of flowback volume from nine females. Figures for male influence are calculated from number of sperm inseminated, equation (1). Figures for female influence are calculated from residuals from equation (3). Other figures are calculated from raw data for numbers ejected and retained.

## Patterns of female orgasm

We now have data on orgasm pattern from two sources: (1) our nationwide survey of the most recent orgasm, if any, reported by 3679 females (Bellis *et al.* 1989); and (2) the details returned by 22 'experimental' females who recorded orgasm pattern in the course of collecting or observing ejaculates and flowbacks. There were no significant differences in orgasm pattern between the two groups of subjects (Table V). The most striking features are that nearly 50% of female orgasms occurred in the absence of a male, around 35% of copulations did not involve a female orgasm (even during foreplay or postplay) and, when copulatory orgasms did occur, the female most often ( >50% of copulatory orgasms) climaxed before the male ejaculated.

In our nationwide sample, 81% of females had experienced at least one non-copulatory orgasm while they were still virgin but only 7% experienced orgasm at their first copulation. After 50 lifetime copulations, 92% had experienced orgasm at some time but only 53% had done so during copulation. After 500 copulations, the figures had increased to 98% and 84%, respectively.

**Table V.** Relative incidence of different types of orgasm reported by two groups of females: nationwide (U.K.) and 'experimental'

| | Relative incidence (%) | | Meddis' non-specific test | |
|---|---|---|---|---|
| | Nationwide ($N$ = 3679) | 'Experimental' ($N$ = 22) | $H_1$ | $P$ |
| **Non-copulatory** | | | | |
| Nocturnal | 3 | 7 | 1.171 | 0.279 |
| Self-masturbation | 52 | 43 | 0.480 | 0.504 |
| Male partner | 44 | 50 | 0.178 | 0.677 |
| Female partner | 1 | 0 | 0.004 | 0.951 |
| **Copulatory** | | | | |
| Before male ejaculates | 55 | 52 | 0.014 | 0.903 |
| Simultaneous | 26 | 24 | 0.074 | 0.782 |
| After male ejaculates | 19 | 24 | 0.204 | 0.657 |
| **Last orgasm with male present** | 56 | 56 | 0.001 | 0.969 |
| **Last copulation not involving female orgasm** | 37 | 41 | 0.114 | 0.735 |

Data refer to the most recent non-copulatory and copulatory orgasms for the nationwide sample and to the first non-copulatory and copulatory orgasms for the 'experimental' sample (i.e. each female contributes up to one non-copulatory and up to one copulatory orgasm to the analysis).

### *Test of the poleaxe theory*

We tested three predictions of the poleaxe theory (using residuals from equation (3), Table I, for tests involving sperm retention). (1) Time from male ejaculation to flowback should be longer if the female has a copulatory orgasm than if she does not ($z$ = 1.643, $N$ = 12 females, 210 in-pair copulations, $P$ = 0.050).

(2) More sperm should be retained if the female has a copulatory orgasm than if she does not ($z$ = 0.841, $N$ = 8 females, 106 in-pair copulations, $P$ = 0.200).

(3) The longer the time-interval from male ejaculation to flowback, the more sperm should be retained ($z = -1.953$, $N = 7$ females, 117 flowbacks, $P = 0.975$).

On the whole, therefore, there is no support for the poleaxe theory in our data. Although time to flowback may be longer after an orgasm, there is no corresponding increase in the number of sperm retained nor is there any positive association between time to flowback and number of sperm retained.

## Female Orgasms and Sperm Retention at $C_2$

Consider two successive in-pair copulations, $C_1$ and $C_2$, hours, days or weeks apart. The female may or may not experience a copulatory orgasm at either of these in-pair copulations and, in addition, may or may not experience one or more non-copulatory orgasms in between. This section is concerned with the residual (equation 3; Table I) number of sperm retained at $C_2$ in relation to events between the beginning of foreplay at $C_1$ to the moment of flowback at $C_2$. As such, therefore, it is concerned specifically with female manipulation of the $C_2$ ejaculate, not with actual numbers of sperm retained or ejected.

We analysed only those 121 flowbacks collected by non-pregnant females.

### Test of the upsuck hypothesis

The upsuck hypothesis (Fox *et al.* 1970) for the function of the female copulatory orgasm would predict that orgasm should be associated with greater sperm retention only if sperm are already present in the female tract.

In this analysis, six categories of female copulatory orgasm were recognized: (1) CX = no orgasm between the beginning of foreplay and ejection of the flowback; (2) BC = orgasm before copulation (i.e. during foreplay; penis not in vagina); BE = orgasm during copulation but before ejaculation (penis in vagina); DE = during ejaculation (i.e. simultaneous climax; penis in vagina); AE = after ejaculation (penis in vagina); AC = after copulation but before flowback (penis not in vagina).

As already shown in testing the poleaxe theory, whether the female has a copulatory orgasm (BC, BE, DE, AE or AC) or not (CX) has no significant influence on residual number of sperm retained. However, if a copulatory orgasm does occur, the timing of the female climax relative to copulation and male ejaculation (i.e. comparison of BC, BE, DE, AE, AC) has a highly significant influence on the residual number of sperm retained ($H_4 = 13.567$, $N = 10$ females, 88 in-pair copulations, $P = 0.009$). The variation is also significant against the specific test that the residual number of sperm retained changes in a linear, rank-order, sequence from the earliest timing for female climax (BC) to the latest (AC) ($z = 3.495$, $N = 10$ females, 88 flowbacks, $P < 0.001$, two-tailed). The direction of the relationship is for the female to retain relatively more sperm when she climaxes later in the sequence than when she climaxes earlier.

In our flowback data, the earliest that a female experienced a copulatory orgasm during an in-pair copulation episode was 45 min before the male ejaculated and the latest was 45 min after (i.e. –45 min to +45 min, using negative numbers to indicate a female climax before the male ejaculates and positive numbers to indicate a climax after the male ejaculates). The frequency distribution of the timing of orgasm for our 11 flowback donors shows the typical pre-ejaculation peak (Fig. 2).

Our data fitted best the model shown in Fig. 3 and are almost entirely consistent with the up suck hypothesis. Essentially, climaxes earlier than 1 min before the male ejaculates were associated with low sperm retention. Maximum sperm retention was associated with climaxes from >0 to 1 min after the male ejaculated onwards. If the female climaxed earlier than –1 min, female influence on sperm retention was no better than if she failed to climax altogether ($z$ = 0.873, $N$ = 8 females, 76 in-pair copulations, $P$ = 0.382, two-tailed). Only climaxes that occurred between –1 min and emergence of the flowback (up to +45 min) were associated with a significant increase in the residual number of sperm retained compared with having no orgasm at all ($z$ = 1.997, $N$ = 10 females, 82 in-pair copulations, $P$ = 0.046, two-tailed).

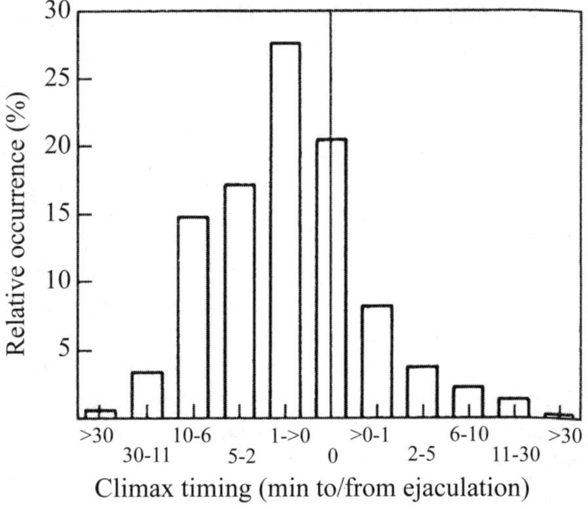


**Figure 2.** Relative frequency of timing of climax of female orgasm in relation to timing of male ejaculation. Histogram bars to the left of this line show the frequency of orgasms that climax before the male ejaculates; bars to the right the same for orgasms that climax after the male ejaculates. Data for 10 females, 88 in-pair copulations.

Essentially, our analysis of female orgasm at $C_2$ identified two levels of sperm retention (henceforth, orgasm regimes) at $C_2$: a higher level associated with climaxes –1 min or later; and a lower level associated with absence of orgasm or orgasms earlier than –1 min.

*$C_1$, inter-copulatory orgasms and sperm retention at $C_2$*

We used residuals from equation (4) (Table I) to test for an influence of $C_1$ and inter-copulatory orgasms on sperm retention at $C_2$. We define an inter-copulatory orgasm as any female orgasm that occurs between the first withdrawal of the penis after male ejaculation at $C_1$ and the last insertion of the penis before male ejaculation at $C_2$. In this paper, we use five categories of inter-copulatory orgasm, two copulatory (A $C_1$ = during post play at $C_1$; B $C_2$ = during foreplay at $C_2$) and three non-copulatory (NO = nocturnal/spontaneous; SS = self-stimulation in absence of male; PS = any stimulation, but not leading to copulation, in the presence of a male or female partner).

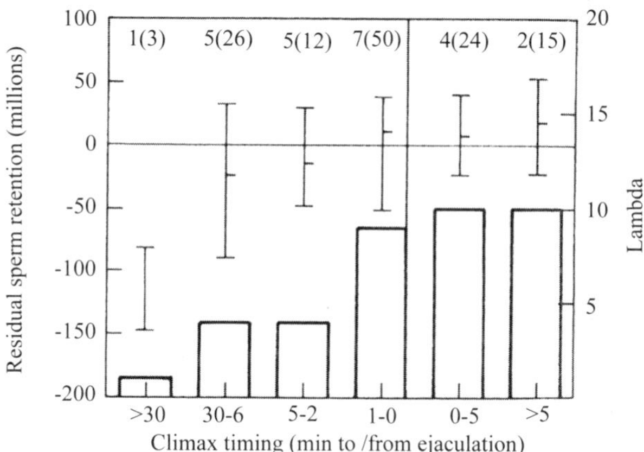


**Figure 3.** Influence of timing of climax of female orgasm on the number of sperm retained. Bars show median and inter-quartile range sperm retention (residuals from equation 3). Numbers above bars: number of females (number of in-pair copulations). Histograms show the lambda coefficients that maximize *z* in a Meddis' specific test when the data are blocked by couple to test for within-couple response. Lambda coefficients are the more accurate-indicators of the form of the relationship between the timing of climax and sperm retention. Vertical line shows the timing of male ejaculation. Histogram bars to the left of this line show lambda coefficients for orgasms that climax before the male ejaculates; bars to the right the same for orgasms that climax after the male ejaculates.

First, we considered whether the different types of inter-copulatory orgasm differ in influence on sperm retention at $C_2$. Analysis was restricted to those occasions on which there was no menstruation and only a single inter-copulatory orgasm between $C_1$ and $C_2$ (four females, 33 flowbacks). We divided the data into five samples according to the type of inter-copulatory orgasm (i.e. $AC_1$, NO, SS, PS or $BC_2$) and tested the residuals from equation (4) for heterogeneity. There was no significant heterogeneity between the five samples ($H_4 = 7.308$, $P = 0.119$), nor between the three types of non-copulatory orgasms (NO, SS, PS; $H_2 = 1.628$, $N = 4$ females, 14 flowbacks, $P = 0.553$). If inter-copulatory orgasms have any influence on the residual number of sperm retained at $C_2$, it is independent of the type of inter-copulatory orgasm. We therefore considered the five types of inter-copulatory orgasm as a single category.

Next, we restricted analysis to occasions on which the female neither menstruated nor had an inter-copulatory orgasm between $C_1$ and $C_2$ (10 females, 47 flowbacks). On such occasions, residual sperm retention at $C_2$ is a highly significant function of time since $C_1$ ($z = 2.757$, $P = 0.006$, two-tailed). Residual sperm retention was higher when inter-copulatory interval was longer, with no clear indication of any plateauing of the relationship up to the limit of our data (i.e. 192 h; = 8 days). The data suggest that $C_1$ sets up some form of block to sperm retention at $C_2$ and that this block persists with declining effectiveness for up to at least 8 days.

In contrast, on occasions when the female has a single inter-copulatory orgasm between $C_1$ and $C_2$, there is no significant association between residual sperm retention at $C_2$ and inter-copulation interval ($z = -0.386$, $N = 4$ females, 33 in-pair copulations, $P = 0.700$, two-tailed). Moreover, the relationship is significantly different from that when there is no inter-copulatory orgasm ($z_1 - z_2 = 2.757 - (-0.386) = 3.143$; $z = 3.143/\sqrt{2} = 2.222$, $P = 0.026$, two-tailed; $z$-difference test). However, time from $C_1$ to first inter-copulatory orgasm does show a significant relationship with level of sperm retention at $C_2$ ($z = 2.580$, $N = 4$ females, 52 in-pair copulations, $P = 0.010$, two-tailed). Thus, both time to $C_2$ (in the absence of an inter-copulatory orgasm) and time to first inter-copulatory orgasm have a positive relationship with sperm retention at $C_2$ (Fig. 4), but an inter-copulatory orgasm removes any further influence of time to $C_2$. The implication is that an inter-copulatory orgasm effectively fixes sperm retention at a level that is a function of time from $C_1$ to the first inter-copulatory orgasm.

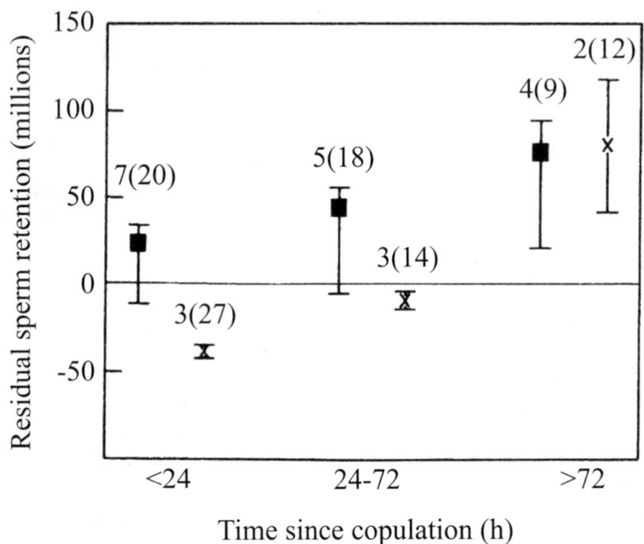


**Figure 4**. Influence of the time from $C_1$ to first inter-copulatory orgasm and from $C_1$ to $C_2$ (in the absence of inter-copulatory orgasms) on level of sperm retention at $C_2$ where $C_1$ and $C_2$ are consecutive copulations. Sperm retention is measured as residuals from equation (4). ■: Median number of sperm retained at $C_2$ for a particular time interval from $C_1$ to $C_2$ (in the absence of intercopulatory orgasms); X: median number of sperm retained at $C_2$ for a particular time interval from $C_1$ to first inter-copulatory orgasm. Error bars show interquartile ranges. Numbers above bars: number of females (number of in-pair copulations).

Crudely dividing time since $C_1$ into four categories ( <2 h; 2–24 h; 25–72 h; >72 h), we compared the level of sperm retention at $C_2$ for occasions when either $C_2$ (in the absence of an inter-copulatory orgasm) or the first inter-copulatory orgasm falls into a particular zone. There is no significant difference between these two types of occasions for either the < 2 h ($H_1 = 1.699$, $N = 1$ female, 12 in-pair copulations, $P = 0.189$), 2–24 h ($H_1 = 0.533$, $N = 7$ females, 35 in-pair copulations, $P = 0.528$), 25–72 h ($H_1 = 0.001$,

$N = 5$ females, 32 in-pair copulations, $P = 0.969$) or >72 h ($H_1 = 0.646$, $N = 5$ females, 21 in-pair copulations, $P = 0.573$) time categories. We conclude, therefore, that when an inter-copulatory orgasm occurs it effectively fixes the level of sperm retention at the next copulation at a level that is more or less the same as the level determined by the efficiency of the block established by the previous copulation.

These analyses combine to suggest that the level of sperm retention at $C_2$ is primarily a function of the time interval from $C_1$ to the next sexual event, either $C_2$ (in the absence of inter-copulatory orgasms) or the first inter-copulatory orgasm. Analysing all flowbacks (excluding those with inter-copulatory menstruation) for level of retention at $C_2$ in relation to time from $C_1$ to the next sexual event reveals a highly significant relationship ($z = 3.142$, $N = 10$ females, 100 in-pair copulations, $P = 0.002$).

In our flowback data, the maximum number of inter-copulatory orgasms between any two successive copulations was six. Using the residuals from equation (4) as the dependent variable, there is a very significant negative relationship between the number of inter-copulatory orgasms between $C_1$ and $C_2$ and the retention of sperm at $C_2$ ($z = -2.744$, $N = 10$ females, 115 in-pair copulations, $P = 0.006$, two-tailed); the more inter-copulatory orgasms, the lower the retention of sperm.

So far, our analysis has shown that copulation ($C_1$) apparently produces some form of block to sperm retention at the next copulation ($C_2$). This block gradually declines in efficiency, at least over the next 8 days, the time that sperm are known to remain alive in the cervix. An inter-copulatory orgasm fixes the level of efficiency of the block at about the level that it has reached during normal decline. The obvious implication is that this putative block is in some way due to the sperm from the $C_1$ ejaculate. Our data allow three tests of this hypothesis.

First, we have five flowbacks (i.e. at $C_2$) for which we can be reasonably certain that no sperm had been transferred at $C_2$ (or at any copulations within the 8 days prior to $C_2$) because the male either wore a condom or withdrew the penis before ejaculation. This limited data set shows no association between sperm retention at $C_2$ and time from $C_1$ to the next sexual event ($z = 0.174$, $N = 2$ females, 5 in-pair copulations, $P = 0.431$).

Second, including these five occasions, we have 54 flowbacks (from $C_2$s) for occasions when we also have flowbacks and relevant details for the corresponding $C_1$ and are thus able to estimate the number of sperm retained by the female at $C_1$. On these occasions there is a significant negative relationship between the number of sperm retained by the female at $C_1$ and the residual (from equation 5) level of sperm retention at $C_2$ ($z = 1.721$, $N = 5$ females, 54 in-pair copulations, $P = 0.042$).

Finally, we used the same 54 in-pair copulations as above and crudely estimated how many sperm might still be present in the female tract from $C_1$. We made no allowance for sperm that may still be present from copulations before $C_1$ on the assumption that most will be dead by the time of $C_2$. We assumed a linear decline from the number of sperm retained at $C_1$ to zero 192 h later. We assumed that if an inter-copulatory orgasm occurs between $C_1$ and $C_2$ it 'fixes' the number of sperm at the level on this decline reached by the time of the first inter-copulatory orgasm. In the absence of an inter-copulatory orgasm, the number of sperm is determined by the inter-copulation time interval. Residual (from equation 4) level of sperm retention at $C_2$ is a significant negative function of the number of sperm remaining in the tract from $C_1$ ($z = -1.996$, $N = 5$ females, 54 in-pair copulations, $P = 0.023$; Fig. 5).

In summary, therefore, our data suggest that sperm transferred during $C_1$ interfere with sperm retention at $C_2$. As their numbers decline with time, the level of interference also declines. Intercopulatory orgasms in some way (see Discussion) use the remaining sperm to fix the level of interference at roughly the point it has reached in its normal decline. The stability of this fixation can be tested by analysing level of retention at $C_2$ in relation to time since inter-copulatory orgasm.

We restricted analysis to occasions when there was only one inter-copulatory orgasm between $C_1$ and $C_2$. Up to 168 h (the longest time-interval in our data) after inter-copulatory orgasm, there is no significant change in residual (from equation 4) retention of sperm at $C_2$ with time since intercopulatory orgasm ($z = -1.752$, $N = 4$ females, 33 in-pair copulations, $P = 0.080$, two-tailed). The non-significant trend for longer time intervals to be. associated with reduced retention is very different from the positive trend from $C_1$ to $C_2$ in the absence of inter-copulatory orgasm.

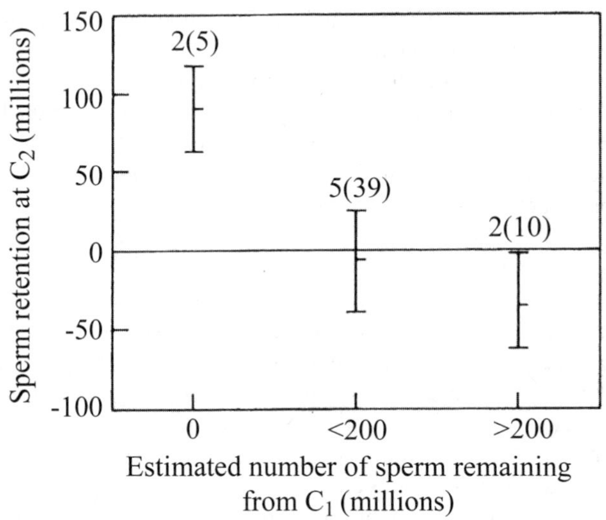


**Figure 5.** Influence of number of sperm remaining from $C_1$ on number of sperm retained at $C_2$ where $C_1$ and $C_2$ are consecutive copulations. Number of sperm retained at $C_2$ is measured as residuals from equation (5). Other conventions as in Fig. 4.

Finally, we used the residuals from equation (4) as the dependent variable and tested for an interaction between the influence of $C_1$ and inter-copulatory orgasms on sperm retention at $C_2$ and the high/low retention orgasm regime at $C_2$. The normal relationship between sperm retention at $C_2$ and time from $C_1$ to first sexual event is present only when there is a low retention orgasm regime at $C_2$ ($z = 2.936$, $N = 6$ females, 61 in-pair copulations, $P = 0.002$). A high retention orgasm regime at $C_2$ , however, seems to remove this relationship ($z = 0.848$, $N = 10$ females, 39 in-pair copulations, $P = 0.198$).

*Orgasm regimes and levels of retention*

In effect, our analyses have led us to recognize four regimes for sperm retention, based on the interaction of inter-copulation interval, time to first inter-copulatory orgasm, and orgasm regime at $C_2$. For ease of description, let FEV = the first event after $C_1$ (i.e. first inter-copulatory orgasm or, in the absence of an inter-copulatory orgasm, $C_2$).The four regimes are: (I) $\leq$24 h to FEV with a low retention orgasm regime at $C_2$; (II) >24 to $\leq$72 h to FEV with a low retention orgasm regime at $C_2$; (III) >72 h to FEV with a low retention orgasm regime at $C_2$; and (IV) any time to FEV with a high retention orgasm regime at $C_2$.

The hypothesis that regimes I–IV show a rank order linear decrease in levels of sperm retention, using residuals from equation (3) as the dependent variable (i.e. before any orgasm factors were controlled) is supported ($z = 3.534$, $N = 11$ females, 100 in-pair copulations, $P < 0.001$). z is maximized (at 3.806), however, by the hypothesis that regimes III and IV produce equal levels of sperm retention (see also Fig. 6) and, when compared directly, there is no significant difference in retention at these two levels ($z = 0.996$, $N = 10$ females, 54 in-pair copulations, $P = 0.320$, two-tailed).

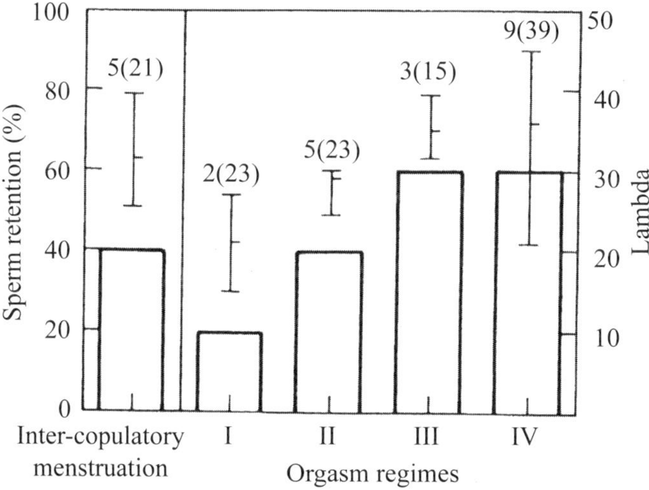


**Figure 6.** Percentage retention of sperm after an intercopulatory menstruation compared with retention after four orgasm regimes. Bars show median and interquartile range sperm retention (residuals from equation 3). Numbers above bars: number of females (number of in-pair copulations). Histograms show the lambda coefficients that maximize z in a Meddis' specific test when the data are blocked by couple to test for within-couple response. Lambda coefficients are the more accurate indicators of the form of the relationship between the timing of climax and sperm retention. Regime I: low retention orgasm regime at $C_2 \leq 24$ h from $C_1$ to first inter-copulatory orgasm or (in absence of an intercopulatory orgasm) $C_2$ where $C_1$ and $C_2$ are consecutive copulations. Regime II: low retention orgasm regime at $C_2$, $\geq$25 – $\leq$72 h from $C_1$ to first inter-copulatory orgasm or (in absence of an inter-copulatory orgasm) $C_2$. Regime III: low retention orgasm regime at $C_2$, >72 h from $C_1$ to first inter-copulatory orgasm or (in absence of an intercopulatory orgasm) $C_2$. Regime IV: high retention orgasm regime at $C_2$ (see Table VI).

**Table VI.** Definition of orgasm regimes and levels of sperm retention (identified in this study and used during discussion in text)

| Orgasm regime | Time to FEV (h) | Retention level at $C_2$ | Level of retention |
|---|---|---|---|
| I | ≤24 | L | I |
| II | >24 – ≤72 | L | II |
| III | > 72 | L | III |
| IV | any | H | III |

$C_1$ and $C_2$ consecutive copulations by a female. FEV, first sexual event after $C_1$ (i.e. either the first inter-copulatory orgasm or, in the absence of an inter-copulatory orgasm, $C_2$ ). L, low retention level at $C_2$ (no copulatory orgasm or a climax earlier than I min before the male ejaculates). H, high retention level at $C_2$ (climax some time between 1 min before the male ejaculates and flowback).

The four orgasm regimes, therefore, impart three levels of sperm retention (levels I–III). These three retention levels are significantly evident even if we restrict analysis to only the first sample produced by each female ($z$ = 2.032, $N$ = 11 females, 11 in-pair copulations, $P$ = 0.021).

In the analyses and discussions that follow, we refer, for ease of description, to regimes I–IV and/or levels I–III. The definitions of these regimes and levels are summarized in Table VI.

*Menstruation and sperm retention*

If the number of sperm retained at $C_2$ is in part a function of the number of sperm remaining in the cervix from $C_1$, we should expect the normal relationship between retention at $C_1$ and $C_2$ to be changed if the female begins to menstruate between $C_1$ and $C_2$. Menstruation not only clears the cervix of the sperm present in the cervical mucus (but not necessarily of all those in the cervical crypts; Discussion), it also, at least temporarily, replaces them with various female tissues and cellular debris. The result should be a change in the normal relationship between $C_1$ and $C_2$ and perhaps a tendency to reduce sperm retention at $C_2$.

Our flowback data set included 21 in-pair copulations (from five females) which were preceded by menstruation between $C_1$ and $C_2$. Using the residuals from equation (5) as the dependent variable, sperm retention at $C_2$ tended to be lower after inter-copulation menstruation, though the difference is not significant ($z$ = 1.343, $N$ = 5 females, 107 in-pair copulations, $P$ = 0.090). However, the normal influence of levels I–III on the residuals from equation (4) is obliterated ($z$ = −1.261, $N$ = 5 females, 21 in-pair copulations, $P$ = 0.896) and changed significantly ($z_1 - z_2 = 3.142 - (-1.261) = 4.403$; $z = 4.403/\sqrt{2} = 3.113$, $P$ = 0.002, two-tailed; $z$-difference test).

The level of sperm retention (residuals from equation 4) after menstruation tended to be greater than level I retention ($z$ = 0.600, $N$ = 7 females, 87 in-pair copulations, $P$ = 0.548, two-tailed) but lower than level III ($z$ = −1.433, $N$ = 6 females, 29 in-pair copulations, $P$ = 0.152, two-tailed). Menstruation thus appears to produce a block to retention that is intermediate in effectiveness between that associated with a cervix relatively full of sperm and a cervix relatively devoid of sperm (i.e. equivalent to level II; Fig. 6).

## Check on Conclusions

Our analyses have used dependent variables (residuals from predictive equations; Table I) that increasingly depart from our raw data (counts of sperm in flowbacks). Moreover, most of our flowbacks were from volunteers, some of whom were more prolific than others, and most were taken while the females concerned were using oral contraceptives. This section checks our conclusions concerning the existence of three levels of sperm retention (Table VI) controlling for these potential sources of error.

All aspects of the female's influence on sperm retention that were identified by our analysis of residuals are also apparent in the variation in actual numbers of sperm ejected and retained (Table VII). Any apparent inconsistencies in Table VII between trends in median values and the results of Meddis' test are due to the latter being blocked by female to test for within-female response.

**Table VII.** Association between actual number of sperm inseminated, ejected and retained at $C_2$ and orgasm regime for sperm retention

| | I | II | III | IV | $z$ | $P$ |
|---|---|---|---|---|---|---|
| Lambda coefficients | 1 | 2 | 3 | 3 | | |
| No. females (flowbacks) | 2 (23) | 5 (23) | 3 (15) | 9 (39) | | |
| **Number inseminated** | | | | | | |
| Median | 190 | 216 | 329 | 135 | 0.120 | 0.453 |
| Inter-quartile range | (176–203) | (214–310) | (270–330) | (76–205) | | |
| **Number retained** | | | | | | |
| Median | 100 | 128 | 167 | 96 | 2.727 | 0.003 |
| Inter-quartile range | (75–125) | (101–181) | (156–226) | (71–105) | | |
| **Number in flowback** | | | | | | |
| Median | 113 | 130 | 62 | 78 | 2.445 | 0.007 |
| Inter-quartile range | (93–133) | (85–173) | (53–111) | (28–85) | | |
| **Volume of flowback** | | | | | | |
| Median | 2.3 | 2.0 | 2.0 | 2.0 | −0.099 | 0.539 |
| Inter-quartile range | (1.0–3.0) | (1.5–2.0) | (1.5–2.0) | (2.0–2.0) | | |

$P$-values are one-tailed against the hypothesis that orgasm regimes with higher lambda coefficients will have larger numbers of sperm inseminated and retained and lower numbers and volume ejected. Positive $z$-values: trend consistent with hypothesis.

We opted at the beginning of this paper, for the reasons given, to measure the female's influence on sperm retention using residuals rather than percentage retention. However, the influence of regimes I–IV (Table VI) and inter-copulatory menstruation are equally applicable (Fig. 6) to the percentage retention of sperm ($z$ = 3.310, $N$ = 11 females, 121 in-pair copulations, $P$ = 0.001), even if we restrict analysis to the first sample provided by each subject ($z$ = 2.159, $N$ = 11 females, 11 ejaculates, $P$ = 0.015).

Finally, but importantly, if we restrict analysis of retention level (Table VI) to those few flowbacks collected by females not using any form of contraceptive, the influence of regimes I–IV is still significant ($z$ = 1.776, $N$ = 3 females, 9 in-pair copulations, $P$ = 0.038).

## Flowback Volume

Flowback volume varied significantly with the % time the pair spent together and time since last copulation (Table IV). There was also a very strong influence of copulatory orgasm on flowback volume. The primary influence is that, when a copulatory orgasm occurred, the volume of the flowback increased significantly ($z$ = 2.980, $N$ = 9 females, 109 in-pair copulations, $P$ = 0.002, two-tailed). $z$ is maximized (at 3.135) by the model that flowback volume was greatest when orgasm occurred during foreplay. No other aspects of the timing of copulatory orgasms or inter-copulatory orgasms were significant and, in general, orgasm regimes that favoured sperm retention did not decrease flowback volume (Table VII).

## Female Orgasm and Sperm Competition

In this section we analyse the way that the pattern of female orgasm varies in relation to socio-sexual situation, with particular reference to in-pair copulation and extra-pair copulation. The results are summarized in Fig. 7.

In our nationwide survey, the majority of subjects provided enough information relating to their last copulation and inter-copulatory orgasm to allow us to estimate the level of sperm retention (in terms of levels I–III; Table VI). In evaluating female behaviour, however, we also determined the extent to which the level shown is a function of behaviour overt to her male partner (i.e. overt copulatory orgasms) or cryptic (inter-copulatory orgasms that are nocturnal, self-masturbatory, or encouraged by a female partner). It is known that these two elements of female sexual behaviour sometimes vary differently (e.g. through the menstrual cycle: Harvey 1987; Baker *et al.* 1989).

To reduce complexity, we present results only for hormonally 'normal' subjects (i.e. we exclude all females who were under an unusual hormonal regime, either taking oral contraception or depo-provera injections, or who had undergone hysterectomy). As shown above, an increase in cryptic inter-copulatory orgasms leads to a decrease in sperm retention at the next copulation unless it is counteracted by a high retention copulatory orgasm.

### *In-pair copulations and monandry*

A total of 1207 hormonally 'normal' females provided information on their last in-pair copulation at times that they claimed they had no other sexual partners. We refer to these as monandrous females.

A slight, but not significant, tendency for monandrous females to show an increase in high retention copulatory orgasms during the main fertile phase of their menstrual cycle (days 6–15) compared with the remainder of the cycle ($z = 1.185$, $P = 0.118$, two-tailed) is counteracted by a very significant increase in cryptic inter-copulatory orgasms ($z = 3.159$, $P = 0.002$, two-tailed; cf. Harvey 1987). The result is no change in retention level during the menstrual cycle ($z = 0.064$, $P = 0.950$, two-tailed).

Twenty-one of our monandrous females were pregnant. These showed no difference in orgasm regime, whether in overt copulatory orgasms, cryptic inter-copulatory orgasms, or overall retention level ($z = 0.427$, $P = 0.668$, two-tailed), relative to non-pregnant monandrous females.

*In-pair copulations and polyandry*

Seventy five hormonally 'normal' females provided information on their last in-pair copulation times that they also had one or more other male sexual partners. We refer to these as polyandrous females.

Compared with monandrous females, polyandrous females showed a significantly lower level (i.e. of levels I–III; Table VI) of sperm retention when copulating with their main partner ($z = 2.464$, $P = 0.014$, two-tailed). They achieved this change, however, not by any change in overt copulatory orgasms ($z = 0.293$, $P = 0.385$) but by a very significant increase in the frequency of cryptic inter-copulatory orgasms ($z = 4.192$, $P < 0.001$).

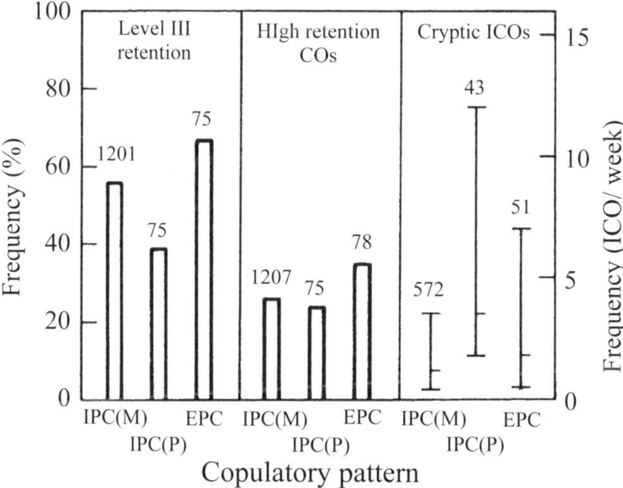


**Figure 7.** Variation in sperm retention, overt copulatory orgasms (COs) and cryptic inter-copulatory orgasms (ICOs) as a function of socio-sexual situation for hormonally 'normal' women. IPC(M), In-pair copulations by monandrous females; IPC(P), in-pair copulations by polyandrous females; EPC, extra-pair copulations (including double-matings). Sperm retention is illustrated as frequency (%) of level III retention; overt copulatory orgasms (COs) illustrated as frequency (%) of high level retention regimes; cryptic inter-copulatory orgasms (i.e. nocturnal, self-masturbatory and lesbian intercopulatory orgasms) illustrated as median (and interquartile ranges) frequency/week. Numbers above bars: number of females providing all necessary information. Data from U.K. nationwide survey.

*Extra-pair copulations*

A further 75 hormonally 'normal', polyandrous females provided information on their last copulation when the latter was an extra-pair copulation. Behaviour during extra-pair copulation for these females can be compared with behaviour during in-pair copulation for the 75 other polyandrous females from the previous section (Fig. 7).

Level of sperm retention was significantly higher during extra-pair copulation than during in-pair copulation ($z = 3.242$, $P = 0.002$, two-tailed). This increase was achieved primarily by a difference in overt copulatory orgasms when females were copulating with their extra-pair male(s) compared with their partner ($z = 1.673$, $P = 0.047$). A slight tendency also to change the frequency of cryptic inter-copulatory orgasms before copulating with the extra-pair male was not significant ($z = 0.720$, $P = 0.236$).

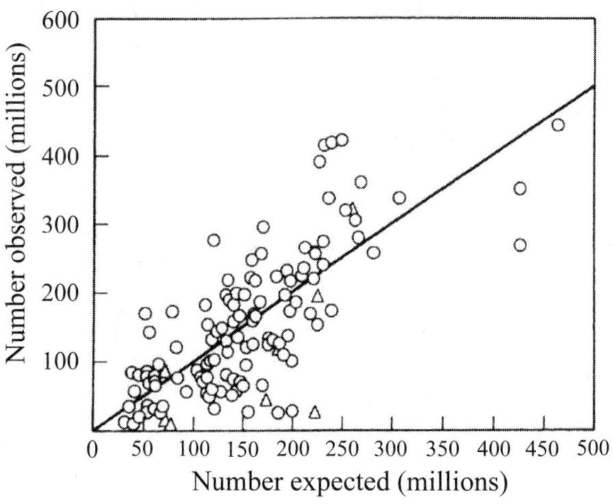


**Figure 8.** Numbers of sperm observed in 127 flowbacks from 11 females compared with the number predicted by equation (6) (Table I). △: First flowback contributed by each couple; ○: subsequent flowbacks. Diagonal line shows the desired 1:1 relationship.

## Predicting the Number of Sperm Retained

During the course of this paper, we have identified a number of factors that significantly influence the retention of sperm during in-pair copulation. These may be combined into equation (6) (Table 1) which explains 71% of the observed variation in number of sperm retained in the first in-pair copulation for which flowbacks were provided by our 11 females and 62% of the entire data set of 127 samples. The difference between number of sperm inseminated (equation 1) and number of sperm retained (equation 6) is the predicted number of sperm in the flowback. Figure 8 shows the relationship between the predicted and observed numbers of sperm in the flowbacks in this study. Our analyses explain 51% of the observed variation in number of sperm in the first flowback provided by each female and 55% of the total data set. Figure 8 shows a fit to the expected 1:1 relationship. The equation now needs to be tested for robustness with other groups of subjects.

## DISCUSSION

### *Methodology*

The methodologies associated with two of the investigations used in this paper (i.e. (1) the collection of whole ejaculates in condoms in the derivation of equation (1); and (2) our nationwide survey of sexual behaviour by questionnaire) have been discussed elsewhere (Baker & Bellis 1989; Bellis & Baker 1990) and need not be debated further here.

Of all of our investigations, the study of flowbacks was the most difficult and, as it was new, the most problematic. In particular, it was naturally impossible to standardize efficiency of collection in any precise way even from sample to sample by the same couple. However, there are several reasons to believe that the methodological problems were not severe enough to mask underlying biological effects. First, we obtained a strong correlation between the number of sperm we calculated to be inseminated by the male partner and the number of sperm collected by the female partner. Within couple variation in collection efficiency was not, therefore, enough to mask biological variation from flowback-to-flowback. Second, the residuals (from equation 3) of sperm in the flowback correlate with female weight (Table IV). Inter-couple variation in collection efficiency was again, therefore, not enough to mask inter-couple biological variation (unless it is proposed that larger females are more efficient at collecting flowbacks). Third, the six flowbacks collected during pregnancy suggest that virtually all of the sperm inseminated were successfully collected in the flowback.

All sperm counts were carried out blind (Baker & Bellis 1989). Estimation of sperm numbers could not, therefore, have been influenced by any preconceptions on the part of the investigators. None of the females who collected the samples or estimated volumes could have had any preconception of the patterns later revealed by the analysis of their samples. In most cases, the volunteers were even unaware of the precise aims of the study. The majority assumed they were simply making some general contribution to the study of human fertility. There was no opportunity, therefore, for subjects to bias results in any particular direction by putting more effort into collecting more of the flowback on some occasions than others. The possibility that the timing of orgasm influenced, not the size of the flowback, but the effort people put into the collection of the flowback, can also be ruled out. The difference between high and low levels of sperm retention was not a simple function of whether or not the female experienced an orgasm. Rather, retention was a fairly precise function of the timing of orgasm relative to the male's ejaculation (Fig. 3). It seems unlikely that any heterogeneity of effort in collection would show such a precise relationship.

Our conclusions concerning the volume of flowback and the number of sperm ejected derived primarily from studies of females using oral contraceptives. However, we had enough information to be able to demonstrate that orgasm regimes and retention levels have the same influence on sperm retention when the female is hormonally 'normal'. Thus, even though orgasm regimes may change in frequency when females take oral contraceptives (Baker *et al.* 1989), any given orgasm regime when it occurs appears to have the same influence on sperm retention whether the female is taking oral contraceptives or not. Our conclusions seem equally applicable, therefore, to hormonally 'normal' females.

Ideally much larger numbers of females should have been used to collect flowbacks. Not surprisingly, however, it was relatively difficult to recruit people to participate in the collection of material that many consider to be highly personal. The result was a relatively large sample of flowbacks but a skewed distribution between only 11 volunteers (Table III). However, for the following reasons, neither skew nor small sample size appears to invalidate our conclusions.

The use of Meddis' blocking technique avoids the worst vagaries of skewed sample sizes. Nevertheless, wherever possible (e.g. when the whole data set was being analysed) we began with the first sample from each female before considering the biologically more meaningful tests of within-couple variation. Most importantly, however, our final conclusions concerning the influence of orgasm regimes on levels of retention significantly predicted variation in sperm retention in the first sample produced by each couple. The generality of our conclusions should not, therefore, be questioned on any grounds of skewed sample sizes and undue bias from the more prolific females. Figures and tables present medians of medians, again avoiding undue visual bias from the more prolific females.

The similarity in pattern shown by our flowback samples and the quite separate volume estimates from a further group of nine subjects (e.g. Table IV) gave some reassurance that our conclusions were not specific to our flowback contributors. Where differences do occur, as discussed below, they are reconciled by the fact that one group was providing sperm numbers, the other group, flowback volume. Also reassuring was the way in which the conclusions from our experimental subjects could then be applied in an apparently meaningful way to our nationwide sample of over 3000 females.

Finally, we were able to show that our main conclusion concerning levels of retention (Table VI) was independent of the analytical parameters and methods chosen. Our conclusions were equally applicable to other parameters (i.e. percentage retention; Fig. 6) and the raw data (i.e. number of sperm in the flowback; Table VII). Moreover, our final predictive equation, although mathematically not necessarily the best and although developed for sperm retention, nevertheless provided a significant fit to our raw flowback data (Fig. 8).

For all of the above reasons, we feel confident that our major conclusions concerning the influence of female orgasm patterns and sperm retention are real and not an artefact of either our sample of volunteers or our method of analysis. The main gap in our study is that, for obvious reasons, we do not yet have direct counts of the sperm content and volume of flowbacks from extra-pair copulations, particularly double-matings. We have been able to make inferences concerning what might happen on such occasions from our nationwide survey, but such indirect evidence is only an interim substitute for actual samples.

## Hypotheses of the Female Copulatory Orgasm

Our flowback data provide no support for the 'poleaxe' hypothesis of the female copulatory orgasm (Morris 1967; Levin 1981). On the other hand, they do provide the first direct evidence in support of the 'upsuck' hypothesis mooted by Fox *et al.* (1970). Highest levels of sperm retention (about 70%) occurred when the seminal pool was present in the upper vagina at the time the female climaxed (Fig. 6). The fact that improvement in sperm retention was observed when the female climaxed up to a minute before the male ejaculated (Fig. 3) suggests that the upsuck mechanism continues to function for at least 1 min after the female subjectively first experiences the climax.

We observed an increase in volume of the flowback if the female experienced an orgasm at any time during a copulation episode, the volume being greatest for orgasms during foreplay. The pattern suggests that the female adds some material, probably cervical mucus, to the seminal fluid in forming the flowback and that more is added if the female climaxes than if she does not. Climaxes during foreplay thus either add more material or add the same amount but without 'sucking' it back up to the same extent as when it is mixed with seminal fluid.

## Inter-copulatory Orgasms: a Hypothesis

The penetrability of the cervical mucus to sperm from the next inseminate is a negative function of the density of cells and debris (Parsons & Sommers 1978; Belsey *et al.* 1987). Pregnancy, characterized by a large population of leucocytes and other cells in the cervical mucus (Davey 1986) irrespective of copulation and inter-copulatory orgasms, should be a time of maximum impenetrability, as our results indicate. Our results for non-pregnancy flowbacks also show that sperm retention seems to be a function of the number of sperm remaining from the previous copulation (Fig. 5).

If the upsuck mechanism applies to copulatory orgasms, it is likely also to apply to inter-copulatory orgasms. The difference would be that instead of sucking-up a mixture of cervical mucus and seminal fluid, inter-copulatory orgasms would suck-up a mixture of cervical mucus and vaginal secretions. This would lower the pH of the cervix and thus have far-reaching repercussions on the mobility and survival of sperm which would be consistent with our data.

Sperm are immediately immobilized in acidic environments at pH levels below 6.5. Even at a pH of 7.0 the ability of sperm to penetrate cervical mucus is minimal. Penetration is 'normal' at pH 7.5 and above normal at pH 8.25 (El-Banna & Hafez 1972). The pH of seminal fluid is normally in the range 7.0–7.8 (Raboch & Skachova 1965) and buffers the sperm from the vaginal environment which is acidic, and thus hostile, having a pH of between 5.8 and 3.5 (Duerden *et al.* 1987). Once sperm are no longer buffered by seminal fluid, they cannot live for more than a maximum of 10–12 h at the hostile pH levels found in the vagina (Vander Vliet & Hafez 1974). The pH of the cervical mucus can vary considerably from a favourable 7.4 to a hostile 4.0 (Kroeks & Kremer 1977), probably depending on the extent to which it has been mixed with material from the vagina.

Mixing cervical mucus and vaginal secretions via inter-copulatory orgasms should lower the pH of the cervix towards the acidic end of its observed range. This will slow down, or even immobilize, any sperm that subsequently enter the cervical mucus from reservoirs such as the cervical crypts. More may also die before reaching the uterus. Any hindrance to passage through the cervical mucus into the uterus may lead to the gradual accumulation of millions of sperm in the cervical mucus (Sagiroglu & Sagiroglu 1970). Here they attract an even larger population of scavenging, female-produced leucocytes (Fig. 9). The result could be that instead of declining with time since the most recent copulation the population of sperm leucocytes and debris in the cervical mucus could remain constant or even increase, at least for as long as sperm continue to leave the crypts.

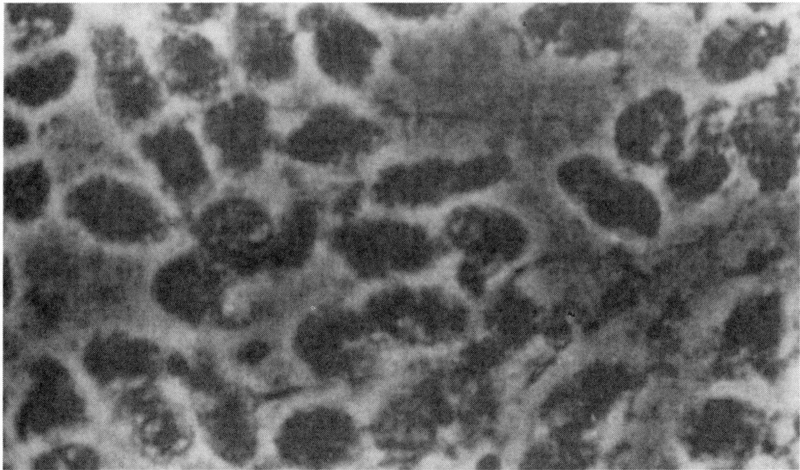

**Figure 9.** Human cervical mucus containing leucocytes and sperm. When cervical mucus contains a dense population of cells, the ease with which sperm can migrate through the channels is reduced.

We suggest, therefore, that an inter-copulatory orgasm lowers the pH of the cervical mucus. In the context of the manipulation of sperm, the function is to slow down or prevent further passage of sperm already present in the cervix, promote a build up of cells and debris in the cervical channel, and thus to reduce sperm retention at the next copulation. As with all such blocks, however, this can perhaps be circumvented by a high retention orgasm at the next copulation which may take sperm directly to their storage sites, bypassing mucus penetrability.

Most medically significant infections of the female tract also prefer a more-alkaline environment (Duerden *et al.* 1987) and any increase in acidity due to inter-copulatory orgasms could have an antibiotic effect. In the absence of copulation (i.e. while still virgin or during temporary periods without insemination) and during pregnancy, the function of nocturnal, masturbatory and other non-copulatory orgasms could be as an 'antibiotic' mechanism aimed at combating cervical infection. Some support for this latter mechanism may be taken from our earlier finding that the frequency of non-copulatory orgasms is greater in females who also report having infection of the reproductive tract (Baker *et al.* 1989).

## Male-Female Conflict and Cooperation over Sperm

Although our data are consistent with the upsuck hypothesis as a mechanism for the female orgasm, both copulatory and non-copulatory, they also suggest a behavioural significance far more complex than simple assistance in sperm retention. The relatively high incidence of non-copulatory orgasms, copulations without orgasm, and copulatory orgasms which climax before the male ejaculates (Table V; Fig. 2) cannot simply be ignored as they have in most previous discussions of the female orgasm. Far from assisting sperm retention, all of these inter-copulatory events reduce sperm uptake at the next copulation, perhaps using the mechanisms just discussed. Yet their influence can be overridden by a high uptake copulatory orgasm at the next copulation. The whole pattern hints strongly at a female strategy to influence sperm retention differently at different copulations.

We are not suggesting, of course, that this strategy is necessarily conscious or that, like any other strategy, it is infallible and omnipotent. As in any other arms race, we should expect male and female strategies to interact and it is by no means inevitable that one sex will always prevail. Nevertheless, the occurrence, pattern and timing of female orgasms emerge from our analysis as part of a female strategy to influence sperm retention from any given copulation. As such, the strategy should influence both the probability of conception and the outcome of sperm competition contests between males.

### *Manipulation in the absence of sperm competition*

In our companion paper (Baker & Bellis 1993), we concluded that, in the absence of sperm competition, both male and female partner may benefit from improved chances of viable conception if fewer sperm are taken into the female tract. The advantage of fewer sperm may either be through reduced risk of debilitating the egg or through reduced risk of pathological polyspermy, eggs fertilized by more than one sperm failing to develop into viable embryos (Englert *et al.* 1986). In this context, females who produce the optimum rate of arrival of sperm at the oviduct by manipulating both the number of sperm that enter the cervical crypts and the rate and length of time that sperm migrate through the cervical mucus will benefit both themselves and their male partner. We assume, however, that the more sperm the male inseminates beyond a certain optimum and/or the more are retained after flowback, the more difficult it is for the female to produce the optimum traffic of sperm.

Lack of orgasm during copulation or a climax more than 1 min before the male ejaculates is associated with low sperm retention (Fig. 3). It has long been known that copulatory orgasm is not essential for conception (Moghissi 1977). In our data we had a situation in which date of conception could be estimated to within +24 h and was attributable to one or other of two in-pair copulations, about 156 and 120 h before conception. The number of sperm inseminated on both occasions could be estimated (equation 1; Table I) and both flowbacks were collected. The amount of information available for this natural conception is probably unique. Neither of the relevant in-pair copulations was associated with a copulatory orgasm and, summed for the two copulations, we estimated –6 ± 48 million sperm (i.e. near zero) to have been retained from a total insemination of 545 million sperm. Two non-copulatory orgasms occurred approximately 28 and 22 h before conception. This unique data point illustrates that neither a copulatory orgasm, nor any more than minimal sperm retention, is necessary for conception.

Sperm retention was significantly reduced during the fertile phase of our three flowback donors who were not taking oral contraceptives, two of whom conceived. This reduction was primarily the result of the timing of inter-copulatory orgasms, menstrual variation in retention no longer being significant once timing of inter-copulatory orgasm was statistically controlled. A significant increase in cryptic inter-copulatory orgasms during the fertile phase was also found not only in our nationwide survey but also by Harvey (1987). Overall, there was no indication in our nationwide survey of any increase in sperm retention (measured in terms of levels I–III; Table VI) by monandrous females during their fertile phase. Finally, there is a tendency for a reduction in sperm retention in hormonally normal females, both in our nationwide survey and in our flowback analysis, compared with those taking oral contraceptives. There is thus no indication in our data that monandrous females favour higher sperm retention when conception is more likely. On the contrary, the impression is that the favoured strategy associated with conception in a monandrous situation is to reduce the number of sperm retained. The results are therefore entirely in line with our previous discussion of ejaculate adjustment by males (Baker & Bellis 1993).

We are not suggesting, of course, that high retention copulatory orgasms never lead to conception. During in-pair copulation, such orgasms are, in any case, more often associated with occasions when the male inseminates relatively few sperm (Table VII). Any negative relationship between the number of sperm retained and the probability of conception (Baker & Bellis 1993) will be quantitative, not absolute (i.e. there is always some chance of conception, no matter how many sperm are retained).

*Manipulation in the presence of sperm competition*

Although a male's probability of fertilization from any given copulation may be increased by decreasing sperm number in the absence of sperm competition (Baker & Bellis 1993), it is generally accepted that it is increased by increasing sperm number in the presence of sperm competition (Parker 1990).

The current explanation for concealed ovulation and continuous receptivity in some female mammals, such as humans, is that it allows the female to confuse the male over which copulations are likely to lead to paternity (Small 1989; Bellis & Baker 1990). In this way, females may manipulate the timing of inseminations from different males to suit their own priorities. Continuing copulation well into pregnancy maintains this deception by concealing the date of conception.

If females use copulation as a form of mate confusion, it is important that no aspect of their behaviour changes predictably with levels of infidelity and chances of conception. We have demonstrated that polyandrous females favour the extra-pair male (or males) not only by the timing of extra-pair copulations (Bellis & Baker 1990) but also by the relative level of sperm retention (Fig. 7). When females switch from a monandrous to a polyandrous situation (but still retaining a main partner), they reduce the level of sperm retention from their partner's inseminates but show a significantly higher level of retention from the extra-pair male's inseminates.

Our data show that females achieve the change in retention with their partner by varying the frequency of inter-copulatory orgasms cryptic to their partner while maintaining the same level of overt copulatory orgasms (Fig. 7). They achieve the difference in level of retention between males by varying the pattern of overt copulatory orgasms. The female response is exactly the one expected if the key factor in her behaviour were the deception of her primary partner. Add to these changes in pattern of real orgasms the possibility of females faking or hiding orgasms and the opportunity for males to detect any underlying female strategy is greatly reduced.

*Sperm retention in the absence of fertilization*

There are three situations in which manipulation of ejaculates by both males and females at first sight seems redundant: (1) during the infertile phase of the female menstrual cycle; (2) when either male or female is using contraceptives; and (3) during pregnancy.

A male's sperm remain fertile for at least 5 days after copulation (Barrett & Marshall 1969). In addition, our data suggest that his sperm, simply by being present in the cervix, can influence retention of sperm at the female's next copulation for up to 8 days (cf. Austin 1975). This in turn could mean that the sperm could indirectly influence retention at the copulation after that, and so on. Thus, sperm from one insemination could in principle have some influence on fertilization for up to at least 13 (8 + 5) days and perhaps even longer. Rarely can males, or even females, predict whether or not the female will become fertile at some time during the next 13 or so days.

In a sense, with this time scale in mind, discussion over why ejaculate manipulation should occur at times that conception is unlikely is largely unnecessary. The general answer is that it is very rare that any copulation has absolutely no chance of influencing fertilization. No form of contraception is perfect (Johnson & Everitt 1988), no phase of the menstrual cycle is absolutely infertile (Barrett & Marshall 1969; Jochle 1975), and even during pregnancy, spontaneous abortion followed by ovulation could occur within the period of influence of sperm.

As far as the male is concerned, a female has to be many months into pregnancy before her symptoms are unequivocal. As discussed above, pregnancy, like the menstrual cycle, is a phase when the female may benefit from confusion. During pregnancy, the vast majority of the sperm are apparently ejected in the flowback, no matter what the orgasm regime. Overt copulatory orgasms, therefore, seem to be functionless for sperm retention and we presume the main reason for overt regimes not to change during pregnancy is to maintain the efficiency of confusion. However, the frequency of cryptic orgasms also remains unchanged during pregnancy. Perhaps, as suggested for virgins, the functions of these cryptic orgasms during pregnancy is largely antibiotic.

*Individual variation in female strategies*

The greatest flexibility of ejaculate manipulation is shown by females who are capable of a varied range of orgasm patterns, both copulatory and inter-copulatory, including at times having no orgasm at all.

According to our nationwide survey, 84% of women, by the time they have had 500 copulations (henceforth, 'experienced' women), are using the whole range of ejaculate manipulation. Their strategy is 'mixed', involving copulatory and intercopulatory orgasms, as well as lack of orgasm. Only 2% of experienced women claimed never to have experienced any orgasm and thus to be using a 'no orgasm' strategy.

Sixteen per cent of experienced women in our survey claimed not to experience copulatory orgasms though the majority of these (88%) were experiencing inter-copulatory orgasms. Women with such an 'inter-copulatory orgasm dominated' strategy are less flexible in ejaculate manipulation than women with a fully 'mixed' strategy only insofar as they cannot convert a level I retention into a level III (see Table VI) during copulation itself.

Our analysis of flowbacks suggests that no matter whether a woman uses a 'no orgasm', 'inter-copulatory orgasm dominated', or a 'mixed' strategy, all are capable of manipulating sperm retention across the whole range from level I to level III (Table VI). Indeed, what 'no orgasm' and 'intercopulatory orgasm dominated' strategies may lose in flexibility, they may recoup, at least partially, through greater crypsis.

## Conclusion

Males adjust the number of sperm in their ejaculates according to the conflicting pressures of an advantage of low numbers for conception in the absence of sperm competition and of high numbers for competitiveness in the presence of sperm competition. On average, partly through female cooperation and perhaps partly through lack of total perfection in the female mechanism, when males inseminate more sperm, more are retained. On anyone occasion, however, a female is capable of negating any male strategy through her implementation of different orgasm regimes to manipulate the male's inseminate. The most flexible strategy involves copulatory and intercopulatory orgasms as well as occasional lack of orgasm. The conception strategy for females in monandrous situations seems primarily to be to favour low retention regimes. In polyandrous situations, females change their orgasm regimes on average to favour the extra-pair male in sperm competition. They do this by varying cryptic orgasms to reduce retention with their partner, then overriding this preparation with overt high retention orgasms should their next copulation be with an extra-pair male.

## ACKNOWLEDGMENTS

We thank Dr Tamsin Peachey for her part in developing our final model for the function of female orgasm; Emma Creighton, Dominic Shaw and Viki Cook for help with our pilot surveys, Gill Hudson and *Company* magazine for publishing our final questionnaire on sexual behaviour, and Jo Bell, Kath Griffin and Phil Wheater for help in transferring the data onto a computer data base. Jo Moffitt organized the recruitment and collection of the subjective estimates of flowback volume. Ray Meddis, Laurence Cook, Mike Hounsome and Kevin McConway gave valuable statistical advice and Professor Tim Halliday and Professor Tim Birkhead read, and greatly improved, an early version of the manuscript. We are particularly grateful to Liz Oram for drawing our attention to the potential value of studying flowbacks and for developing the techniques of flowback collection.

## REFERENCES

Austin, C. R. 1975. Sperm fertility, viability and persistence in the female tract. *J. Reprod. Fert., Suppl.*, **22**, 75–89.

Baker, R. R. & Bellis, M. A. 1989. Number of sperm in human ejaculates varies in accordance with sperm competition theory. *Anim. Behav.*, **37**, 867–869.

Baker, R. R. & Bellis, M. A. 1993. Human sperm competition: ejaculate adjustment by males and the function of masturbation. *Anim. Behav.*, **46**, 861–885.

Baker, R. R., Bellis, M. A., Hudson, G., Oram, E. & Cook, V. 1989. *Company*, **September**, 60–62.

Barrett, J. C. & Marshall, J. 1969. The risk of conception on different days of the menstrual cycle. *Pop. Studies*, **23**, 455–461.

Bellis, M. A. & Baker, R. R. 1990. Do females promote sperm competition: data for humans. *Anim. Behav.*, **40**, 997–999.

Bellis, M. A., Baker, R. R., Hudson, G., Oram, E. & Cook, V. 1989. *Company*, **April**, 90–92.

Belsey, M. A., Eliason, R., Gallegos, A. J., Moghissi, K. S., Paulsen, C. A. & Prasad, M. R. N. 1987. *WHO Laboratory Manual for Examination of Human Semen and Semen–Cervical Mucus Interaction.* 2nd Edn. Cambridge: Cambridge University Press.

Birkhead, T. R. & Hunter, F. M. 1990. Mechanisms in sperm competition. *Trends Ecol. Evol.*, **5**, 48–52.

Davey, D. A. 1986. Normal pregnancy: physiology and antenatal care. In: *Dewhurst's Textbook of Obstetrics and Gynaecology for Postgraduates* (Ed. by C. R. Whitfield), pp. 126–158. London: Blackwell Scientific Publications.

Davies, N. B. 1983. Polyandry, cloaca-pecking and sperm competition in dunnocks. *Nature, Lond.*, **302**, 334–336.

Duerden, B. I., Reid, T. M. S., Jewsbury, J. M. & Turk, D. C. 1987. *A New Short Textbook of Microbial and Parasitic Infection.* London: Hodder & Stoughton.

El-Banna, A. A. & Hafez, E. S. E. 1972. The uterine cervix in mammals. *Am. J. Obstet. Gynecol.*, **112**, 145–164.

Englert, Y, Puissant, F., Camus, M., Degueldre, M. & Leroy, F. 1986. Factors leading to tripronucleate eggs during human *in-vitro* fertilization. *Human Reprod.*, **1**, 117–119.

Evans, E. I. 1933. The transport of spermatozoa in the dog. *Am. J. Physiol.*, **105**, 287–293.

Fisher, S. 1973. The Female Orgasm. London: Allen Lane.

Ford, C. S. & Beach, F. A. 1952. *Patterns of Sexual Behaviour.* London: Eyre & Spottiswoode.

Fox, C. A., Wolff, H. S. & Baker, J. A. 1970. Measurement of intra-vaginal and intra-uterine pressures during human coitus by radio-telemetry. *J. Reprod. Fert.*, **22**, 243–251.

Fugle, G. N., Rothstein, S. I., Osenberg, C. W. & McGinley, M. A. 1984. Signals of status in wintering white-crowned sparrows, *Zonotrichia leucophrys gambelii. Anim. Behav.*, **32**, 86–93.

Ginsberg, J. R. & Huck, U. W. 1989. Sperm competition in mammals. *Trends Ecol. Evol.*, **4**, 74–79.

Ginsberg, J. R. & Rubenstein, D. I. 1990. Sperm competition and variation in zebra mating behaviour. *Behav. Ecol. Sociobiol.*, **26**, 427–434.

Hartman, C. G. 1957. How do sperms get into the uterus? *Fert. Steril.*, **8**, 403–427.

Harvey, S. M. 1987. Female sexual behaviour: fluctuations during the menstrual cycle. *J. Psychosom. Res.*, **31**, 101–110.

Howarth, B. 1971. Transport of spermatozoa in the reproductive tract of turkey hens. *Poult. Sci.*, **50**, 84.

Jochle, W. 1975. Current research in coitus-induced ovulation: a review. *J. Reprod. Fert., Suppl.*, **22**, 165–207.

Johnson, M. & Everitt, B. 1988. *Essential Reproduction.* London: Blackwell Scientific Publications.

Kroeks, M. V. A M. & Kremer, J. 1977. The pH in the lower third of the genital tract. In: *The Uterine Cervix in Reproduction* (Ed. by V. Insler & G. Bettendorf ), pp. 109–118. Stuttgart: Thieme.

Levin, R. J. 1981. The female orgasm: a current appraisal. *J. Psychosom. Res.*, **25**, 119–133.

LoPiccolo, J. & Lobitz, W. J. 1972. The role of masturbation in the treatment of orgasmic dysfunction. *Arch. Sex. Behav.*, **2**, 163–171.

Meddis, R. 1984. Statistics Using Ranks: a Unified Approach. Oxford: Blackwell.

Mock, D. W. & Fujioka, M. 1990. Monogamy and longterm pair bonding in vertebrates. *Trends Ecol. Evol.*, **5**, 39–42.

Moghissi, K. S. 1977. Sperm migration through the human cervix. In: *The Uterine Cervix in Reproduction* (Ed. by V. Insler & G. Bettendorf), pp. 146–165. Stuttgart: Thicme.

Morris, D. 1967. *The Naked Ape.* New York: McGrawHill.

Morton, D. B. & Glover, T. D. 1974. Sperm transport in the female rabbit: the role ofthe cervix. *J. Reprod. Fert.*, **38**, 131–138.

Overstreet, J. W. 1983. Transport of gametes in the reproductive tract of the female mammal. In: *Mechanism and Control of Animal Fertilization* (Ed. by J. F. Hartmann), pp. 499–543. London: Academic Press.

Parker, G. A. 1970. Sperm competition and its evolutionary consequences in the insects. *Biol. Rev.*, **45**, 525–567.

Parker, G. A. 1990. Sperm competition: games, raffles and roles. *Proc. R. Soc. Lond. B.*, **242**, 120–126.

Parsons, L. & Sommers, S. C. 1978. *Gynecology.* London: W. B. Saunders.

Raboch, J. & Skachova, J. 1965. The pH of human ejaculate. *Fert. Steril.*, **16**, 252–256.

Sagiroglu, N. & Sagiroglu, E. 1970. Biological mode of action of the Lippes Loop in intrauterine contraception. *Am. J. Obstet. Gynecol.*, **106**, 506–515.

Singer, I. 1973. Fertility and the female orgasm. In: *The Goals of Human Sexuality* (Ed. by I. Singer), pp. 159–197. London: Wildwood House.

Small, M. F. 1989. Aberrant sperm and the evolution of human mating patterns. *Anim. Behav.*, **38**, 544–546.

Sumption, L. J. 1961. Multiple sire mating in swine; evidence of natural selection for mating. *J. Agric. Sci.*, **56**, 31–37.

Tilbrook, A. J. & Pearce, D. T. 1986. Patterns of loss of spermatozoa from the vagina of the ewe. *Austral. J. Biol. Sci.*, **39**, 295–303.

Vander Vliet, W. L. & Hafez, E. S. E. 1974. Survival and aging of spermatozoa: a review. *Am. J. Obstet. Gynecol.*, **118**, 1006–1105.

Wells, B. L. 1986. Predictors of female nocturnal orgasms: a multivariate analysis. *J. Sex Res.*, **22**, 421–437.

# PART III: CONTEMPORARY READINGS IN HUMAN SPERM COMPETITION

# 12. NO EVIDENCE FOR KILLER SPERM OR OTHER SELECTIVE INTERACTIONS BETWEEN HUMAN SPERMATOZOA IN EJACULATES OF DIFFERENT MALES *IN VITRO*

H. D. M. Moore[1,2*], M. Martin[2] and T. R. Birkhead[3]

## ABSTRACT

This study examines one of the possible mechanisms of sperm competition, i.e. the kamikaze sperm hypothesis. This hypothesis states that sperm from different males interact to incapacitate each other in a variety of ways. We used ejaculates from human donors to compare mixes of semen *in vitro* from the same or different males. We measured the following parameters: (i) the degree of sperm aggregation, velocity and proportion of morphologically normal sperm after 1 and 3 h incubation in undiluted semen samples, (ii) the proportion of viable sperm plus the same parameters as in (i) in 'swim-up' sperm suspensions after 1 and 3 h incubation, (iii) the degree of self and non-self sperm aggregation using fluorescent dyes to distinguish the sperm of different males, and (iv) the extent of sperm capacitation and acrosome-reacted sperm in mixtures of sperm from the same and different males. We observed very few significant changes in sperm aggregation or performance in mixtures of sperm from different males compared with mixtures from the same male and none that were consistent with previously reported findings. The incapacitation of rival sperm therefore seems an unlikely mechanism of sperm competition in humans.

---

Departments of [1]Obstetrics and Gynaecology, [2]Molecular Biology and Biotechnology and [3]Animal and Plant Sciences, University of Sheffield.

## 1. INTRODUCTION

Sperm competition is the competition between the ejaculates of different males for fertilization of a female's ova (Parker 1970). Among species with internal fertilization sperm competition occurs when a female copulates with and is inseminated by two or more males during a single breeding cycle. Throughout a wide range of animal taxa, including reptiles, birds and mammals, sperm competition is widespread and is a powerful selective force shaping the morphology, behaviour and physiology of reproduction (Smith 1984*a*; Birkhead & Møller 1998). In most cases females appear to seek multiple copulation partners actively and the types of benefit females might obtain have been extensively discussed (Birkhead & Parker 1997). The mechanistic aspects of sperm competition have been studied in detail in only a small number of species. In insects and birds, the last male to inseminate a female before ovulation usually fertilizes the majority of ova. Last male sperm precedence, as it is called, occurs through a variety of processes. In insects, incoming spermatozoa may displace those of previous inseminations (e.g. Simmons *et al.* 1999) or substances in the male's seminal fluid may incapacitate spermatozoa from previous inseminations stored in the female (Harshman & Prout 1994; Chapman *et al.* 1995; Clark *et al.* 1995; Price *et al.* 1999). In birds, last male sperm precedence arises largely as a consequence of passive sperm loss between successive inseminations (Birkhead 1998). In mammals with induced ovulation, the first male usually fertilizes the ova but in spontaneous ovulators there are no consistent insemination order effects (Gomendio *et al.* 1998). Instead, the outcome of inseminations by different males is thought to be determined by interactions between the order of copulation, the interval between copulations and the timing of the insemination relative to when the female ovulates (Ginsberg & Huck 1989; Huck *et al.* 1989). In addition to these basic mechanisms in all species, other factors may also influence the outcome of sperm competition. These include differences between males in the rate at which spermatozoa are capacitated (Moore & Bedford 1983), the velocity of spermatozoa (Holt *et al.* 1996; Froman & Feltman 1998), the relative fertilizing capacity of spermatozoa (Dzuik 1996; Birkhead *et al.* 1999) and the compatibility between spermatozoa and the female tract and ova (Zeh & Zeh 1997; Clark *et al.* 1999).

Baker & Bellis (1988) proposed an additional mechanism of sperm competition in mammals. Their kamikaze sperm hypothesis states that the variation which exists in sperm morphology within an ejaculate is not the result of production errors during spermatogenesis, as other studies have indicated (Cohen 1967, 1973; see also Manning & Chamberlain 1994), but instead is an adaptation to sperm competition. This idea was previously introduced by Sivinski (1980) and Silberglied *et al.* (1984) to explain the variation in insect sperm. Specifically, Baker & Bellis (1988) proposed that, within an ejaculate, there are two broad categories of spermatozoa. So-called 'egg getters' were spermatozoa which were capable of fertilizing the oocyte and 'kamikaze' spermatozoa were those which would sacrifice themselves in the 'warfare' between spermatozoa of different males when a female is inseminated by two or more males. Baker & Bellis (1988) suggested that only a small proportion of spermatozoa (with large-headed morphology) were capable of fertilization and that the role of the other spermatozoa was 'warfare'. They asserted that certain spermatozoa with oval-headed morphology had evolved to kill or incapacitate the spermatozoa of rival males (Baker & Bellis 1988,

1989). Using data from humans, Baker & Bellis (1995, p. 274) subsequently reported that, when they mixed semen from two different males *in vitro*, they observed the following statistically significant changes: (i) a reduction in the proportion of oval-headed sperm (otherwise known as morphologically normal sperm; World Health Organization 1992), (ii) an increase in agglutination, (iii) an increase in mortality, and (iv) a reduction in sperm velocity compared with controls in which self-spermatozoa had been mixed. Their explanation for these effects was that, on making contact with a spermatozoon from another male, oval-headed spermatozoa released their acrosomal enzymes (which include proteases) thereby killing rival spermatozoa.

Baker & Bellis's (1995) kamikaze sperm hypothesis has received a great deal of uncritical media attention even though their results leave a large number of questions unanswered, particularly regarding their methods (see also Harcourt 1991; Birkhead *et al.* 1997; Gomendio *et al.* 1998; Short 1998). For example, they stated that they were apparently able to distinguish normal and acrosome-reacted human spermatozoa readily without the use of specific supravital stains or fluorescent dyes (Kohn *et al.* 1997), yet no other student of human spermatozoa has been able to achieve this. Indeed, most researchers agree that definitive identification of the acrosomal status of a human spermatozoon requires a specific label and that light microscopy alone (bright field or phase contrast) is unreliable (De Jonge *et al.* 1989). Baker & Bellis (1995) did not label spermatozoa from different males nor did they assess either sperm viability or morphology in an objective or quantitative manner.

The aim of the present study was to investigate the kamikaze sperm hypothesis using standard andrology protocols and quantitative assessment by computer assisted sperm analysis (CASA) to address the question of whether spermatozoa from one human ejaculate can specifically influence the spermatozoa from the ejaculate of another male *in vitro*. Specifically, we tested the following predictions of the kamikaze sperm hypothesis: compared with self-sperm mixtures, non-self mixtures of sperm show a reduction in sperm velocity and the proportion of morphologically normal sperm and an increase in agglutination, mortality and the incidence of capacitated and acrosome-reacted sperm.

## 2. MATERIAL AND METHODS

### (a) Semen samples

Ejaculates were obtained by masturbation following three days sexual abstinence from 15 donors of proven fertility attending the Jessop Hospital for Women, Sheffield, who had been screened regularly for sexually transmitted disease. The samples were processed within 1 hour of collection. All samples used were in the normal range for sperm concentration, motility and morphology as determined by standard protocols (World Health Organization 1992): sperm concentration $46 \pm 23 \times 10^6$ ml$^{-1}$, sperm motility $54 \pm 18\%$ progressively motile sperm (average path velocity $44 \pm 12$ μms$^{-1}$) and sperm morphology $36 \pm 14\%$ normal morphology (World Health Organization (1992) criteria) and $19 \pm 5\%$ normal morphology (strict criteria; Kruger *et al.* 1995). The World Health Organization (1992) criteria for normal morphology comprise sperm head length

4–5.5 μm, head width 2.5–3.5 μm, acrosome 60–70% head area and no neck, midpiece or tail defect.

All tests were done blind by one of the authors (M.M.) with no prior knowledge of the hypothesis or its predictions. CASA was used to measure the sperm motility parameters objectively as previously reported (Mohammad *et al.* 1996). Sperm morphology was determined using a Hobson sperm morphology analyser (Sense and Vision Ltd, Sheffield, UK) with sperm smears which had been fixed in methanol (100%) and stained by the Papanicolaou staining procedure (World Health Organization 1992). Sperm mixtures were examined after 1 and 3 h incubation as this is the time that sperm normally remain in the human vagina (Tredway *et al.* 1978; Insler *et al.* 1980) and, hence, the most likely duration of sperm mixing following copulations with different males. Baker & Bellis (1995, p. 273) provided few details of their experimental protocol, but apparently examined sperm mixtures after 3–6 h. We considered that, with our more sensitive assays, any effects should have been apparent after the more biologically realistic interval of 3 h.

Semen from either the same male or pairs of males was mixed and duplicate aliquots of each mixture were prepared with the order of mixing changed (i.e. A + B and B + A). For each order of mixing we used paired *t*-tests to check that mixing order had no effect and we used repeatability values (Lessells & Boag 1987) to assess the similarity of the two samples. In 12 out of 14 comparisons, the values from the different mixing orders did not differ significantly and seven out of the 14 measures of repeatability were significant. It is important to note that the non-significant repeatability values were obtained, as expected, from the swim-up samples (seven out of eight samples; see below) since the sperm used in these tests were a more uniform, selected sample from the whole ejaculate. Since in the majority of cases mixing order had no effect on the values obtained, we used the means from the two mixes in subsequent analyses.

### (b) Tests with diluted semen

To test the most basic prediction of the kamikaze sperm hypothesis, that mixtures of semen from different men (non-self mixtures) resulted in greater aggregation of sperm than samples mixed from the same man (self mixtures), we conducted the following experiments. Ejaculates from ten different donors were mixed both with themselves and pairwise with all other donors. Aliquots (100 μl) of the two semen samples were mixed together in a microcentrifuge tube diluted in 200 μl of modified Earle's medium (MEM) and incubated for 1 or 3 at 37°C in 5% $CO_2$ in air.

A sperm smear was made for each mixture after 1 and 3 h incubation. No adjustment was made for sperm concentration (22–140 $\times 10^6$ sperm $ml^{-1}$). After incubation, the sperm suspension was gently agitated by a single finger flick of the tube and a 7 μl aliquot was pipetted onto a microscope slide and a cover-slip carefully placed on top. Cell aggregation was assessed at × 200 magnification by phase-contrast microscopy. Ten fields were selected at random and the number of fields containing an aggregation of two or more spermatozoa counted after 1 and 3 h. Aggregation was considered to have occurred when more than two fields contained two or more spermatozoa bound to each other. Sperm motility (200 sperm tracks, 1.5–3.0s duration per track) was assessed by CASA after 1 and 3 h incubation. Sperm morphology was determined for non-aggregated sperm only since it was not possible to assess the morphology of aggregated sperm either subjectively or by CASA.

### (c) Tests with swim-up sperm suspensions

In order to simulate a situation resembling that when sperm have left the vagina and are 'selected' on the basis of their motility and morphology (Overstreet & Katz 1990; Mortimer 1995), we mixed sperm, as in §2(b), in suspensions prepared from aliquots (0.5 ml) of the same ejaculates by the swim-up method. In this, progressively motile sperm are tested after they migrate out of seminal plasma into an overlying layer of MEM (World Health Organization 1992) and are adjusted to a concentration of $10^6$ sperm $ml^{-1}$ in MEM. The sperm aggregation test was performed as described above. In addition, a sperm viability test was performed, which measures the percentage of viable sperm, on the mixture of sperm suspensions after 1 and 3 h incubation using the 'Fertilyte' cell viability test (Molecular Probes Inc., Oregon, USA) as reported elsewhere (Mohammad *et al.* 1996).

### (d) Tests with fluorochrome-labelled sperm

In order to assess whether aggregations were random with respect to the donor or, as the kamikaze sperm hypothesis predicts, more likely to comprise sperm from different males, we conducted the following experiment. A separate batch of five donors was used to prepare sperm suspensions labelled with fluorochrome. This was effective in distinguishing the origin of the spermatozoa and did not adversely affect sperm motility over the 3 h incubation period as assessed by CASA. Swim-up sperm suspensions were prepared as described previously and samples adjusted with MEM medium to $10^6$ sperm $ml^{-1}$. Aliquots (500 μl) were labelled with either fluorescein isothiocyanate (FITC) or tetramethyl rhodamine isothiocyanate (TRITC) as described by Lui *et al.* (1989) and incubated in combinations for 1 or 3 h. A fluorochrome-labelled aliquot (100 ml) was mixed with either a non-labelled or a differently labelled aliquot. There was no evidence that labelling had any effect on the incidence of sperm aggregations in mixtures comprising (i) FITC-labelled and TRITC-labelled sperm, (ii) FITC-labelled and unlabelled sperm, and (iii) TRITC-labelled and unlabelled sperm ($\chi^2 = 4.18$, $df = 2$ and $p > 0.05$). Labelled sperm suspensions were kept away from light (which can adversely affect cell viability) by wrapping the tubes in aluminium foil. Sperm motility and morphology was assessed after 1 and 3 h incubation as described above. Aggregation of spermatozoa was observed under phase-contrast and epifluorescent microscopy (Olympus (UK) Ltd) with appropriate filters for FITC (498 nm) and TRITC (525 nm) fluorescence.

### (e) Induction of capacitation and spontaneous acrosome reaction

To test the hypothesis that non-self mixtures of semen resulted in a greater incidence of capacitated and acrosome-reacted sperm than self mixtures, we conducted the following experiment. The same swim-up preparations of donor ejaculates as in §2 (c) were used to assess the phosphorylation of sperm proteins and, hence, the capacitation of sperm (Emiliozzi & Fenichel 1997; Aitken *et al.* 1998; Brewis *et al.* 1998). Aliquots (0.5ml) from different donors were mixed together in combinations and incubated for 1 or 3 h at 37°C in 5% $CO_2$ in air in capacitating MEM containing human serum albumin as described elsewhere (Brewis *et al.* 1996). At the end of the incubation, 100 µl aliquots of sperm suspension were removed, washed by gentle centrifugation at 600 *g* for 5 min, resuspended in phosphate-buffered saline and each smeared onto several clean microscope slides for assessment of the sperm acrosome reaction using a monoclonal antibody (mAb 18.6) probe (Moore *et al.* 1987). The remainder of each aliquot was centrifuged at 5000 *g* for 5 min and the supernatant removed. The sperm pellet was immediately resuspended in sodium-dodecacyl sulphate sample buffer and protein analysed for phosphotyrosine residues by gel electrophoresis and immunoblotting using specific monoclonal antibody and chemiluminescent detection. The spontaneous sperm acrosome reaction was measured as a defined end-point of capacitation (Burks *et al.* 1995; Moore 1995; Brewis *et al.* 1996). The intensity of specific phosphoprotein bands, which is a measure of the extent of acrosome activation, was determined as pixel points using image analysis software (NIH Image, 1.52).

Unless stated otherwise, chemicals were from Sigma Chemical Co., Poole, Dorset, UK. Means and s.d.s are expressed. All statistical tests are two-tailed.

## 3. RESULTS

### (a) Tests with diluted semen

The results of mixing combinations of ejaculates from ten donors are summarized in figure l*a*. Non-self semen samples were no more likely to show any aggregation (31%) than self samples (50%; see table 1). There was no significant difference in the velocity of sperm after 1 or 3 h incubation in self and non-self mixtures (table 1). There was a significantly lower proportion of morphologically normal sperm in self mixtures (30%) than in non-self mixtures (35%) after 1 h, although after 3 h the difference was not significant (table 1).

### (b) Tests with swim-up sperm suspensions

The results of mixing combinations of swim-up sperm suspensions from the same ten donors are summarized in figure l*b*. The swim-up procedure significantly reduced the incidence of aggregation (ten cases of aggregation out of 100 possible cases) compared with the diluted semen (see above) (26 cases out of 100, $\chi^2 = 8.6$, $df = 1$ and $p < 0.02$; compare figure l*a*,*b*). As expected, the sperm in swim-up suspensions were an improved subset compared with those in diluted semen (see above). For example, after 1 h of incubation virtually all (99–100%) sperm in the swim-up suspension were viable. There was no evidence that swim-up suspensions from different males were any more likely to show aggregation than self mixtures (table 1). Nor were there any significant differences in the proportion of morphologically normal sperm and the velocity or viability of sperm in self and non-self mixtures after either 1 or 3 h incubation (table 1).

(*a*)

| donor | A | B | C | D | E | F | G | H | I | J |
|---|---|---|---|---|---|---|---|---|---|---|
| A | X | | | | | | | | X | |
| B | | X | | | | | X | | | |
| C | X | X | | | | X | | | | |
| D | | | | X | | | | | | X |
| E | X | X | | | | X | | | X | |
| F | | | | | X | | | | | |
| G | | X | | | | | X | X | | X |
| H | | | X | | | | X | | | |
| I | | | | | X | | | | | X |
| J | | | | X | | | X | | X | X |

(*b*)

| donor | A | B | C | D | E | F | G | H | I | J |
|---|---|---|---|---|---|---|---|---|---|---|
| A | | | | | | | | | | |
| B | | X | | | | | | | | |
| C | | | | | | | | | | |
| D | | | | X | | X | | | X | |
| E | X | | | | | | | | | |
| F | | | | X | | | X | | | |
| G | | | | | | | | | | |
| H | | | | | | | | | | |
| I | | | | | | | | | | |
| J | | | | | | X | | X | | X |

(*c*)

| donor | R | S | T | U | V |
|---|---|---|---|---|---|
| R | | | X | | |
| S | | | X | X | |
| T | X | X | X | X | |
| U | | | | | X |
| V | X | | | | |

**Figure 1.** Summary of the results of three experiments in which semen from different donors (A–J and R–V) was mixed either with themselves (shaded boxes) or with another donor. Cells below the self-self diagonal represent cases where sperm were mixed B–A, C–A, etc. and cells above the diagonal indicate mixings A–B, A–C, etc. (*a*) Diluted semen only from ten donors (A–J). (*b*) Swim-up sperm using the same ten donors (A–J). (*c*) Swim-up sperm labelled with fluorochromes (see the text) from five different donors (R–V). A cross in a cell indicates that aggregation of sperm was recorded at either 1 or 3 h after mixing (see 2).

### (c) Tests with fluorochrome-labelled sperm

There was no difference in the incidence of aggregation between self and non-self sperm labelled with fluorochrome, although the sample size was relatively small (figure 1c and table 1). Sperm suspensions from donor T had a tendency to aggregate with samples from other donors as well as with itself (figure 1*c*). There was no differences in the velocity of sperm in self and non-self mixtures after 1 or 3 h incubation (table 1), but there was a tendency for there to be fewer morphologically normal sperm in nonself mixtures (table 1). This result is consistent with the kamikaze sperm hypothesis, but since the same effect did not occur in either the diluted semen or the swim-up test (table 1), it provides little support for Baker & Bellis's (1995) hypothesis.

**Table 1.** *Summary of results of three types if test involving self-self and non-self mixtures of semen from ten different donors.* (The values are means and s.d.s. The tests are either $\chi^2$-tests, Mann-Whitney U-tests (*z*-statistic) or Fisher exact tests. The units are as follows: velocity (μms$^{-1}$), morphology (percentage of normal head morphology) and viability (percentage viable).)

| test | self | non-self | statistic | *P* |
|---|---|---|---|---|
| **diluted semen** | | | | |
| incidence of aggregation | 50% (5/10) | 31% (14/45) | $\chi^2 = 1.29$ | 0.25 |
| velocity (1 h) | 38.2 ± 1.9 | 39.3 ± 0.6 | $z = 0.35$ | 0.73 |
| velocity (3 h) | 30.7 ± 2.5 | 33.2 ± 0.8 | $z = 0.88$ | 0.37 |
| morphology (1 h) | 30.1 ± 2.1 | 35.2 ± 0.9 | $z = 2.16$ | 0.03 |
| morphology (3 h) | 31.6 ± 1.9 | 31.0 ± 0.6 | $z = 0.17$ | 0.86 |
| **swim-up test** | | | | |
| incidence of aggregation | 30% (3/10) | 13% (6/45) | $\chi^2 = 3.18$ | 0.02 |
| velocity (1 h) | 52.2 ± 1.5 | 51.8 ± 0.6 | $z = 0.16$ | 0.87 |
| velocity (3 h) | 46.1 ± 1.3 | 45.0 ± 0.6 | $z = 0.84$ | 0.40 |
| morphology (1 h) | 41.9 ± 2.5 | 38.7 ± 0.6 | $z = 1.23$ | 0.22 |
| morphology (3 h) | 40.1 ± 1.8 | 39.4 ± 0.6 | $z = 0.18$ | 0.85 |
| viability (1 h) | 99–100 | 99–100 | no test | – |
| viability (3 h) | 95.5 ± 0.9 | 94.3 ± 0.4 | $z = 1.03$ | 0.29 |
| **fluororochrome test** | | | | |
| incidence of aggregation | 20% (1/5) | 60% (6/10) | Fisher exact | 0.20 |
| velocity (1 h) | 60.6 ± 1.7 | 56.3 ± 1.7 | $z = 1.46$ | 0.14 |
| velocity (3 h) | 53.6 ± 2.8 | 51.0 ± 1.2 | $z = 0.67$ | 0.50 |
| morphology (1 h) | 52.8 ± 3.1 | 47.3 ± 1.1 | $z = 1.90$ | 0.06 |
| morphology (3 h) | 51.8 ± 1.7 | 46.6 ± 1.2 | $z = 2.21$ | 0.04 |

We compared the observed proportion of sperm aggregations containing sperm from both males with those expected from a null model for sperm aggregations of sizes two, three to five and six or more sperm (the maximum clump size observed was 12 sperm). If sperm aggregate at random with respect to male then for aggregations of size two sperm the expected proportion of aggregations containing sperm from one or both males is predicted to be 25% that both are male A, 25% that both are male B and 50% are mixed. For larger clumps binomial probabilities provide the predicted proportions of the various possible mixtures. As clump size increases the proportion of clumps containing sperm from both males increases, for example 87.5% for clumps of four sperm and 96.9% for clumps of six sperm. The observed proportion of sperm clumps containing sperm from both males for aggregations of two sperm did not differ significantly ($\chi^2$-test $p > 0.05$) from chance (table 2), but for larger aggregations there were significantly fewer clumps containing sperm from both males than expected by chance ($\chi^2$-tests $p < 0.01$). This suggests that sperm from the same male were significantly more likely to form aggregations.

**Table 2.** Number (and percentage) of sperm aggregations containing sperm of two males in non-self mixtures with different combinations of labels in swim-up preparations from five donors. (The label combinations were green (FITC) plus red (TRITC) (g + r), green plus unlabelled (g + u) and red plus unlabelled sperm (r + u) (see the text).)

| | number (percentage) of sperm aggregations with sperm from two males in different labelling combinations | | |
|---|---|---|---|
| aggregation size | g + r | g + u | r + u |
| two | 37/64 (50) | 34/70 (49) | 35/84 (42) |
| three-five | 23/34 (68) | 27/40 (68) | 31/45 (69) |
| six or more | 18/27 (67) | 22/25 (88) | 22/31 (71) |

### (d) Induction of capacitation and spontaneous acrosome reaction

There was no evidence that non-self sperm mixtures resulted in any greater incidence of capacitated or acrosome-reacted sperm than self mixtures. In terms of capacitation, a three-way ANOVA (with male pair, treatment (self or non-self mixture) and time (1 or 3 h) as factors) revealed that neither the effect of male pair nor treatment was significant, but, as expected, there was a significant increase in capacitation (i.e. phosphotyrosine residues in sperm proteins) with time ($p < 0.001$). The results are summarized in figure 2*a*. The same samples were also analysed in the same way for the spontaneous acrosome reaction using mAb 18.6. There was no effect of treatment (self versus non-self $p = 0.83$), but there was a significant (and expected) effect of time ($p < 0.001$; figure 2*a*). In addition, there was also a significant interaction between time and male pair ($p < 0.001$), suggesting a difference between individuals in the rate at which their sperm underwent the acrosome reaction (figure 2*b*).

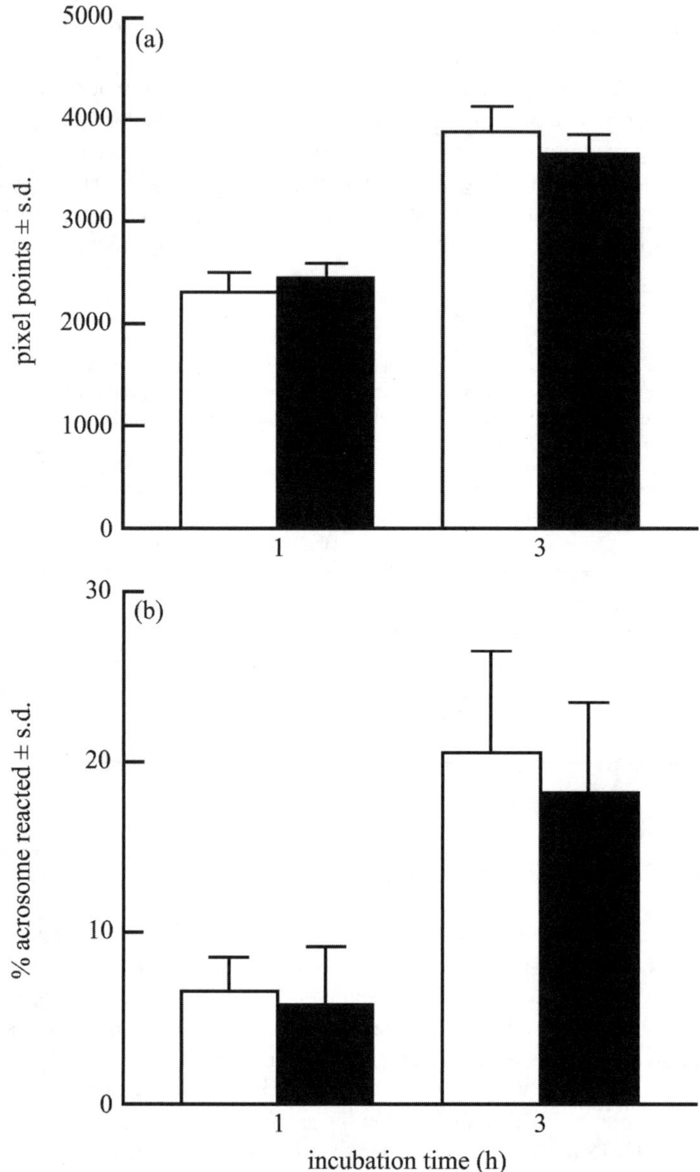


**Figure 2**. (a) Phosphotyrosine expression of the 95 kDa epitope as determined by image analysis (mean ± s.d. pixel points) for self (unshaded) and non-self (shaded) mixtures of sperm. This assay provides an index of the incidence of capacitation (see the text). The difference with incubation time was significant (ANOVA $p < 0.001$), but there was no difference between self and non-self mixtures. (*b*) The difference with incubation time was significant (ANOVA $p < 0.001$), but there was no difference between self and nonself mixtures of semen.

## 4. DISCUSSION

On the basis of several different *in vitro* assays our results provide no support for Baker & Bellis's (1995) hypothesis that human sperm selectively make contact with sperm from a different male and then incapacitate them.

In several of our experiments some aggregation of spermatozoa was observed. However, in all our experiments agitation of the samples was kept to a minimum to allow any potential interactions between cells to occur. In the clinical laboratory, sperm clumping or agglutination of a few sperm in undiluted semen is common (World Health Organization 1992) and normally an ejaculate is agitated by vortexing before analysis (World Health Organization 1992). Indeed, for all the samples in the present study, agitation of an aliquot by vortexing eliminated sperm clumps indicating that the sperm interactions observed during the incubation period were weak. Within the female tract, smooth muscle action may cause considerable agitation of spermatozoa so that only strong agglutination may be maintained. Aggregation of sperm in an ejaculate may be due to incomplete liquefaction, leaving groups of spermatozoa in microscopic seminal clots. This may explain our observation that aggregations tended to be more common in self mixtures. Sperm may also adhere unselectively to cellular debris, prostatic droplets and other insoluble material in seminal plasma (World Health Organization 1992). Many previous studies have also shown that autoantibodies can be present in seminal plasma and differentially affect sperm aggregation from different ejaculates whether from the same or a different male (Shulman 1995; Sigman *et al.* 1997). This is probably due to sperm-coating antibodies being present in variable amounts in an ejaculate. At low titres (and occasionally at high titres), autoantibodies in seminal plasma may cause sperm aggregation without any influence on the overall fertilizing capacity of an ejaculate (Turek 1997), although individual sperm in aggregations may be less likely to participate in fertilization. Seminal plasma may also be directly detrimental to spermatozoa in contact with it for a prolonged period due to the presence of reactive oxygen molecules (Sigman *et al.* 1997) and over the 3 h incubation period in the present study we recorded a reduction in sperm velocity in several combinations of ejaculates. However, we found no difference in velocity in self and non-self mixtures of sperm.

Mixing undiluted semen from different ejaculates *in vitro* to some extent mimics sequential copulation with different partners over a short period of time when ejaculates could intermingle within the vagina before any selection processes occur in the female tract. This situation might arise in primates such as the chimpanzee *Pan troglodytes* where sperm competition is intense, copulation frequency is high and females are routinely inseminated by several males and often in rapid succession (e.g. Gagneux *et al.* 1997, 1999). In humans sperm competition is less intense (Smith 1984*b*; Harcourt *et al.* 1995; Gomendio *et al.* 1998), female copulation frequency lower and the interval between successive inseminations likely to be longer. As a result it is less likely that undiluted semen samples and more likely that selected sperm from different ejaculates would encounter each other in the female tract. Within a few hours of insemination, spermatozoa involved in fertilization pass into reservoirs in the cervical mucus and enter the uterine lumen and the isthmus region of the oviduct (Overstreet & Katz 1990; Mortimer 1995). Spermatozoa in these reserves are probably selected on the basis of their progressive motility and morphology characteristics and occur at much lower concentrations ($\times 10^{-3}$ to $10^{-6}$) than those deposited initially in the vagina at coitus. There is therefore much less opportunity for direct sperm-sperm contact once spermatozoa leave the vagina. *In vitro* swim-up sperm preparations represent such selected cells since this method allows spermatozoa with good progressive motility and possibly more normal morphology (World Health Organization 1992) to swim from the remainder of the ejaculate (Yovich 1995). Although we found a significant improvement in the mean sperm morphology after the swim-up procedure, spermatozoa with an oval head morphology, which Baker & Bellis (1995) indicated were involved in 'warfare', were not excluded by this procedure. Nonetheless, mixing experiments involving equal numbers of swim-up spermatozoa from different donors provided no evidence in support of selective interactions between spermatozoa from the ejaculates of different men.

One of the major criticisms of the study by Baker & Bellis (1995) is the lack of validation of their results. It is impossible to determine whether sperm agglutinations are due to self or non-self interactions or a combination of both on the basis of light microscopy alone. By fluorescent labelling of spermatozoa of one or both ejaculates the composition of a clump was determined unequivocally. The mixing experiments with labelled spermatozoa indicated that, in clumps comprising two sperm, self and non-self clumps were equally likely to occur but, with larger clumps, self aggregations were more likely than non-self aggregations. It could be argued that this result is consistent with the kamikaze sperm hypothesis if, as Baker & Bellis (1995, p. 268) proposed, self sperm aggregate to form a block to subsequent inseminations. However, had we found that sperm from different males were more likely to form aggregations, that could also have been construed as evidence that sperm from one male had blocked the progress of the other. Consequently, at best this result provides equivocal support for the kamikaze sperm hypothesis.

If a kamikaze spermatozoon acts by undergoing a spontaneous acrosome reaction in the vicinity of a foreign spermatozoon then one would expect non-self sperm mixtures to show higher rates of capacitation and a higher proportion of acrosome-reacted spermatozoa compared with self mixtures (Baker & Bellis 1995). A number of previous reports have shown that the amount of tyrosine-phosphorylated epitopes in human spermatozoa is related to their capacitation status (Brewis & Moore 1997). Our results clearly indicated that tyrosine phosphorylation increased during incubation but that this was not related to the mixture of sperm samples. Similarly, the proportion of sperm that had undergone the acrosome reaction did not differ between self and non-self mixtures.

Baker & Bellis (1995) used additional circumstantial evidence from the literature to support their kamikaze sperm hypothesis, including the phenomenon of differential fertilizing capacity. When semen pooled from several different males is used for artificial insemination (as in breeding in many domestic animals) one male usually fertilizes a disproportionate number of offspring (Dziuk 1996). Baker & Bellis (1995, p. 302) suggested that the differential fertilizing capacity 'will only make sense when viewed in the perspective of the KSH' (kamikaze sperm hypothesis). However, recent studies on nonhumans, while not explicitly testing the kamikaze sperm hypothesis, have shown that the differential fertilizing capacity can be adequately explained by inherent differences in the attributes of the sperm of different males (Birkhead *et al.* 1999; Donoghue *et al.* 1999).

In summary, there is no evidence from this study to support the conclusions reached by Baker & Bellis (1995) that particular spermatozoa of an ejaculate act in a 'kamikaze' manner against spermatozoa of a different ejaculate. In a mixture of ejaculates, there may be interactions between spermatozoa but *in vitro* most of these interactions appear to be random and do not involve specific recognition processes.

## ACKNOWLEDGEMENTS

We are grateful to the staff of the Andrology Laboratory, Jessop Hospital for Women, for organization of the donor semen samples used in this study. We also wish to thank Professor J. D. Biggins for statistical advice and Dr B. J. Hatchwell, Dr F. M. Hunterand, Dr E. R. S. Roldan and four anonymous referees for their constructive comments on the manuscript.

## REFERENCES

Aitken, R. J., Harkiss, D., Knox, W., Paterson, M. & Irvine, D. S. 1998 A novel signal transduction cascade in capacitating human spermatozoa characterised by a redox-regulated CAMP-mediated induction of tyrosine phosphorylation. *J. Cell Sci.* **111**, 645–656.

Baker, R. R. & Bellis, M. A. 1988 Kamikaze sperm in mammals? *Anim. Behav.* **36**, 936–939.

Baker, R. R. & Bellis, M. A. 1989 Elaboration of the kamikaze sperm hypothesis: a reply to Harcourt. *Anim. Behav.* **37**, 865–867.

Baker, R. R. & Bellis, M. A. 1995 *Human sperm competition.* London. Chapman & Hall.

Birkhead, T. R. 1998 Sperm competition in birds. *Rev. Reprod.* **3**, 123–129.

Birkhead, T. R. & Møller, A. P. 1998 *Sperm competition and sexual selection.* London. Academic Press.

Birkhead, T. R. & Parker, G. A. 1997 Sperm competition and mating systems. In *Behavioural ecology: an evolutionary approach* (ed. J. R. Krebs & N. B. Davies), pp. 121–145. Oxford, UK: Blackwell.

Birkhead, T. R., Moore, H. D. M. & Bedford, J. M. 1997 Sex, science and sensationalism. *Trends Ecol. Evol.* **12**, 121–122.

Birkhead, T. R., Martinez, J. G., Burke, T. & Froman, D. P. 1999 Sperm mobility determines the outcome of sperm competition in the domestic fowl. *Proc. R. Soc. Lond.* B **266**, 1–6.

Brewis, I. & Moore, H. D. M. 1997 Molecular mechanisms of gamete recognition and fusion at fertilization. *Hum. Reprod.* **12** (Suppl. 2), 156–165.

Brewis, I. A., Clayton, R., Barratt, C. L. R., Hornby, D. P. J. & Moore, H. D. M. 1996 Recombinant human zona pellucida glycoprotein 3 induces calcium influx and acrosome reaction in human spermatozoa. *Mol. Hum. Reprod.* **2**, 583–589.

Brewis, I. A., Clayton, R., Browes, C. E., Martin, H., Barratt, C. L. R., Hornby, D. P. & Moore, H. D. M. 1998 Tyrosine phosphorylation of a 95 kDa protein and induction of the acrosome reaction in human spermatozoa by recombinant human zona pellucida glycoprotein 3. *Mol. Hum. Reprod.* **12**, 1136–1144.

Burks, D. J., Carballada, R., Moore, H. D. M. & Saling, P. M. 1995 A novel tyrosine kinase from human sperm interacts with the zona pellucida at fertilization. *Science* **269**, 83–86.

Chapman, T., Liddle, L. F., Kalb, J. M., Wolfner, M. F. & Partridge, L. 1995 Cost of mating in *Drosophila melanogaster* females is mediated by male accessory gland products. *Nature* **373**, 241–244.

Clark, A. G., Aguade, M., Prout, T., Harshman, L. G. & Langley, C. H. 1995 Variation in sperm displacement and its association with accessory gland protein in *Drosophila melanogaster*. *Genetics* **139**, 189–201.

Clark, A. G., Begun, D. J. & Prout, T. 1999 Female × male interactions in *Drosophila* sperm competition. *Science* **283**, 217–220.

Cohen, J. 1967 Correlation between chiasma frequency and sperm redundancy. *Nature* **215**, 862–863.

Cohen, J. 1973 Crossovers, sperm redundancy, and their close associations. *Heredity* **31**, 408–415.

De Jonge, C. J., Mack, S. R. & Zaneveld, L. J. D. 1989 Synchronous assay for human sperm capacitation and acrosome reaction. *J. Androl.* **10**, 232–239.

Donoghue, A. M., Sonstegard, T. S., King, L. M., Smith, E. J. & Burt, D. W. 1999 Turkey sperm mobility influences paternity in the context of competitive fertilization. *Biol. Reprod.* **61**, 422–427.

Dzuik, P. J. 1996 Factors that influence the proportion of offspring sired by a male following heterospermic insemination. *Anim. Reprod. Sci.* **43**, 65–88.

Emiliozzi, C. & Fenichel, P. 1997 Protein tyrosine phosphorylation is associated with capacitation of human sperm *in vitro* but is not sufficient for its completion. *Biol. Reprod.* **56**, 674–679.

Froman, D. P. & Feltman, A. J. 1998 Sperm mobility: a quantitative trait in the domestic fowl (*Gallus domesticus*). *Biol. Reprod.* **58**, 379–384.

Gagneux, P., Woodruff, D. S. & Boesch, C. 1997 Furtive mating in female chimpanzees. *Nature* **387**, 358–359.

Gagneux, P., Boesch, C. & Woodruff, D. S. 1999 Female reproductive strategies, paternity and community structure in wild West African chimpanzees. *Anim. Behav.* **57**, 19–32.

Ginsberg, J. R. & Huck, U. W. 1989 Sperm competition in mammals. *Trends Ecol. Evol.* **4**, 74–79.

Gomendio, M., Harcourt, A. H. & Roldan, E. R. S. 1998 Sperm competition in mammals. In *Sperm competition and sexual selection* (ed. T. R. Birkhead & A. P. Møller). London: Academic Press.

Harcourt, A. H. 1991 Sperm competition and the evolution of nonfertilizing sperm in mammals. *Evolution* **45**, 314–328.

Harcourt, A. H., Purvis, A. & Liles, L. 1995 Sperm competition: mating system, not breeding season, affects testes size of primates. *Funct. Ecol.* **9**, 468–476.

Harshman, L. G. & Prout, T. 1994 Sperm displacement without sperm transfer in *Drosophila melanogaster*. *Evolution* **48**, 758–766.

Holt, C., Holt, W. V., Moore, H. D. M., Reed, H. C. B. & Curnock, R. M. 1996 Objectively measured boar sperm motility parameters correlate with the outcome of on-farm inseminations. Results of two fertility trials. *J. Androl.* **18**, 312–323.

Huck, U. W., Tonias, B. A. & Lisk, R. D. 1989 The effectiveness of competitive male inseminations in golden hamsters, *Mesocricetus auratus*, depends on an interaction of mating order, time delay between males and the time of mating relative to ovulation. *Anim. Behav.* **37**, 674–680.

Insler, V., Glezerman, M., Ziedel, L., Bernstein, D. & Misgar, N. 1980 Sperm storage in the human cervix: a quantitative study. *Fertil. Steril.* **33**, 288–293.

Kohn, F. M., Mack, S. R., Schill, W. B. & Zaneveld, L. J. D. 1997 Detection of human sperm acrosome: comparison between methods using double staining, *Pisum sativum* agglutinin, concanavalin A and transmission electron microscopy. *Hum. Reprod.* **12**, 714–721.

Kruger, T. F., Dutoit, T. C., Franken, D. R., Menkveld, R. & Lombard, C. J. 1995 Strict morphology-assessing the agreement between the manual method (strict criteria) and the sperm morphology analyzer IVOS. *Fertil. Steril.* **63**, 134–141.

Lessells, C. M. & Boag, P. T. 1987 Unrepeatable repeatabilities: a common mistake. *Auk* **104**, 116–121.

Lui, D. Y., Clarke, G. N., Lopata, A., Johnston, W. I. H. & Baker, H. W. G. 1989 A sperm-zona pellucida binding test and *in vitro* fertilization. *Fertil. Steril.* **52**, 281–287.

Manning, J. T. & Chamberlain, A. T. 1994 Sib-competition and sperm competitiveness: an answer to 'Why so many sperms?' and the recombination/sperm number correlation. *Proc. R. Soc. Lond.* B **256**, 177–182.

Mohammad, S. N., Barratt, C. L., Cooke, I. D. & Moore, H. D. M. 1996 Continuous assessment of human spermatozoa viability during cryopreservation. *J. Androl.* **18**, 43–50.

Moore, H. D. M. 1995 Modification of sperm membrane antigens during capacitation. In *Human acrosome reaction* (ed. P. Fenichel & J. Parinaud), pp. 35–43. Paris: John Libbey.

Moore, H. D. M & Bedford, J. M. 1983 The interaction of mammalian gametes in the female. Sperm/egg interactions *in vivo*. In *Mechanism and control of animal fertilization* (ed. J. F. Hartmann), pp. 453–497. New York: Academic Press.

Moore, H. D. M., Hartman, T. D. & Smith, C. A. 1987 A monoclonal antibody for marking the human acrosome reaction. *Gamete Res.* **17**, 245–259.

Mortimer, D. 1995 Sperm transport in the female tract. In *Gametes: the spermatozoon* (ed. J.G. Grudzinsksa & J. Yovich), pp. 157–174. Cambridge University Press.

Overstreet, J. W. & Katz, D. F. 1990 Interaction between the female reproductive tract and spermatozoa. In *Controls of sperm motility: biological and clinical aspects* (ed. C. Gagnon), pp. 63–75. Boca Raton, FL: CRC Press.

Parker, G. A. 1970 Sperm competition and its evolutionary consequences in the insects. *Biol. Rev.* **45**, 525–567.

Price, C. S. C., Dyer, K. A. & Coyne, J. A. 1999 Sperm competition between *Drosophila* males involves both displacement and incapacitation. *Nature* **400**, 449–452.

Short, R. V. 1998 Review of human sperm competition: copulation, masturbation and infidelity by R. R. Baker and M. A. Bellis. *Eur. Sociobiol. Soc. Newslett.* **47**, 20–23.

Shulman, S. 1995 Immunological reactions and infertility. In *Immunobiology of human reproduction* (ed. M. Kurpisz & N. Fernandez), pp. 53–78. Oxford, UK: Bios Scientific Publishers Ltd.

Sigman, M., Lipshultz, L. I. & Howards, S. S. 1997 Evaluation of the subfertile male. In *Infertility in the male* (ed. L. I. Lipshultz & S. S. Howards), pp. 173–193. St Louis, MO: Mosby-Year Book Inc.

Silberglied, R. E., Shepherd, J. G. & Dickinson, J. L. 1984 Eunuchs: the role of apyrene sperm in Lepidoptera? *Am. Nat.* **123**, 255–265.

Simmons, L. W., Parker, G. A. & Stockley, P. 1999 Sperm displacement in the yellow dungfly, *Scatophaga stercoraria*: an investigation of male and female processes. *Am. Nat.* **153**, 302–314.

Sivinski, J. 1980 Sexual selection and insect sperm. *Fl. Entomol.* **63**, 99–111.

Smith, R. L. 1984*a* *Sperm competition and the evolution of animal mating systems*. Orlando, FL: Academic Press.

Smith, R. L. 1984*b* Human sperm competition. In *Sperm competition and the evolution of animal mating systems* (ed. R. L. Smith), pp. 601–659. Orlando, FL: Academic Press.

Tredway, D. R., Buchanan, G. C. & Drake, T. S. 1978 Comparison of the fractional postcoital test and semen analysis. *Am. J. Obstet. Gynecol.* **130**, 647–652.

Turek, P. J. 1997 Immunopathology and infertility. In *Infertility in the male* (ed. L. I. Lipshultz & S. S. Howards), pp. 305–325. St Louis, MO: Mosby-Year Book Inc.

World Health Organization 1992 *Laboratory manual for the examination of human semen and sperm-cervical mucus interactions*. Cambridge University Press.

Yovich, J. L. 1995 Sperm preparation for assisted conception. In *Gametes: the spermatozoon* (ed. J. G. Grudzinsksa & J. Yovich), pp. 268–281. Cambridge University Press.

Zeh, J. A. & Zeh, D. W. 1997 The evolution of polyandry. II. Post-copulatory defences against genetic incompatibility. *Proc. R. Soc. Lond.* B **264**, 69–75.

# 13. PSYCHOLOGICAL ADAPTATION TO HUMAN SPERM COMPETITION

Todd K. Shackelford[1], Gregory J. LeBlanc[1], Viviana A. Weekes-Shackelford[1], April L. Bleske-Rechek[2], Harald A. Euler[3], and Sabine Hoier[3]

## ABSTRACT

Sperm competition occurs when the sperm of two or more males simultaneously occupy the reproductive tract of a female and compete to fertilize an egg. We used a questionnaire to investigate psychological responses to the risk of sperm competition for 194 men in committed, sexual relationships in the United States and in Germany. As predicted, a man who spends a greater (relative to a man who spends a lesser) proportion of time apart from his partner since the couple's last copulation reported (a) that his partner is more attractive, (b) that other men find his partner more attractive, (c) greater interest in copulating with his partner, and (d) that his partner is more sexually interested in him. All effects were independent of total time since the couple's last copulation and the man's relationship satisfaction. Discussion addresses two failed predictions and directions for future work.

## 1. INTRODUCTION

Sperm competition occurs when the sperm of two or more males simultaneously occupy the reproductive tract of a female and compete to fertilize an egg (Baker & Bellis, 1995; Parker, 1970*a*, 1970*b*; Parker, 1984). Sperm Competition Theory provides the theoretical framework for a body of work investigating adaptations in males and in

---

[1] Division of Psychology, Florida Atlantic University.

[2] Vanderbilt University.

[3] University of Kassel.

Reprinted from *Evolution and Human Behavior,* 23 (2), Shackelford, T. K., LeBlanc, G. J., Weekes-Shackelford, V. A., Bleske-Rechek, A. L., Euler, H. A., & Hoier, S., Psychological adaptation to human sperm competition. pp. 123–138, Copyright (2002) with permission from Elsevier.

females designed to solve problems posed by sperm competition (Parker, 1970*a*, 1970*b*; Smith, 1984*a*). Current research on the evolutionary causes and consequences of sperm competition focuses primarily on birds (Birkhead & Møller, 1992) and insects (Gage, 1995; Thornhill & Alcock, 1983). The research on sperm competition in birds is particularly interesting for students of human mating.

Over 90% of bird species practice social monogamy, the mating system most common to humans (Birkhead & Møller, 1992). Social monogamy is a mating system in which a male and a female form a long-term pair bond. Within this mating system, a male benefits by gaining sexual access to the reproductive value of a female, whereas a female benefits by gaining access to the paternal investment of the male in her offspring (Birkhead & Møller, 1992; Trivers, 1972). Only rarely is this exchange of reproductive resources between a male and a female exclusive, however. Recent work, including DNA fingerprinting to identify paternity, documents that social monogamy in birds does not necessarily indicate sexual and genetic monogamy (e.g., Burke, Davies, Bruford, & Hatchwell, 1989). Female infidelity in socially monogamous birds is widely documented and is the primary context for sperm competition (Birkhead & Møller, 1992).

When a female bird in a socially monogamous mateship is sexually unfaithful to her partner, she places her partner at risk of investing his limited time and resources in offspring to whom he is genetically unrelated. Female infidelity places selective pressures on males to detect the infidelity and to inflict costs on an unfaithful female. These costs include abandoning, raping, and physically assaulting an unfaithful female (Birkhead & Møller, 1992). The costs inflicted by a male bird on a partner suspected of infidelity reveal psychological features designed to motivate behaviors that might "correct" the infidelity by an immediate or forced copulation, for example. These actions might be corrective in the sense that the in-pair male is placing his sperm in the female's reproductive tract, providing his sperm an opportunity to compete with the sperm of a rival male for the fertilization of his partner's eggs.

The socially monogamous mating system of birds – and the female infidelity that is the primary context for sperm competition – is characteristic of several primates, including humans. Humans are socially monogamous, with both males and females engaging in extra-pair mating under certain predictable conditions (Baker & Bellis, 1995; Buss, 1994). Smith (1984*a*) and Baker and Bellis (1995) provided empirical and theoretical arguments that female infidelity is the primary context for sperm competition in humans. Work by evolutionary psychologists, in turn, has identified the conditions under which female infidelity occurs and, furthermore, the costs that a male inflicts on a female partner who is suspected of infidelity. Shackelford and Buss (Buss & Shackelford, 1997*a*, 1997*b*, 1997*c*; Shackelford & Buss, 1997*a*, 1997*b*, 2000), for example, documented that female infidelity is associated with her personality characteristics, her partner's personality characteristics, and the discrepancy in mate value between her and her partner. Shackelford and Buss also identified some of the costs that a male inflicts on a partner that he suspects of infidelity. These costs include physical and psychological abuse, rape, and divorce (Buss & Shackelford, 1997*a*, 1997*b*; Shackelford & Buss, 1997*a*, 1997*b*).

Human sperm competition has become a recent focus of evolutionary biologists and evolutionary psychologists (Baker & Bellis, 1988, 1989*a*, 1989*b*, 1993*a*, 1993*b*, 1995; Bellis & Baker, 1990; Gangestad & Thornhill, 1997, 1998; Pound, 1998; Singh, Meyer,

Zambarano, & Hulbert, 1998; Thornhill, Gangestad, & Comer, 1995). Baker and Bellis (1993*a*, 1995), for example, documented that male humans, like male birds, males insects, and other male non-human primates, appear to be physiologically designed to solve the adaptive problems of sperm competition. Studying couples in committed, sexual relationships, Baker and Bellis assessed (decreased) risk of sperm competition as the proportion of time a couple has spent together since their last copulation, controlling for the total time since their last copulation. As hypothesized, and consistent with Sperm Competition Theory, Baker and Bellis documented a large positive correlation between the risk of sperm competition and the number of sperm a male inseminates into his partner at the couple's next copulation.

Baker and Bellis investigated physiological reactions to the risk of sperm competition. There must be psychological mechanisms linked to these physiological adjustments. Male psychology may include evolved mechanisms that motivate behavior that would have increased the probability of success in sperm competition for ancestral males. The present research seeks to identify some of these evolved mechanisms and their associated design features.

Baker and Bellis operationalized risk of sperm competition as the proportion of time a couple has spent together since their last copulation. As this proportion decreases, the risk that a rival male will inseminate a woman increases (Baker & Bellis, 1995). The proportion of time that a couple has spent together since their last copulation is perfectly negatively correlated with the proportion of time that a couple has spent apart since their last copulation. The proportion of time spent apart since the couple's last copulation is arguably a more intuitive index of the risk of sperm competition, and we therefore refer to the proportion of time spent apart rather than the proportion of time spent together. We propose that the proportion of time that a couple has spent apart since their last copulation is information processed by male psychological mechanisms that subsequently motivates a man to want to inseminate his partner as soon as possible, to combat the increased risk of sperm competition.

Total time since last copulation is not clearly linked to the risk of sperm competition. Instead, it is the proportion of time a couple has spent apart since their last copulation – time during which a man cannot account for his partner's activities – that is linked to the risk of sperm competition (Baker & Bellis, 1995). Nevertheless, it might be argued that total time since last copulation is the relevant variable. As the total time since last copulation increases, a man might feel increasingly "sexually frustrated" and, therefore, increasingly interested in copulation with his partner, independent of the proportion of time he and his partner have spent apart since their last copulation. We could not locate any research that investigated the relationship between a man's sexual interest in his partner and the total time since the couple's last copulation. To address the potential confound, however, we assessed the relationships between male sexual psychology and behaviors predicted to be linked to the risk of sperm competition (as assessed by the proportion of time spent apart since last copulation), independent of the total time since a couple's last copulation.

A man's sexual interest in his partner, for example, might be higher with a greater proportion of time spent apart from his partner since their last copulation, but only for a man that is satisfied with his relationship. A man who is more satisfied with or invested in his relationship may have more to lose in the event of cuckoldry. More generally, a man's satisfaction with the relationship might be related to his reported sexual interest in and attraction to his partner. Unlike a greater proportion of time spent apart since the couple's last copulation, a man's satisfaction with the relationship is not clearly linked to the risk of sperm competition. To address this possible confound, however, we investigated the relationships between male sexual psychology and behaviors predicted to be linked to the risk of sperm competition (as assessed by the proportion of time spent apart since last copulation), independent of a man's relationship satisfaction.

In summary, this research tested the hypothesis that human male psychology includes evolved mechanisms designed to attend to the risk of sperm competition as assessed by the proportion of time a couple has spent apart since their last copulation, independent of (a) the total time since the couple's last copulation and (b) the man's relationship satisfaction. A key component of this hypothesis is that the output of these evolved mechanisms includes increased motivation to copulate with one's partner. We generated six predictions from this hypothesis. A man may be more likely to want to copulate with his partner when he perceives his partner as more attractive. This leads to the first prediction:

*Prediction 1*: A man who spends a greater (relative to a man who spends a lesser) proportion of time apart from his partner since the couple's last copulation will rate his partner as more attractive, independent of (a) the total time since last copulation and (b) relationship satisfaction.

A man who perceives that other men find his partner attractive might, as a consequence, perceive his partner as more attractive. This effect might occur because the perception that other men find his partner attractive is a cue to increased risk of sperm competition, should these other men pursue copulation with his partner. Or this effect might occur because the perception that other men find his partner attractive draws his attention to his partner's attractiveness. Following Prediction 1, the greater perceived attractiveness of his partner might motivate a desire to copulate with his partner as soon as possible.

*Prediction 2*: A man who spends a greater (relative to a man who spends a lesser) proportion of time apart from his partner since the couple's last copulation will provide higher ratings of other men's assessments of his partner's attractiveness, independent of (a) the total time since last copulation and (b) relationship satisfaction.

A man's self-reported desire to copulate with his partner presumably predicts his active pursuit of copulation with his partner. This assumption leads to the third prediction:

*Prediction 3*: A man who spends a greater (relative to a man who spends a lesser) proportion of time apart from his partner since the couple's last copulation will report greater interest in copulating with his partner, independent of (a) the total time since last copulation and (b) relationship satisfaction.

A man's perception that his partner is sexually interested in *him* may motivate him to pursue copulation with her. A man's perception that his partner is sexually interested in *other men* might also motivate him to pursue copulation with her. The perception that his partner is sexually attracted to other men may signal an increased risk of sperm competition or may highlight his partner's sexual needs, desires, and attractiveness. This focused attention on his partner's sexuality might motivate him to pursue copulation with her. Regardless of the proximate causal processes, we state the fourth and fifth predictions as follows:

*Prediction 4*: A man who spends a greater (relative to a man who spends a lesser) proportion of time apart from his partner since the couple's last copulation will provide higher ratings of his partner's sexual attraction to *him*, independent of (a) the total time since last copulation and (b) relationship satisfaction.

*Prediction 5*: A man who spends a greater (relative to a man who spends a lesser) proportion of time apart from his partner since the couple's last copulation will provide higher ratings of his partner's sexual attraction to *other men*, independent of (a) the total time since last copulation and (b) relationship satisfaction.

According to Strategic Interference Theory (Buss, 1989), negative emotions such as anger, frustration, and upset signal interference with a plan of action that was reproductively successful for our ancestors. These negative emotions call attention to the strategic interference and motivate removal of the source of interference. The intensity of the negative emotions corresponds to the intensity or severity of strategic interference. A woman's denial of her partner's request for copulation is predicted to cause feelings of distress in her partner. These feelings of distress might, first, call her partner's attention to the strategic interference and, second, motivate him to remove the source of interference − in this case, to negotiate when the next copulation will occur. A woman's denial of her partner's request for copulation is expected to generate a level of distress in her partner that corresponds to the current risk of sperm competition he faces and which is indexed by the proportion of time the couple has spent apart since their last copulation. This leads to Prediction 6:

*Prediction 6*: A man who spends a greater (relative to a man who spends a lesser) proportion of time apart from his partner since the couple's last copulation will report greater distress following his partner's denial of a request for copulation, independent of (a) the total time since last copulation and (b) relationship satisfaction.

To summarize, we predict that a man who spends a greater (relative to a man who spends a lesser) proportion of time apart from his partner since the couple's last copulation will rate his partner as more attractive, will provide higher ratings of other men's assessments of his partner's attractiveness, will report greater interest in copulating with his partner, will provide higher ratings of his partner's sexual attraction to himself and to other men, and will report greater distress following his partner's denial of a request for copulation. These findings are predicted to be independent of the total time since the couple's last copulation and independent of the man's relationship satisfaction. To test these predictions, we collected self-report data in the United States and in Germany from men in committed, sexual relationships.

## 2. METHODS

### 2.1 Participants

Participants were 194 men currently in a committed, sexual, heterosexual relationship. Participants were drawn from universities and surrounding communities in two Western countries, the United States ($n$ = 126) and Germany ($n$ = 68). Across samples, the mean age of participants was 26.3 years, and the mean age of participants' partners was 25.2 years. The mean length of participants' relationships, across samples, was 55.9 months. The predictions assume that participants are involved in a committed relationship. We included in analyses data provided by men who were currently in a relationship that had lasted at least one year. Although this particular minimum criterion is somewhat arbitrary, it ensures that all participants were involved in a "committed" relationship in the sense that the relationship has lasted at least one year. Additional descriptive statistics are provided in Table 1.

### 2.2. Materials

Each participant completed a three-page survey titled "Short Survey About Men's Sexual Behavior." Each participant was instructed to answer the questions honestly and was reminded that his responses were anonymous. The first section of the survey collected demographic information, including the participant's age, his partner's age, and the length of their relationship. The second section of the survey addressed the key sexuality variables. First, each participant was asked how many hours had passed since he last had sexual intercourse with his partner. Next, each participant was asked how many hours he had spent together with his partner, including sleeping time, since he last had sexual intercourse with his partner. The next set of questions asked each participant to think about his feelings "at this moment in time" and to rate the following items, from 0 = *Not at all* to 9 = *Extremely*: How attractive does the participant think his partner is? How attractive do other men think the participant's partner is? How interested is the participant in having sex with his partner? How sexually attracted to the participant is the participant's partner? How sexually attracted to other men is the participant's partner? Finally, each participant was asked to indicate, in hours, when he would next like to have sexual intercourse with his partner.

The third section of the survey asked each participant to answer questions about how he would feel "at that moment in time" if his partner denied a request to have sexual intercourse with him. Each participant was asked to rate his resulting anger, frustration, and upset on a scale anchored by 0 = *Not at all* and 9 = *Extremely*. The fourth section of the survey asked each participant to rate his commitment to his partner and his overall, sexual, and emotional satisfaction with his partner. These ratings were made on a 10-point scale anchored by 0 = *Not at all* and 9 = *Extremely*.

The translation of the survey from English to German proceeded as follows: A bilingual speaker translated the English language survey into German. A second bilingual speaker unaware of the contents of the original English language survey back-translated the German language survey into English. The two bilingual speakers consensually resolved the few resulting discrepancies between the original English language survey and the back-translated English language survey.

### 2.3. Procedures

Data were collected from two sources. First, students in social science courses were asked to voluntarily complete the survey during the last few minutes of a class session. Second, trained research assistants approached prospective participants at various locations on and around university campuses. These locations included coffee shops, shopping malls, and airports. The research assistant approached a man and asked if he would be interested in completing a short survey about sexuality. If the prospective participant was interested, the research assistant then asked if he was currently in a committed, sexual, heterosexual relationship. If these criteria were met, the research assistant handed the participant a consent form, the survey, and a 9-inch × 12-inch brown security envelope. The participant was instructed to read and sign the consent form, complete the survey, place the completed survey in the envelope, and then seal the envelope. The participant was instructed not to seal the consent form inside the envelope to maintain anonymity. Finally, the participant was instructed to place the sealed envelope in a box that contained other sealed envelopes. The participant was asked to place the signed consent form in a separate envelope that contained other signed consent forms. The research assistant explained to the participant the purpose of the study, answered any questions, and thanked the participant for his participation.

The procedure for data collection in class settings began with a trained research assistant explaining that men were needed for a study on sexuality in committed, sexual, heterosexual relationships. This request for participants occurred during the last few minutes of a class session. The research assistant explained that men in a committed, sexual relationship that were interested in voluntarily completing a short survey on sexuality in their relationship should remain in the classroom. Students who did not fit the criteria or who did not want to participate were given several minutes to exit the classroom. The remainder of the procedure for classroom data collection was identical to the procedure described above for the public approach method. About half the participants in the United States and in Germany were obtained through the public approach method. Most participants completed the survey in three to five minutes.

## 3. RESULTS

We used multiple regression analyses to test the six predictions. An important assumption of multiple regression and other parametric statistical analyses is that the independent variables are approximately normally distributed (see Tabachnick & Fidell, 2001). Non-normal distribution or skewness can be corrected in several ways, the most common of which is logarithmic transformation of the skewed variable (Tabachnick & Fidell, 2001). Three variables were log-transformed prior to analyses to correct substantial skew: (a) total time since the couple's last copulation, (b) desired time to next copulation with partner, and (c) interest in sex with partner. The proportion of time the couple had spent apart since their last copulation was calculated by subtracting the number of hours the couple had spent together since their last copulation from the total number of hours since the couple's last copulation and dividing this difference by the total number of hours since the couple's last copulation.

Several variables examined were composites. Relationship satisfaction ($\alpha$ = .78) is the mean of four variables: (a) participant's overall satisfaction, (b) sexual satisfaction and (c) emotional satisfaction with his partner, and (d) participant's commitment to his partner. Own rating of partner's attractiveness ($\alpha$ = .86) is the mean of two variables: Own rating of partner's (a) sexual attractiveness and (b) physical attractiveness. Own rating of other's assessment of partner's attractiveness ($\alpha$ = .88) is the mean of two variables: Own rating of other's assessment of partner's (a) sexual attractiveness and (b) physical attractiveness. Disinterest in copulating with partner ($\alpha$ = .65) is the mean of two log-transformed variables: (a) desired time to next copulation with partner and (b) interest in sex with partner (reverse-coded). Finally, hypothetical distress following partner's denial of a request for copulation ($\alpha$ = .87) is the mean of three variables: (a) anger, (b) frustration, and (c) upset following partner's denial of a request for copulation.

Table 1 presents descriptive statistics for the key variables, across both samples and separately for the American and German samples. Table 1 also displays the results of independent means tests of country-of-origin differences on these variables. The German men and their partners were older than the American men and their partners. Relative to the German men, the American men reported greater relationship satisfaction, provided higher ratings of other men's assessments of their partners' attractiveness, reported greater interest in copulating with their partners, and provided higher ratings of their partners' sexual attraction to them (the participants). The German and American participants did not differ significantly in relationship duration, the total time since last copulation with their partners, the proportion of time spent apart from their partners since last copulation, assessments of partners' attractiveness, ratings of partners' sexual attraction to other men, or distress following their partners' denial of a request for copulation.

To increase the power of the statistical analyses, we collapsed the data across country. These results are presented in Table 2 and discussed below. In addition, we analyzed the data separately by country. The pattern of results when analyzed separately by country was the same as when analyzed across countries. For one of the country-level analyses (testing Prediction 4, regarding partner's sexual attraction to the participant), the significant effects found in the across-country analyses held for the American sample but not the German sample. For the remaining country-level analyses, significant effects identified in the across-country analyses were in the same direction but did not reach statistical significance. For reportorial efficiency and to capitalize on greater statistical power, we present only the results of analyses conducted on data collapsed across country (country-level analyses are available from the first author upon request).

We tested each of the six predictions with multiple regression. For each prediction, three predictors were entered into a regression equation predicting a single dependent variable. These three predictors were (a) the proportion of time the couple spent apart since their last copulation, (b) the total time since the couple's last copulation (log-transformed), and (c) the participant's relationship satisfaction. Each prediction states that the proportion of time spent apart since the couple's last copulation will predict the target dependent variable, independent of the total time since the couple's last copulation and independent of the participant's relationship satisfaction. One of the reviewers of this article inquired as to whether we could analyze the results as a function of method of data collection (classroom versus public approach). Unfortunately, we did not code the surveys for method of data collection (although we estimate that about half the data were collected by each method), and so we cannot conduct these analyses.

**Table 1:** Descriptive statistics for target variables

| | Sample | | | | | | | |
|---|---|---|---|---|---|---|---|---|
| | Overall | | United States | | Germany | | | |
| Variable | Mean | SD | Mean | SD | Mean | SD | t | |
| Relationship duration (months) | 55.9 | 72.6 | 50.6 | 72.7 | 65.7 | 71.9 | –1.38 | |
| Own age (years) | 26.3 | 10.1 | 24.6 | 8.8 | 29.4 | 11.4 | –3.23 | ** |
| Partner's age (years) | 25.2 | 9.4 | 23.6 | 8.3 | 28.2 | 10.6 | –3.32 | ** |
| Total time since last copulation (hours) | 120.6 | 160.3 | 137.2 | 182.3 | 89.1 | 101.4 | 1.94 | |
| Proportion of time apart since last copulation | 0.6 | 0.3 | 0.6 | 0.3 | 0.6 | 0.3 | 0.78 | |
| Relationship satisfaction[a] | 7.4 | 1.3 | 7.5 | 1.3 | 7.1 | 1.3 | 2.3 | * |
| Own assessment of partner's attractiveness[b] | 7.2 | 1.5 | 7.4 | 1.5 | 6.9 | 1.5 | 1.93 | |
| Own rating of other men's assessments of partner's attractiveness[b] | 6.8 | 1.6 | 7.2 | 1.4 | 6.2 | 1.7 | 4.48 | *** |
| Disinterest in copulating with partner[c] | 1.5 | 0.9 | 1.3 | 0.9 | 1.7 | 0.9 | –2.7 | ** |
| Estimate of how sexually attracted partner is to participant[d] | 7.0 | 1.6 | 7.4 | 1.5 | 6.4 | 1.6 | 4.0 | *** |
| Estimate of how sexually attracted partner is to other men[d] | 4.0 | 2.4 | 4.2 | 2.4 | 3.6 | 2.3 | 1.59 | |
| Hypothetical distress following partner's denial of request for copulation[e] | 3.6 | 2.2 | 3.7 | 2.4 | 3.4 | 1.9 | 0.99 | |

For the combined (United States and Germany) sample, N = 194; for the United States sample, n = 126; for the Germany sample, n = 68. The t values were generated by independent means tests of the differences between the United States and Germany. [a]Composite variable (see text), with scale anchored by 0 = not at all satisfied, 9 = extremely satisfied. [b]Composite variable (see text), with scale anchored by 0 = not at all attractive, 9 = extremely attractive. [c]Composite variable constructed using log-transformed variables (see text). [d]Scale anchored by 0 = not at all sexually attracted, 9 = extremely sexually attracted. [e]Composite variable (see text), with scale anchored by 0 = not at all distressed, 9 = extremely distressed. * $p < 0.05$, ** $p < 0.01$, *** $p < 0.001$ (two-tailed).

Table 2 presents the results of the six multiple regressions conducted to test the six predictions. The table presents, for each multiple regression, the standardized regression coefficient and the associated *t*-value for each of the three predictors. The first row of Table 2 shows that the greater the proportion of time a couple has spent apart since their last copulation, the more attractive a man rated his partner, independent of the total time since last copulation and independent of the man's relationship satisfaction. Prediction 1 therefore was supported. Total time since last copulation did not independently predict a man's rating of his partner's attractiveness. Relationship satisfaction positively predicted own ratings of partner's attractiveness.

**Table 2:** Results of multiple regressions of proportion of time spent apart since last copulation, total time since last copulation, and relationship satisfaction on target dependent variables.

| | **Predictor variable** | | | | | |
|---|---|---|---|---|---|---|
| | **Proportion of time spent apart since last copulation** | | **Total time since last copulation (log transformed)** | | **Relationship satisfaction** | |
| **Dependent variable** | ***B*** | ***t*** | ***B*[a]** | ***t*** | ***B*** | ***t*** |
| Own assessment of partner's attractiveness (Prediction 1) | 0.70 | 2.01* | 0.92 | 1.02 | 0.66 | 8.68*** |
| Own rating of other men's assessments of partner's attractiveness (Prediction 2) | 0.95 | 2.34* | 0.38 | 0.36 | 0.52 | 5.78*** |
| Disinterest in copulating with partner (Prediction 3) | −0.56 | −2.24* | 0.6 | 0.92 | −0.16 | −2.93** |
| Estimate of how sexually attracted partner is to participant (Prediction 4) | 1.23 | 2.95** | −0.9 | −0.82 | 0.42 | 2.95** |
| Estimate of how sexually attracted partner is to other men (Prediction 5) | −0.01 | −0.02 | 2.40 | 1.36 | −0.02 | −0.02 |
| Hypothetical distress following partner's denial of request for copulation (Prediction 6) | 0.43 | 0.69 | −0.82 | −0.51 | −0.48 | −3.49** |

[a]For reportorial efficiency, the *B* value shown for "Total time since last copulation" was produced by multiplying the original *B* value by 10. *Note.* $N = 194$ men. *B* = unstandardized regression coefficient, *t* = test statistic associated with *B*. * $p < 0.05$; ** $p < 0.01$; *** $p < 0.001$ (two-tailed).

The second row of Table 2 shows that the greater the proportion of time a couple spent apart since their last copulation, the more attractive a man thinks other men find his partner, independent of the total time since last copulation and independent of the man's relationship satisfaction. Prediction 2 therefore was supported. Total time since last copulation did not independently predict a man's rating of other men's perceptions of his partner's attractiveness. Relationship satisfaction positively predicted own ratings of others' assessments of partner's attractiveness.

The third row of Table 2 shows that the greater the proportion of time a couple spent apart since their last copulation, the more interested (or the less disinterested) he is in copulating with his partner, independent of the total time since last copulation and independent of the man's relationship satisfaction. Prediction 3 therefore was supported. Total time since last copulation did not independently predict a man's interest in copulating with his partner. Relationship satisfaction positively predicted copulatory interest in partner.

The fourth row of Table 2 shows that the greater the proportion of time a couple has spent apart since their last copulation, the more sexually interested in him a man thinks his partner is, independent of the total time since last copulation and independent of the man's relationship satisfaction. Prediction 4 therefore was supported. Total time since last copulation did not independently predict a man's assessment of his partner's sexual interest in him. Relationship satisfaction positively predicted assessments of partner's sexual interest in the participant.

The fifth row of Table 2 shows that that neither the proportion of time a couple has spent apart since their last copulation nor the total time since the couple's last copulation predicted a man's assessment of his partner's sexual interest in other men. Prediction 5 therefore was not supported. Additionally, relationship satisfaction did not predict a man's assessment of his partner's sexual interest in other men.

Finally, the sixth row of Table 2 shows that neither the proportion of time a couple has spent apart since their last copulation nor the total time since the couple's last copulation predicted a man's distress following his partner's denial of his request for copulation. Prediction 6 therefore was not supported. Relationship satisfaction, however, negatively predicted distress following partner's denial of a request for copulation. One of the reviewers of this article suggested that a man who was in a relationship of longer duration might have had experiences in that relationship that would temper distress reactions following a partner's denial of a request for copulation. This reviewer suggested that we might include relationship length as a covariate in our test of Prediction 6. We conducted a second test of Prediction 6, following the analysis presented above. In this second analysis, however, we included relationship length as a covariate. The results did not change, either at the county-level or across country. Distress following a partner's denial of a request for copulation was not predicted by the proportion of time apart since last copulation, total time since last copulation, the man's relationship satisfaction, or the length of the couple's relationship (analyses are available from the first author upon request).

## 4. DISCUSSION

This research tested the hypothesis that human male psychology includes mechanisms designed to solve the adaptive problems confronted when a rival might inseminate a partner. Over human evolutionary history, males risked squandering reproductive resources on offspring to whom they were genetically unrelated. A male equipped with psychological mechanisms that motivated him to inseminate his partner as soon as possible following his partner's sexual infidelity would have out-reproduced a male that lacked the requisite mechanisms and therefore took no such "corrective" action. The present research provides empirical evidence that male psychology may include mechanisms designed to solve the adaptive problem of a rival's insemination of a long-term partner.

We assessed the risk of sperm competition as the proportion of time a couple has spent apart since their last copulation. When a couple is apart, a man cannot know for certain his partner's activities. The time a woman spends apart from her partner may be time spent copulating with another man. A woman is increasingly likely to be inseminated by another man with an increasing proportion of time spent apart from her partner since the couple's last copulation (Baker & Bellis, 1995). We tested six predictions derived from the hypothesis that male psychology includes mechanisms designed to attend to the proportion of time a couple has spent apart since their last copulation as an indicator of the risk of sperm competition. We found support for four of these predictions. A man who spends a greater (relative to a man who spends a lesser) proportion of time apart from his partner since the couple's last copulation rates his partner as more attractive, reports that other men find his partner more attractive, reports greater interest in copulating with his partner, and reports that his partner is more interested in copulating with him.

These effects of the proportion of time spent apart since the couple's last copulation are independent of the total number of hours since the couple's last copulation. It is not, therefore, that these men are simply "sexually frustrated" or "sexually pent-up." Total time since the couple's last copulation has no independent predictive effect on any of the dependent variables we examined. A man's motivation to inseminate his partner apparently is not contingent on the time that has passed since he last inseminated his partner, once the proportion of time the couple has spent apart since their last copulation is taken into account. One possibility noted by Daly and Wilson (personal communication, September 2, 2001) for the null effects of total time since last copulation on the target variables is the presence of bi-directional effects between the variables. For a man who copulates with his partner relatively frequently, greater time since last copulation might produce an increase in copulatory interest in his partner, for example. For this same man, an effect opposite in causal direction and in sign also might be operating: Greater interest in copulating with his partner might cause him to initiate copulation with her sooner rather than later, resulting in a shorter duration of time since last copulation. Future research will need to assess a couple's copulatory frequency to investigate the possibility of these opposing effects.

The effects of the proportion of time spent apart since the couple's last copulation also are independent of the man's relationship satisfaction. Regardless of his satisfaction with the relationship, a man who has spent a greater proportion of time apart from his partner since the couple's last copulation appears to be motivated to inseminate his partner sooner than a man who has spent a greater proportion of time together with his partner since the couple's last copulation. Consistent with previous research (see Buss, 1994 for a partial review), a man's relationship satisfaction does independently predict five of the six dependent variables investigated. A man who is more satisfied with the relationship (relative to a man who is less satisfied with the relationship) rates his partner as more attractive, reports that other men find his partner more attractive, is more interested in copulating with his partner, reports that is partner is more interested in copulating with him, and is less distressed by his partner's denial of a request for copulation.

Two of the six predictions tested in this research were not supported. The proportion of time a man spent apart from his partner since the couple's last copulation does not predict his assessment of his partner's sexual interest in other men. The failure of this prediction may have occurred for several reasons. One possibility is that we did not test the prediction we intended to test. Perhaps, for example, we did not validly or reliably measure a man's assessments of his partner's sexual interest in other men. Rather than immediately discarding the failed prediction, it might be worth securing multiple assays of a man's assessments of his partner's sexual interest in other men. Assuming a composite measure that is valid and reliable, we can re-examine the independent predictive influence of proportion of time spent apart since the couple's last copulation on a man's assessments of his partner's sexual interest in other men. One other dependent measure, a man's assessments of his partner's sexual attraction to him, was a single-item measure. Although the prediction for this variable was supported, future work also should include multiple-item assessments of this variable.

An intriguing possibility noted by Martin Daly and Margo Wilson (personal communication, September 2, 2001) for the failure of Prediction 5 is that there may be bi-directional effects between the target variables. Prediction 5 stated that a man who spends a greater (relative to a man who spends a lesser) proportion of time apart from his partner since the couple's last copulation will provide higher ratings of his partner's sexual attraction to other men. Daly and Wilson suggested that an effect opposite in causal direction and in sign also might be operating, with the result that this prediction was not supported. According to Daly and Wilson, a man who provides higher ratings of his partner's sexual interest in other men might, as a consequence, engage in more intense mate guarding, with the result that the couple spends a smaller proportion of time apart since their last copulation. These two effects, opposite in causal direction and opposite in sign, might produce the null relationship identified in this research.

One other prediction failed. The proportion of time a man spent apart from his partner since the couple's last copulation does not predict his distress following his partner's denial of his request for copulation. A possible reason for the failure of this prediction is that we did not test it appropriately. The prediction states that a man who spends a greater (relative to a man who spends a lesser) proportion of time apart from his partner since the couple's last copulation will report greater distress following his partner's denial of a request for copulation, independent of the total time since last copulation and independent of the participant's relationship satisfaction. The test of the prediction, however, asked each man to *imagine* a scenario in which his partner denied his request for copulation. A better test of this prediction may require that a participant report his distress following his partner's most recent denial of a request for copulation since the couple's last copulation. Distress following a partner's *actual* denial of a request for copulation then could be regressed on the proportion of time spent apart since the couple's last copulation, total time since the couple's last copulation, and relationship satisfaction. This analysis strategy may have problems, however. The analysis would need to be restricted to men whose partners had denied them copulation in the interim since the last copulation. There might exist factors unknown to the researcher that differentiate a man who does and a man who does not report that his partner has denied him sexual access.

The present research has several important limitations. One limitation of this research is a sample limitation. The mechanisms that may have evolved in response to the adaptive problems of sperm competition are proposed to be universally present in male psychology. The present samples included mostly young men spending time at or around large universities in one of two Western countries. Future work must sample men from different socioeconomic, age, educational, ethnic, and cultural groups.

A second limitation of the present research is a design limitation. We secured one-shot assessments of feelings and desires that were predicted to covary with the proportion of time spent apart since the couple's last copulation. The nature of this design precluded a causal analysis of the relationships between the *risk* of sperm competition and the psychological *reactions* to risk of sperm competition. It would be valuable to assess these relationships over an extended period of time, such as in a one-month daily diary study. A daily diary study of couples involved in a committed sexual relationship would allow for a causal analysis of predicted relationships such as the relationship between the proportion of time the couple has spent apart since their last copulation and a man's interest in copulating with his partner.

This research can be extended in several different directions. One such direction, for example, is identifying whether mate retention tactics used by a man in a committed relationship are linked to characteristics of each partner and to characteristics of the relationship (Buss, 1988; Buss & Shackelford, 1997; Flynn, 1988). Buss and Shackelford (1997; see also Buss, 1988) identified 19 different tactics used by people to prevent their spouse from becoming romantically involved with someone else. These tactics varied from vigilance about a partner's whereabouts to violent reprisal for a suspected infidelity. Of particular interest for the present research is whether some of the more devastating male mate retention tactics, such as psychological abuse, physical violence, and sexual coercion, may be linked to the risk of sperm competition. Identifying such links may be a first step to helping women who suffer at the hands of abusive and controlling men (see Daly & Wilson, 1988; Dobash & Dobash, 1979; Dutton, 1995; Jacobson & Gottman, 1998).

In conclusion, this research suggests that human male psychology may include mechanisms designed to solve the adaptive problems of sperm competition. We are not aware of a theory other than Sperm Competition Theory that can account for the predictive utility of the proportion of time spent apart since the couple's last copulation, independent of the total time since last copulation and independent of relationship satisfaction. This research suggests that male humans, like the males of other socially monogamous but not sexually exclusive species, may have psychological mechanisms that are designed to solve the adaptive problem of a partner's sexual infidelity.

## ACKNOWLEDGEMENTS

The authors thank Dave Bjorklund, David Buss, Bram Buunk, Martin Daly, Todd DeKay, Barry Friedman, Steve Gangestad, Martie Haselton, Erika Hoff, Lee Kirkpatrick, Craig LaMunyon, Rick Michalski, Nick Pound, Percy Rhode, Virgil Sheets, Randy

Thornhill, Margo Wilson, and an anonymous reviewer for many helpful suggestions that improved this article.

## REFERENCES

Baker, R. R. & Bellis, M. A. (1988). "Kamikaze" sperm in mammals? *Animal Behaviour,* **36**, 937–980.

Baker, R. R. & Bellis, M. A. (1989*a*). Number of sperm in human ejaculates varies in accordance with sperm competition theory. *Animal Behaviour,* **37**, 867–869.

Baker, R. R. & Bellis, M. A. (1989*b*). Elaboration of the kamikaze sperm hypothesis: A reply to Harcourt. *Animal Behaviour,* **37**, 865–867.

Baker, R. R. & Bellis, M. A. (1993*a*). Human sperm competition: Ejaculate adjustment by males and the function of masturbation. *Animal Behaviour,* **46**, 861–885.

Baker, R. R. & Bellis, M. A. (1993*b*). Human sperm competition: Ejaculate manipulation by females and a function for the female orgasm. *Animal Behaviour,* **46**, 887–909.

Baker, R. R. & Bellis, M. A. (1995). *Human sperm competition*. London: Chapman & Hall.

Bellis, M. A. & Baker, R. R. (1990). Do females promote sperm competition: Data for humans. *Animal Behavior,* **40**, 197–199.

Birkhead, T. R. & Møller, A. P. (1992). *Sperm competition in birds*. London: Academic Press.

Burke, T., Davies, N. B., Bruford, M. W. & Hatchwell, B. J. (1989). Parental care and mating behaviour of polyandrous dunnocks, *Prunella modularis*, related to paternity by DNA fingerprinting. *Nature,* **338**, 249–251.

Buss, D. M. (1989). Conflict between the sexes: Strategic interference and the evocation of anger and upset. *Journal of Personality and Social Psychology,* **56**, 735–747.

Buss, D. M. (1994). *The evolution of desire*. New York: Basic Books.

Buss, D. M. & Shackelford, T. K. (1997*a*). From vigilance to violence: Mate retention tactics in married couples. *Journal of Personality and Social Psychology,* **72**, 346–361.

Buss, D. M. & Shackelford, T. K. (1997*b*). Human aggression in evolutionary psychological perspective. *Clinical Psychology Review,* **17**, 605–619.

Buss, D. M. & Shackelford, T. K. (1997*c*). Susceptibility to infidelity in the first year of marriage. *Journal of Research in Personality,* **31**, 193–221.

Daly, M. & Wilson, M. (1988). *Homicide*. Hawthorne, NY: Aldine de Gruyter.

Davies, N. B. (1992). *Dunnock behaviour and social evolution*. Oxford: Oxford University Press.

Dobash, R. E. & Dobash, R. P. (1979). *Violence against wives*. New York: Free Press.

Dutton, D. G. (1995). *The domestic assault of women* (rev. ed.). Vancouver: University of British Columbia Press.

Flinn, M. V. (1988). Mate guarding in a Caribbean village. *Ethology and Sociobiology,* **9**, 1–28.

Gage, M. J. G. (1995). Continuous variation in reproductive strategy as an adaptive response to population density in a moth, *Plodia interpunctella*, *Proceedings of the Royal Society of London: B,* **261**, 25–30.

Gangestad, S. W. & Thornhill, R. (1997). The evolutionary psychology of extrapair sex: The role of fluctuating asymmetry. *Evolution and Human Behavior,* **18**, 69–88.

Gangestad, S. W. & Thornhill, R. (1998, July). *The scent of symmetry: Evidence for a human sex pheromone.* Paper presented at 10th Annual Human Behavior and Evolution Society Conference, Davis, CA.

Jacobson, N. & Gottman, J. (1998). *When men batter women*. New York: Simon & Schuster.

Parker, G. A. (1970*a*). Sperm competition and its evolutionary consequences in the insects. *Biological Review,* **45**, 525–567.

Parker, G. A. (1970*b*). The reproductive behaviour and the nature of sexual selection in *Scatophaga stercoraria* V. The female's behaviour at the oviposition site. *Behaviour,* **37**, 140–168.

Parker, G. A. (1984). Sperm competition and the evolution of animal mating strategies. In R. L. Smith (Ed.), *Sperm competition and the evolution of animal mating systems* (pp. 1–60). London: Academic Press.

Pound, N. (1998, July). *Polyandry in contemporary pornography*. Paper presented at 10th Annual Human Behavior and Evolution Society Conference, Davis, CA.

Shackelford, T. K. & Buss, D. M. (1997*a*). Anticipation of marital dissolution as a consequence of spousal infidelity. *Journal of Social and Personal Relationships,* **14**, 793–808.

Shackelford, T. K. & Buss, D. M. (1997*b*). Spousal esteem. *Journal of Family Psychology,* **11**, 478–488.

Shackelford, T. K. & Buss, D. M. (2000). Marital satisfaction and spousal cost-infliction. *Personality and Individual Differences,* **28**, 917–928.

Singh, D., Meyer, W., Zambarano, R. J. & Hulbert, D. F. (1998). Frequency and timing of coital orgasm in women desirous of becoming pregnant. *Archives of Sexual Behavior,* **27**, 15–29.

Smith, R. L. (1984*a*). (Ed.). Sperm competition and the evolution of animal mating systems. New York: Academic Press.

Smith, R. L. (1984*b*). Human sperm competition. In R. L. Smith (Ed.), *Sperm competition and the evolution of animal mating systems* (pp. 601–660). New York: Academic Press.

Tabachnick, B. G. & Fidell, L. S. (2001). *Using multivariate statistics* (4th ed.). Boston: Allyn & Bacon.

Thornhill, R. & Alcock, J. (1983). *Evolution of insect mating systems.* Cambridge, MA: Harvard University Press.

Thornhill, R., Gangestad, S. W. & Comer, R. (1995). Human female orgasm and mate fluctuating asymmetry. *Animal Behaviour,* **50**, 1601–1615.

Trivers, R. L. (1972). Parental investment and sexual selection. In B. Campbell (Ed.), *Sexual selection and the descent of man* (pp. 139–179). London: Aldine.

# 14. SEMEN DISPLACEMENT AS A SPERM COMPETITION STRATEGY IN HUMANS

Gordon G. Gallup, Jr.[1], and Rebecca L. Burch[2]

## ABSTRACT

We examine some of the implications of the possibility that the human penis may have evolved to compete with sperm from other males by displacing rival semen from the cervical end of the vagina prior to ejaculation. The semen displacement hypothesis integrates considerable information about genital morphology and human reproductive behavior, and can be used to generate a number of interesting predictions.

## THE HUMAN PENIS AS A SEMEN DISPLACEMENT DEVICE

The penis evolved as an internal fertilization device. There are, however, striking differences in penis morphology between different species (see Birkhead, 2000). In addition to the ostensible impact of female choice on the evolution of more elaborate male genitalia (Eberhard, 1996), there is reason to believe that sperm competition played a role in shaping the human penis. The human penis, with a relatively larger glans and more pronounced coronal ridge than is found in many other primates, may function to displace seminal fluid from rival males in the vagina by forcing it back over/under the glans. During intercourse the effect of repeated thrusting would be to draw out and displace foreign semen away from the cervix. As a consequence, if a female copulated with more than one male within a short period of time this would allow subsequent males to "scoop out" semen deposited by others before ejaculating (Baker and Bellis, 1995).

To test this hypothesis, Gallup, Burch, Zappieri, Parvez, Stockwell, and Davis (2003) simulated sexual encounters using artificial models and measured the magnitude of

---

[1] Department of Psychology, State University of New York at Albany.

[2] Department of Psychology, State University of New York at Oswego.

Reprinted from *Evolutionary Psychology,* 2, Gallup, G. G. Jr., and Burch, R. L. Semen Displacement as a Sperm Competition Strategy in Humans, pp. 12–23, Copyright (2004) with permission from Evolutionary Psychology.

artificial semen displacement as a function of phallus configuration, depth of thrusting, and semen viscosity. The displacement of simulated semen was robust across different prosthetic phalluses, different artificial vaginas, different semen recipes, and different semen viscosities. The magnitude of semen displacement was directly proportional to the depth of thrusting and inversely proportional to semen viscosity. By manipulating different characteristics of artificial phalluses, the coronal ridge and frenulum were identified as key morphological features involved in mediating the semen displacement effect.

Under conditions that raise the possibility of females engaging in extra-pair copulations (i.e., periods of separation from their partner, allegations of female infidelity), Gallup *et al.* (2003) also found that males appear to modify the use of their penis in ways that are consistent with the displacement hypothesis. Based on anonymous surveys of over 600 college students, many sexually active males and females reported deeper and more vigorous thrusting when in-pair sex occurred under conditions related to an increased likelihood of female infidelity.

## IMPLICATIONS FOR SPECIES DIFFERENCES IN PENIS LENGTH AND MORPHOLOGY

Semen displacement as a means of competing with sperm from rival males is not an uncommon strategy in animals. The males of some species possess penile barbs, hooks, combs, or a textured glans to remove copulatory plugs and semen from the female reproductive tract (for a review see Baker and Bellis, 1995). In humans, the distinctive characteristics of the penis, relative to other primates, are its length, circumference, glans, and coronal ridge. In order for the human penis to serve as an efficient semen displacement device, it needs to be of sufficient size to fill the vagina and supplant foreign semen. The typical erect human penis ranges from 127mm to 178mm in length (Masters and Johnson, 1966), with an average circumference of 24.5mm (Wessells, Lue, and McAninch, 1996). In contrast with our closest living relative, the human penis is roughly twice as long and wide as that of the common chimpanzee (Short, 1980). The glans and coronal ridge of the human penis are also uniquely configured (Izor, Walchuk and Wilkins, 1981). The posterior portion of the human glans is larger in diameter than the penis shaft, and at the interface between the glans and the shaft the coronal ridge is positioned perpendicular to the shaft. Common chimpanzees have no clearly differentiated glans or coronal ridge (Kinzey, 1974).

As evidence that the human penis may have been shaped by the recurrent adaptive problem posed by sperm competition consider the following. Magnetic resonance imaging studies show that during coitus, the typical penis fills and expands the human vagina, and with complete penetration often pushes up against the cervix (Weijmar Schultz, van Andel, Sabelis, and Mooyart, 1999). When ejaculation occurs, thrusting diminishes and vaginal penetration reaches its maximum point (Masters and Johnson, 1966). Not only does this serve to release semen in close proximity to the cervix, but data on ejaculatory pressure shows that the first several ejaculatory contractions project seminal fluid with such force that it can be expelled at a distance of 30–60cm if not contained in a vagina (Masters and Johnson, 1966). Thus, there appear to have been a series of adaptations that serve to confine or focus the release of semen to the uppermost portion of the vaginal tract, possibly as a means of making it less vulnerable to displacement by other males. A longer penis would not only have been an advantage for leaving semen in a less accessible part of the vagina, but by filling and expanding the vagina it also would aid and abet the displacement of semen left by other males as a means of maximizing the likelihood of paternity.

In addition to competing with sperm from rival males, there may be other benefits of deep semen placement. In contrast to organisms that walk on all fours, the assumption of an upright posture and the emergence of bipedalism brought the human female reproductive tract, and the vagina in particular, into a perpendicular orientation with gravity that is poorly suited to semen retention. Copulation in the ventral-ventral mode with the female in a supine position brings the female reproductive tract back into a more primitive parallel orientation with gravity, and enhances the likelihood that semen will be retained. However, due to the effects of gravity, the resumption of an upright posture following coitus has the potential to endanger semen retention. Consistent with this hypothesis, there are several mechanisms that appear to postpone getting up after a sexual encounter, such as post-copulatory petting, patterns of nocturnal copulation, and the sedative-like effects of orgasm (Gallup and Suarez, 1983). Likewise, a long penis that provides for the release of semen deep in the vagina could also serve as a hedge against semen loss.

## ORDINAL EJACULATION EFFECTS

It follows from the semen displacement hypothesis that sperm competition among humans ought to involve ordinal ejaculation/mating order effects. Under conditions in which several males copulate with a female in close temporal proximity to one another the male who mated with the female last would have an advantage, known as last male precedence (see Birkhead, 2000). In the case of humans, the last male to copulate would be in a position to displace the semen left by previous males before inseminating the female with his own semen. This assumes, of course, that other factors such as sperm quality, ejaculate size, penis length, and female control of mating are constant. As yet, little research has been conducted on this topic, but in the sections that follow we elaborate a number of potentially testable implications of the semen displacement hypothesis.

## DOUBLE MATING

For semen displacement to be adaptive it presupposes situations in which human females have sex with multiple (two or more) males in fairly close succession/temporal proximity to one another. Situations that satisfy this criterion include 1) consensual sex with multiple concurrent partners, 2) nonconsensual sex with multiple concurrent partners, and 3) multiple successive consensual and/or nonconsensual sexual encounters that occur within a relatively brief period of time. Examples include, group sex, gang rape, extra pair copulations, promiscuity, prostitution, and resident male insistence on sex in response to suspected infidelity (see subsequent section on female reproductive strategy). Instances of human heteroparity, or heteropaternal superfecundation, where members of a pair of fraternal twins are actually half sibs as a consequence of being conceived by different fathers, are well documented (e.g., Ambach, Parson, and Brezinka, 2000; Wenk, Houtz, Brooks, and Chiafari, 1992), and testify to the existence of double mating by females. It is also worth noting that patterns of consensual concurrent mating with multiple males by female chimpanzees, our closest living relatives, are common (Tutin, 1979), and as such may have been prevalent during earlier phases of human evolutionary history as well.

## EFFECTS OF CIRCUMCISION

What effect, if any, does the practice of surgically removing the foreskin have on the semen displacement properties of the human penis? For intact males, when the penis is fully erect the foreskin is pulled back over the glans and down the shaft of the penis. As a result, whether a man has been circumcised is often only apparent when the penis is flaccid. However, depending on how thick the foreskin is and how far it extends over the end of the glans, circumcision could affect the magnitude of semen displacement. During circumcision the foreskin is cut away from the shaft of the penis immediately behind the glans (Holman and Stuessi, 1999). As a consequence of removing the foreskin the circumference of the shaft posterior to the glans may be slightly reduced, causing the coronal ridge to be more pronounced and creating a larger area for semen to collect where it could be scooped back away from the cervix.

Laumann, Masi, and Zuckerman (1997) found that circumcised men masturbate more often and engage in more elaborate sexual behaviors. Anecdotal reports of adult circumcision by affected males and their partners also suggest that the procedure leads to changes in sexual behavior. Money and Davison (1983) and Fink, Carson, and De Vellis (2002) found that among males who underwent circumcision as adults, the majority reported a loss of penile sensitivity and a prolongation of sexual intercourse, and some reported less sexual gratification.

Anecdotal reports from females also bear on the semen displacement properties of the circumcised penis. In a study of 139 women who had experienced intercourse with a number of both circumcised and uncircumcised partners, O'Hara and O'Hara (1999) found most (73%) reported that circumcised men thrust harder and deeper, and used more elongated strokes than their uncircumcised counterparts. The majority of the respondents preferred sex with uncircumcised males, citing greater displacement of vaginal secretions and resulting vaginal dryness, increased friction, and physical discomfort during intercourse with men that were circumcised. Among the minority of respondents who preferred circumcised partners ($N = 20$), the most common reason given was prolonged intercourse. But complaints about the loss of vaginal secretions, friction, and discomfort were still prevalent in this group. Perhaps due to reduced penile sensitivity, circumcised men thrust deeper and withdraw farther and thereby displace more vaginal fluids. O'Hara and O'Hara conclude that the loss of vaginal lubrication and discomfort is "because of the tight penile skin, the corona of the glans, which is configured like a one way valve, pulls the vaginal secretions out of the vagina when the shaft is withdrawn" (p. 82).

Therefore, although practiced primarily for religious and/or hygienic reasons, an unintended consequence of circumcision may be to enhance the semen displacement properties of the human penis.

Another, albeit indirect, way to examine differences in the effectiveness of semen displacement would be to compare the incidence of cuckoldry between men that have and have not been circumcised. It follows from the displacement hypothesis that the risk of extra-pair paternity might vary with variation in semen displacement effectiveness. In addition to DNA testing to assess paternity, a less obtrusive (but far from perfect) measure of paternity would be to examine the extent to which children ostensibly sired by circumcised males exhibit a higher incidence of paternal resemblance (Platek, Burch, Panyavin, Wasserman, and Gallup, 2002).

## THE PIGGY BACKING HYPOTHESIS

Another intriguing implication of the difference between circumcised and intact males is the question of self-cuckoldry. Put another way, is it possible (short of artificial insemination) for a women to become pregnant by a man she never had sex with? We think the answer is "yes."

If an uncircumcised man (Male B) were to have sex with a women (Female A) who recently had sex with another man (Male A), in the process of thrusting his penis back and forth in her vagina some of Male A's semen would be forced under Male B's frenulum, collect behind his coronal ridge, and be displaced from the area proximate to the cervix. After Male B ejaculates and substitutes his semen for that of the other male, as he withdraws from the vagina some of Male A's semen will still be present on the shaft of his penis and behind his coronal ridge (see Ordinal Ejaculation Effects above). As his erection subsides the glands penis will withdraw under the foreskin, raising the possibility that some of Male A's semen could be captured underneath the foreskin and behind the coronal ridge in the process. Were Male B to then have sex with Female B several hours later, it is possible that some of the displaced semen from Male A would still be present under his foreskin and thus may be unwittingly transmitted to Female B who, in turn, could then be impregnated by Male A's sperm. Were Male B circumcised, this would be a far less likely outcome because the residual foreign sperm on his penis would not be afforded the protection by the foreskin from desiccation, light, and cooling and would likely perish during the interim separating sexual encounters with different partners.

The transfer of another male's semen from one female to the next as a consequence of genitals specialized for sperm removal, is also known as fertilization by proxy, and has been documented in insects (Haubruge, Arnaud, Mignon, and Gage, 1999). If the foreskin makes the human penis a vector for fertilization by proxy, why is the foreskin still there? We assume that during human evolutionary history the incidence of self-cuckoldry was not high enough to offset either the advantages of semen displacement or the advantages of the foreskin, which affords protection of the glans. Indeed, it is possible that the adaptive problems posed by the existence of piggybacking semen from rival males led to compensatory adaptations that incapacitate foreign sperm. For example, it would be interesting to determine if smegma, a glandular discharge that collects under the foreskin and lubricates the glans, has spermicidal properties. Alternatively, because smegma tends to be sticky and viscous it may entrap piggybacking sperm from rival males and minimize self-cuckoldry.

## ADAPTATIONS TO SELF-SEMEN DISPLACEMENT

A potential pitfall of the displacement hypothesis is the problem posed by self-semen displacement. If the human penis evolved to displace semen left by other males, what is to prevent this adaptation from displacing the male's own semen? The data derived from artificial genitals (Gallup *et al.*, 2003) strongly suggest that continued thrusting beyond the point of ejaculation would lead to displacement of the male's own semen. Therefore, the tenability of the displacement hypothesis is predicated on identifying putative collateral mechanisms that serve to minimize the likelihood of self-semen displacement.

Obvious candidate mechanisms that appear to preclude or at least diminish self-semen displacement include the following post-ejaculatory changes: 1) penile hypersensitivity, 2) loss of an erection, and 3) the refractory period. Due to ensuing penile hypersensitivity, continued thrusting for many males can become unpleasant and even mildly aversive following ejaculation (e.g., Aversa, Mazzilli, Rossi, Delfino, Isidori, and Fabbri, 2000). Post-ejaculatory thrusting may also be diminished as a consequence of an inability to sustain an erection. Typically within the first minute after ejaculation half of the erection is lost, and many males experience complete penile tumescence (Byer, Shainberg and Galliano, 1999). The refractory period, as measured by the inability to achieve another erection following ejaculation, varies with age, lasts from 30 minutes to 24 hours (Rathus, Nevid and Fichner Rathus, 2000), and also qualifies as an obvious adaptation that would serve to minimize self-semen displacement. It is interesting that the "Coolidge effect" as measured by an abbreviated refractory period, is usually a consequence of an opportunity to mate with a different female, and as such precludes the problem of self-semen displacement.

As still another corollary adaptation to self-semen displacement, we predict that males who continue to thrust past the point of ejaculation will show post-ejaculatory thrusting that is noticeably shallower and less vigorous. In contrast to deep thrusting, Gallup *et al.* (2003) found that shallow thrusting with prosthetic genitals failed to produce semen displacement.

Finally, one practical implication of this analysis of self-semen displacement would be to advise couples with infertility problems to refrain from engaging in post-ejaculatory thrusting. Indeed, it would be interesting to see if patterns of robust post-ejaculatory thrusting are more common among couples experiencing fertility problems.

## SEMEN COAGULATION

Semen coagulates within seconds after ejaculation and then liquefies or decoagulates about 15–30 minutes later (Mandal and Bhattacharyya, 1985; Robert and Gagnon, 1999). While semen hyperviscosity is associated with infertility (Gonzales, Kortebani and Mazzolli, 1993), the first part of the ejaculate does not typically coagulate, only the last fraction (Baker and Bellis, 1995). Baker and Bellis speculate that this keeps the semen in place while sperm travel to the cervix, and at the same time prevents the passage of rival sperm from subsequent males.

As evidence that semen coagulation may have emerged (in part) as a sperm competition tactic that functions to block sperm from rival males, Dixon and Anderson (2002) examined semen coagulation and copulatory plugs in 40 species of primates. Coagulation rates were highest in species where females commonly mate with multiple partners, and lowest in those where females are primarily monogamous or belong to polygynous groups. Likewise, Mandel and Bhattacharyya (1986) measured semen coagulation in humans, and found that if the male had not ejaculated in the previous two days liquefication times were significantly decreased. Thus, by implication, men who copulate frequently (which may include multi-partner matings) deposit semen that coagulates for longer periods of time.

The data derived from artificial genitals (Gallup *et al.*, 2003) also show that viscous semen is more difficult to displace, and as a consequence another function of semen coagulation may be to minimize self-semen displacement and/or displacement by other males.

## IMPLICATIONS FOR PREMATURE EJACULATION

The latency between insertion of the penis into the vagina and the occurrence of ejaculation in humans ranges from 2 minutes to an hour (Michael, Gagnon, Laumann, and Kolata, 1994). The average duration of coitus is 7.9 minutes (Grenier and Byers, 2001), with 100 to 500 thrusts per encounter (Hrdy and Whitten, 1987). Premature ejaculation is one of the most common forms of male "sexual dysfunction," affecting as many as one in four men (Laumann, Gagnon, Michael, and Michaels, 1994). Premature ejaculation takes two forms. The least common is when the male ejaculates prior to achieving intromission. The other is when ejaculation occurs upon or shortly after insertion of the penis into the vagina. Men who suffer from this form of premature ejaculation have an average ejaculation latency of 1.1 minutes following intromission (Spiess, Geer and O'Donohue, 1984).

Ejaculation that occurs outside the vagina is reproductively nonfunctional. However, after achieving intromission, selection may have operated to minimize the amount of time it takes to inseminate a female; i.e., premature ejaculation may have been ancestrally adaptive. Indeed, one article that champions this hypothesis is entitled "Survival of the Fastest" (Hong, 1984). This thesis is based on the fact that there are a number of potential costs associated with extended bouts of copulation. For example, the longer it takes to ejaculate the greater the risk of predation, the greater the likelihood of detection by jealous mates or offended kin, and the more one might have to contend with competition or interference from other sexually aroused males.

On the other hand, these potential costs have to be weighed against the existence of several compensatory benefits that might accrue to sexual encounters of longer duration. As the duration of coitus increases, the likelihood of female orgasm also increases, and it has been theorized that the vaginal and uterine contractions that accompany orgasm in females may be conducive to sperm uptake, transport, and retention (Baker and Bellis,

1993). From the standpoint of sperm competition, another benefit of extended periods of copulation would be more effective displacement of rival semen from the female reproductive tract. Indeed, premature ejaculation can be thought of as a failure to achieve semen displacement. This line of reasoning leads us to predict that among males with premature ejaculation, jealousy induction procedures (such as watching pornography which features infidelity) might antagonize such symptoms.

It is interesting that Spiess, Geer, and O'Donohue (1984) found men who suffer from premature ejaculation had fewer sexual encounters. Indeed, the longer they went without intercourse, the more prone they were to premature ejaculation. Thus, sexually disadvantaged males appear to be at greater risk of premature ejaculation. Perhaps premature ejaculation functions as an adaptive mechanism that enables subordinate males to minimize the risk of detection and retaliation by dominant/rival males during opportunistic sexual encounters.

## IMPACT OF SEMEN DISPLACEMENT ON FEMALE REPRODUCTIVE STRATEGY

On the basis of retrospective reports from both males and females, Gallup *et al.* (2003) found that human males often modify the use of their penis under conditions in which their long-term female partner may have been unfaithful. Using anonymous surveys of sexually active college students, it was determined that when males accused their partner of cheating, or when the couple had been separated for a period of time, many males thrust deeper, quicker, and more vigorously than during a typical sexual encounter.

Like many other features of the displacement hypothesis, this has interesting implications for future research. For example, we would predict that the sexual behavior of chronically or pathologically jealous men would feature strategies that might produce greater displacement; e.g., deeper and more vigorous thrusting. It would also be interesting to determine whether displacement behaviors are sensitive to contextual cues. Are there fluctuations in displacement behaviors that reflect cyclic changes in female receptivity? Do parameters of female attractiveness (facial features, waist to hip ratio, age, fecundity) affect displacement behaviors? Does variation in paternal resemblance among a male's ostensible offspring affect semen displacement behavior?

If females use extra-pair copulations with alpha males to cuckold their mates, the resident male's capacity for competing with and displacing the interloper's semen puts the reproductive best interests of the resident male and female at odds with one another. The effectiveness of sperm competition strategies in general, and semen displacement in particular, is time dependent (i.e., related to the elapsed time since the extra-pair copulation). Therefore, if semen displacement and other sperm competition strategies have been featured prominently during human evolutionary history, we would expect the desired timing of in-pair copulations by males and females following a female extra-pair encounter to be very different. Other things being equal, the effectiveness of sperm competition ought to be inversely proportional to the amount of time that has elapsed since insemination by the extra-pair male. Therefore, following an extra-pair encounter we would expect females to attempt to postpone copulation with the resident male, as an evolutionary strategy for minimizing sperm competition and increasing the likelihood of impregnation by the extra-pair male. Just the opposite strategy would hold true for males. We predict resident males who have reason to suspect female infidelity would attempt copulation immediately with their partner following a possible extra-pair encounter, as an adaptation to displacing and substituting their semen for the interloper's. In support of both these predictions, Goetz and Shackelford (in press) have uncovered preliminary evidence that a substantial proportion of wife rapes involve husbands who suspect their wives had been unfaithful.

Thus, as illustrated by the different predictions we derive in this paper, the possibility that semen displacement may function as a sperm competition strategy among human males serves to both integrate many diverse features of human sexuality, and it can be used to generate a number of testable hypotheses.

## ACKNOWLEDGEMENTS

The authors thank Todd K. Shackelford and Aaron Goetz for helpful comments on an earlier draft of this paper.

## REFERENCES

Ambach, E., Parson, W., & Brezinka, C. (2000). Superfecundation and dual paternity in a twin pregnancy ending with placental abruption. *Journal of Forensic Science*, **45**, 181–183.

Aversa, A., Mazzilli, F., Rossi, T., Delfino, M., Isidori, A. M., & Fabbri, A. (2000). Effects of Sildenafil (Viagra) administration on seminal parameters and post-ejaculatory refractory time in normal males. *Human Reproduction,* **15**, 131–134.

Baker, R. R., & Bellis, M. A. (1993). Human sperm competition: Ejaculation manipulation by females and a function for the female orgasm. *Animal Behavior*, **46**, 887–909.

Baker, R. R., & Bellis, M. A. (1995). *Human sperm competition: Copulation, Masturbation, and Infidelity*. Chapman and Hall, London.

Birkhead, T. R. (2000). *Promiscuity: An Evolutionary History of Sperm Competition*. Harvard University Press, Cambridge.

Busse, C. D., & Estep, D. Q. (1984). Sexual arousal in male pigtailed monkeys (*Macaca nemestrina*): Effects of serial matings by two males. *Journal of Comparative Psychology,* **98**, 227–231.

Byer, C. O., Shainberg, L. W., & Galliano, G. (1999). *Dimensions of Human Sexuality*. Boston, McGraw Hill.

Dixson, A. L., & Anderson, M. J. (2002). Sexual selection, seminal coagulation and copulatory plug formation in primates. *Folia Primatologica; International Journal of Primatology*, **73**, 63–69.

Fink, K. S., Carson, C. C., & DeVellis, R. F. (2002). Adult circumcision outcomes study: Effect on erectile function, penile sensitivity, sexual activity and satisfaction. *Journal of Urology,* **167**, 2113–2116.

Gallup, G. G. Jr., Burch, R. L., Zappieri, M. L., Parvez, R., Stockwell, M., & Davis, J. A. (2003). The human penis as a semen displacement device. *Evolution and Human Behavior,* **24**, 277–289.

Gallup, G. G., Jr., & Suarez, S. D. (1983). Optimal reproductive strategies for bipedalism. *Journal of Human Evolution,* **12**, 193–196.

Gonzales G. F., Kortebani G., & Mazzolli A. B. (1993). Hyperviscosity and hypofunction of the seminal vesicles. *Archives of Andrology,* **30**, 63–68.

Grenier, G., & Byers, E. S. (2001). Operationalizing premature or rapid ejaculation. *Journal of Sex Research,* **38**, 369–378.

Haubruge, E., Arnaud, L., Mignon, J. and Gage, M. J. G. (1999). Fertilization by proxy: Rival sperm removal and translocation in a beetle. *Proceedings of the Royal Society of London B*, **266**, 1183–1187.

Holman, J. R., & Stuessi, K. A. (1999). Adult circumcision. *American Family Physician,* **1559**, 1514–1518.

Hong, L. K. (1984). Survival of the fastest: On the origin of premature ejaculation. *The Journal of Sex Research*, **20**, 109–122.

Hrdy, S. B., & Whitten, P. L. (1987). Patterning of sexual activity. In Smuts, B. B., Cheney, D. L., & Seyfarth, R. M. (Eds) *Primate Societies* (pp. 370–384). University of Chicago Press, London.

Izor, R., Walchuk, S., & Wilkins, L. (1981). Anatomy and systematic significance of the penis of the pygmy chimpanzee, *Pan paniscus*. *Folia Primatologica,* **35**, 218–224.

Kinzey, W. G. (1974). Male reproductive systems and spermatogenesis. *Comparative Reproduction of Nonhuman Primates*, Academic Press, London, pp. 85–114.

Kvist, U. (1991). Can disturbances of the ejaculatory sequence contribute to male infertility? *International Journal of Andrology*, **14**, 389–393.

Laumann, E. O., Gagnon, J. H., Michael, R. T., & Michaels, S. (1994). The social organization of sexuality: Sexual practices in the United States. Chicago: University of Chicago Press.

Laumann, E. O., Masi, C. M., & Zuckerman, E .W. (1997). Circumcision in the United States: Prevalence, prophylactic effects, and sexual practice. *Journal of the American Medical Association*, **277**, 1052–1057.

Mandal A., & Bhattacharyya A. K. (1985). Physical properties and non-enzymic components of human ejaculates. Relationship to spontaneous liquefaction. *International Journal of Andrology,* **8**, 224–231.

Mandal, A., & Bhattacharyya, A. K. (1986). Grouping of human ejaculates according to the degree of coagulation and the relationship to the levels of choline and cholinesterase. *International Journal of Andrology,* **9**, 407–415.

Masters, W. H., & Johnson, V. E. (1966). *Human Sexual Response*. Little, Brown and Company, Boston.

Michael, R. T., Gagnon, J. H., Laumann, E. O., & Kolata, G. (1994). *Sex in America: A definitive survey.* Boston: Little, Brown.

Money, J. and Davison, J. (1983). Adult penile circumcision: Erotosexual and cosmetic sequelae. *Journal of Sex Research,* **19**, 289–292.

O'Hara, K., & O'Hara, J. (1999). The effect of male circumcision on the sexual enjoyment of the female partner. *British Journal of Urology, International*, **83**, 79–84.

Platek, S. M., Burch, R. L., Panyavin, I. S., Wasserman, B. H. and Gallup, G. G. Jr. (2002). Reactions to children's faces: Resemblance affects males more than females. *Evolution and Human Behavior*, **23**, 159–166.

Rathus, S. A., Nevid, J. S., & Fichner Rathus, L. (2000). *Human Sexuality in a World of Diversity*. Boston: Allyn and Bacon.

Robert, M., & Gagnon, C. (1999). Semenogelin I: a coagulum forming, multifunctional seminal vesicle protein. *Cellular and Molecular Life Sciences,* **55**, 944–60.

Short, R. V. (1979). Sexual selection and its component parts somatic and genital selection, as illustrated by man and the great apes. *Advances in the Study of Behavior*, **9**, 131–158.

Short, R. V. (1980). The origins of human sexuality. In Austin, C. R and Short, R. V. (Eds.) *Reproduction in Animals, Book 8, Human Sexuality*, Cambridge University Press, Cambridge.

Smith, R. L. (1984). *Sperm Competition and the Evolution of Animal Mating Systems*, Academic Press, London.

Spiess, W. F. J., Geer, J. H., & O'Donohue, W. T. (1984). Premature ejaculation: Investigations of factors in ejaculatory latency. *Journal of Abnormal Psychology*, **93**, 242–245.

Tutin, C. E. G. (1979). Mating patterns and reproductive strategies in a community of wild chimpanzees. *Behavioral Ecology and Sociobiology*, **6**, 29–38.

Weijmar Schultz, W., van Andel, P., Sabelis, I., & Mooyart, E. (1999). Magnetic resonance imaging of male and female genitals during coitus and female sexual arousal. *British Medical Journal*, **319**, 18–25.

Wenk, R. E., Houtz, T., Brooks, M., & Chiafari, F. A. (1992). How frequent is heteropaternal superfecundation? *Acta Geneticae Medicae et Gemellologiae*, **41**, 43–47.

Wessells, H., Lue, T. F., & McAninch, J. W. (1996). Penile length in the flaccid and erect states: Guidelines for penile augmentation. *Journal of Urology,* **156**, 995–997.

# 15. HUMAN FEMALE ORGASM AND MATE FLUCTUATING ASYMMETRY

Randy Thornhill[1*], Steven W. Gangestad[2], and Randall Comer[2]

## ABSTRACT

Human, *Homo sapiens*, female orgasm is not necessary for conception; hence it seems reasonable to hypothesize that orgasm is an adaptation for manipulating the outcome of sperm competition resulting from facultative polyandry. If heritable differences in male viability existed in the evolutionary past, selection could have favoured female adaptations (e.g. orgasm) that biased sperm competition in favour of males possessing heritable fitness indicators. Accumulating evidence suggests that low fluctuating asymmetry is a sexually selected male feature in a variety of species, including humans, possibly because it is a marker of genetic quality. Based on these notions, the proportion of a woman's copulations associated with orgasm is predicted to be associated with her partner's fluctuating asymmetry. A questionnaire study of 86 sexually active heterosexual couples supported this prediction. Women with partners possessing low fluctuating asymmetry and their partners reported significantly more copulatory female orgasms than were reported by women with partners possessing high fluctuating asymmetry and their partners, even with many potential confounding variables controlled. The findings are used to examine hypotheses for female orgasm other than selective sperm retention.

---

The human female orgasm has attracted great interest from many evolutionary behavioural scientists. Several hypotheses propose that female orgasm is an adaptation. First, human female orgasm has been claimed to create and maintain the pair bond between male and female by promoting female intimacy through sexual pleasure (e.g. Morris 1967; Eibl-Eibesfeldt 1989). Second, a number of evolutionists have suggested

[1] Department of Biology, University of New Mexico.

[2] Department of Psychology, University of New Mexico.

Reprinted from *Animal Behaviour,* 50 (6), Thornhill, R., Gangestad, S. W., & Comer, R. Human female orgasm and mate fluctuating asymmetry. pp. 1601–1615, Copyright (1995) with permission from Elsevier.

that human female orgasm functions in selective bonding with males by promoting affiliation primarily with males who are willing to invest time or material resources in the female (Alexander 1979; Alcock 1987) and/or males of genotypic quality (Smith 1984; Alcock 1987). Third, female orgasm has been said to motivate a female to pursue multiple males to prevent male infanticide of the female's offspring and/or to gain material benefits from multiple mates (Hrdy 1981). Fourth, Morris (1967) proposed that human female orgasm functions to induce fatigue, sleep and a prone position, and thereby passively acts to retain sperm.

Additional adaptational hypotheses suggest a more active process by which orgasm retains sperm. The 'upsuck' hypothesis proposes that orgasm actively retains sperm by sucking sperm into the uterus (Fox *et al.* 1970; see also Singer 1973). Smith (1984) modified this hypothesis into one based on sire choice; he argued that the evolved function of female orgasm is control over paternity of offspring by assisting the sperm of preferred sires and handicapping the sperm of non-preferred mates. Also, Baker & Bellis (1993; see also Baker *et al.* 1989) speculated that timing of the human female orgasm plays a role in sperm retention. Baker & Bellis (1993) showed that orgasm occurring near the time of male ejaculation results in greater sperm retention, as does orgasm up to 45 min after ejaculation. Orgasm occurring more than a minute before male ejaculation appears not to enhance sperm retention. Baker & Bellis (1993) furthermore argued that orgasms occurring at one time may hinder retention of sperm from subsequent copulations up to 8 days later.

In addition, a number of theorists have argued that human female orgasm has not been selected for because of its own functional significance and hence is not an adaptation, Rather, these theorists claim that female orgasm is an incidental by-product of male orgasm, which is an adaptation (e.g. Symons 1979; Gould 1987; Fox 1993).

The primary criterion by which features can be identified as adaptations is purposeful design (Williams 1966; Thornhill 1990). Is the feature designed to solve a particular problem posed by selective pressures? The functional hypotheses of human female orgasm suggest that it ought to have certain special design features. The by-product hypothesis predicts the absence of functional design in female orgasm. Human female orgasm clearly is not necessary for conception; hence, its function could not have arisen simply to ensure conception (Smith 1984; Baker & Bellis 1993). Perhaps the leading functional hypothesis is that human female orgasm is a female choice adaptation, designed to manipulate sperm competition and promote conception with males of high quality (Smith 1984; Baker & Bellis 1993). In hominid evolution, the primary context in which sperm competition has taken place is probably facultative polyandry, or copulation with an extra-pair male (Smith 1984). Female orgasm as a selective response thus may have evolved as a means by which females could favour sperm of an extra-pair male over that of an in-pair male, or vice versa.

Although extra-pair mating in humans may have evolved because of a number of potential benefits, a primary theory concerns good genes (Benshoof & Thornhill 1979; Smith 1984; Gangestad 1993). A woman who perceives her mate to be of low genetic quality may employ a strategy of garnering resources from her primary mate but having extra-pair sex with a male who is of higher genetic quality. These circumstances could have selected female design favouring retention of sperm from men who possess phenotypic markers of good genes. Baker & Bellis (1993) reported provisional support for these notions. In particular, women in multiple mating situations appear more likely to orgasm during copulation with extra-pair partners than with in-pair partners (see also Bellis & Baker 1990).

Currently, the leading theory of good genes sexual selection is the pathogen theory. Host-parasite coevolution maintains heritable pathogen resistance and, hence, viability in host populations. Thereby, sexual selection favours preferences for mates who possess honest indicators of pathogen resistance (Hamilton & Zuk 1982). A marker of male mating advantage in a variety of species may be fluctuating asymmetry, which is asymmetry of the two sides of bilateral characters (e.g. wings, fins, hands, feet, ears) for which the signed differences between the two sides have a population mean of zero and are normally distributed (Van Valen 1962). Because the two sides of such characters are not controlled by different genes, it is thought that fluctuating asymmetry represents imprecise expression of underlying developmental design because of developmental perturbations (developmental instability). Also, although a variety of factors can cause developmental perturbation (e.g. extreme environmental conditions, genetic mutations, toxins; see Parsons 1990), potentially important causes in natural populations are pathogens (e.g. Bailit *et al.* 1970; Parsons 1990; Møller 1992*a*). Within populations, fluctuating asymmetry can vary considerably across individual organisms. In a range of species, individuals' fluctuating asymmetry negatively predicts their fecundity, growth rate and survival (Mitton & Grant 1984; Palmer & Strobeck 1986; Parsons 1990; Thornhill 1992*a*,*b*; Thornhill & Sauer 1992). Reduced fluctuating asymmetry is associated with male mating success in a variety of species, including barn swallows, *Hirundo rustica* (Møller 1992*b*), scorpionflies, *Panorpa* spp. (Thornhill 1992*a*,*b*; Thornhill & Sauer 1992), *Drosophila* (Markow & Ricker 1992), non-human primates (Manning & Chamberlain 1993) and others (for reviews, see Møller & Pomiankowski 1993; Watson & Thornhill 1994). A meta-analysis of 29 studies of 13 species revealed a mean heritability of fluctuating asymmetry of 0.27 and an overall statistically significant effect size of 0.15 (Møller & Thornhill, in press). This heritability of fluctuating asymmetry is consistent with the good genes hypothesis of sexual selection.

Fluctuating asymmetry may be a marker of male mating success in humans as well. Men who possess low fluctuating asymmetry tend to be judged more attractive than other men (Gangestad *et al.* 1994; Grammer & Thornhill 1994; Thornhill & Gangestad 1994). Moreover, they tend to have had relatively many sexual partners (Thornhill & Gangestad 1994), more extra-pair sexual relations and have had sex with women after shorter courtship (i.e. period of 'dating'; Gangestad & Thornhill, in press). No evidence suggests that these men invest more in their relationships than do others and, in fact, in certain ways, they invest less. In particular, they appear to sexualize (e.g. flirt with) women other than their partner more than do other men (Gangestad & Thornhill, in press) and therefore may engage in greater efforts to mate outside the primary relationship.

If (1) low fluctuating asymmetry is a marker of male quality, independent of any investment he might provide, and (2) human female orgasm is a conditional response that has been designed to favour sperm of men who possess markers of quality, then women should orgasm during copulation with men who possess low fluctuating asymmetry more often than with men who do not. If Baker & Bellis (1993) are correct that orgasms occurring near the time of or after male ejaculation particularly favour sperm retention, then the frequency of female orgasms at these times should particularly correlate with partner fluctuating asymmetry. The current study attempts to test these predictions using data from reports of female orgasmic response from women and their male sexual partners. In examining the role of male fluctuating asymmetry in women's orgasms, we also consider a number of potential confounding variables (e.g. socioeconomic status, perceived future income, relationship investment behaviour, social potency, relationship length, women's and men's sexual experience and attitudes towards casual sex). Also, the data are used to examine the various functional hypotheses for female orgasm mentioned above.

## METHOD

### Subjects

Subjects were 86 heterosexual adult couples involved in sexually active, heterosexual romantic relationships. Subjects were recruited from introductory psychology courses at the University of New Mexico. This source is often used in scientific studies conducted in academic psychology research departments. At the University of New Mexico, introductory psychology students must meet a research requirement, and one way to meet the requirement is to be a subject in studies of the student's own choosing. Our subjects were self-selected, therefore, for participation in a study of physical features and romantic relationships for individuals who had been romantically involved for 1 month or more. Furthermore, they knew at the time they signed up for our study that they would be asked questions about sexual behaviour and history on a questionnaire. Average (± s.d.) ages of men and women were 21.38 ± 3.66 and 20.28 ± 3.38, respectively (ranges 17–40 and 17–37). Relationships had a mean duration of 24 ± 20 months (range 2–108). Subjects were 53% Caucasian ($N$ = 46), 36% Hispanic ($N$ = 31), 6% native American ($N$ = 5), 2% Black ($N$ = 2), 1% Asian ($N$ = 1), and 1% other ($N$ = 1). Eight couples were married, four had had children together, and another four men and two women had had children with other partners. (An additional 18 couples were used in this study, but either had not engaged in sexual intercourse or did not fully complete the orgasm questionnaire; see Procedure.)

### Procedure

Subjects reported for a questionnaire study in groups of one to four couples. Our study was approved by the Human Subjects Committee of the University of New Mexico. We guaranteed anonymity of their responses on the questionnaire, including that their partner would not know any of their responses. We told subjects of the sensitive nature of some of the questions (e.g. sexual history) and that if for any reason they wished to not answer certain questions, they could do so without penalty. Of individuals in couples who claimed that the relationship involved sexual intercourse, 100% of the women and 99% of the men answered all questions about actual female orgasm during intercourse. All of the women and 99% of the men answered a question about faked female orgasm. After reading and signing a consent form, each subject was escorted to his or her own room where they privately completed the questionnaire. Meanwhile, the experimenters (two at each data-collection session) interrupted each individual, one at a time and took him or her to a different room where they measured fluctuating asymmetry and took a photograph. Subjects did not see their partner until both partners had turned in their questionnaires. After measurement and questionnaire completion, we told subjects the purposes, measures and procedures of the study, and answered subjects' questions. Finally, they were told that they could return the next semester after results were analysed to receive an overview of the patterns in the data, though they would not be told of their individual results.

## Questionnaire measures

We recorded the following measures.

(1) Basic information sheet, including age, height, weight, ethnicity and socioeconomic status of home of origin (upper class, upper-middle class, middle class, lower-middle class, lower class), marital status, duration of current relationship, number of offspring, and the number of offspring with the current partner. This basic information sheet also asked whether the subject had ever broken a foot, ankle, hand, wrist or elbow, or sprained any of these body parts, within the past 3 months. These reports were taken into account when we calculated our measure of fluctuating asymmetry (see below).

(2) Self-and partner-estimated future earnings. Each person was asked to estimate the yearly earnings they and their partners would achieve in 10 years. A small proportion (6%, N = 5) put down outlying values (over \$100,000, up to \$500,000); hence we truncated the measure at \$100,000. Self- and partner-reports correlated 0.44 for men and 0.35 for women. We averaged the two reports to estimate perceived future earnings.

(3) Social potency scale of the adjective rating form of the multidimensional personality questionnaire (Tellegen & Waller, in press; A. Tellegen, unpublished data). A three-item adjective rating measure of social potency, patterned after a 26-item, true-false format scale. The three adjectives (accompanied by descriptions of the high and low scorers) were 'dominant', 'persuasive' and 'socially visible'. Internal consistency reliability was moderate (Cronbach's alpha = 0.61).

(4) Factor-analytically derived measures of investment behaviour. Each partner completed a series of measures about their own and their partner's behaviour in their relationship, including the relationship-specific investment inventory (B.Ellis, unpublished data), a series of measures of mate-retention tactics derived from Buss (1988) and a report of extra-pair sex. Self-partner composite measures were factor-analysed in a larger sample of 104 heterosexual couples within each sex to yield three factors: Nurturance/Commitment, marked by measures indicating a willingness to spend time with and money on the relationship, commitment to the relationship, concern for the partner's well-being; Non-exclusivity, marked by measures indicating a tendency to sexualize (for example, flirt with or attend to) individuals of the opposite sex other than the partner, deceive the partner or neglect the partner at social functions; Proprietariness, marked by measures indicating a tendency to monopolize the partner, be vigilant to the partner's interactions with potential rivals or deter potential rivals (S. W. Gangestad & R. Thornhill, unpublished data).

(5) The Rubin love scale (Rubin 1970). A 10-item face-valid measure of self-professed love for a romantic partner (Cronbach's alpha = 0.83).

(6) Sexual behaviour and attitudes questionnaire. Subjects were asked a series of questions about their sexual history and attitudes. Embedded in this questionnaire were the five components of the sociosexual orientation inventory (SOI; Simpson & Gangestad 1991), including a brief three-item measure of attitudes toward sex without commitment: 'Sex without love is okay'; 'I can imagine myself being comfortable and enjoying casual sex with a different partner'; 'I would have to be closely attached to someone (both emotionally and psychologically) before I could feel comfortable and fully enjoy sex with him or her'. (Subjects responded to these items on a 1–9 scale, with 1 = 'I strongly disagree' and 9 = 'I strongly agree'; the last item was reverse-keyed.) In addition, we asked subjects their number of lifetime sex partners to date. We explicitly told subjects to count only heterosexual episodes and to exclude occasions in which sex was either exchanged for money or forced. Finally, we asked subjects their frequency of sexual intercourse in the last month. Men's and women's reports correlated 0.49 (Pearson's $r$); we treated the mean as a composite measure.

(7) Female orgasm questionnaire. Each partner was asked the percentage of overall time (in 10% increments) that the female partner orgasmed (a) during any sexual bout with the partner (including masturbation and oral sex); (b) during penile-vaginal sexual intercourse in particular; (c) before the partner during sexual intercourse; (d) after the partner during sexual intercourse; (e) at the same time as the partner during sexual intercourse. In addition, each partner was asked how often the female partner faked orgasm. In cases in which (c)–(e) summed to a value greater than (b), we transformed the values by taking the proportion of each of (c)–(e) over their total and multiplying times (b). Using both partners' reports of orgasm increases the validity of our estimates.

Women's and men's reports correlated moderately for (a)–(e): Pearson's $r$ = 0.51, 0.47, 0.49, 0.52 and 0.24, respectively. Certain male-female discrepancies were due to cases in which women reported substantial faking that their partners did not detect. For instance, one woman reported that she never orgasmed but faked 100% of the time; her partner reported that she orgasmed 100% of the time and never faked. When we eliminated the 10 cases in which the proportion of total copulations the woman claimed to fake orgasm was more than 20% higher than that reported by the man (for example, 50% faked orgasm reported by the woman and 20% reported by the man would be a difference of 30%), the male-female correlations for (a)–(e) improved: Pearson's $r$ = 0.62, 0.58, 0.54, 0.60 and 0.28. To estimate these variables, then, we computed male–female composite measures as means, except in these 10 cases, for which we used the woman's report only. To the extent that each partner's report is somewhat valid (as reflected by the sizeable correlation between partners' reports) yet fallible, the composite measure should be more valid than either measure alone (Anastasi 1988).

Based on Baker & Bellis' (1993) work, we combined probability of copulatory orgasm coinciding with and following male ejaculation into a composite of (relatively) high sperm retention (HSR) orgasms. Orgasms occurring before male ejaculation were designated (relatively) low sperm retention (LSR) orgasms. Inter-rater reliabilities (Cronbach's alpha) for proportions of copulations associated with any orgasm, HSR orgasm, and LSR orgasm were 0.73, 0.74 and 0.70, respectively.

(8) Contraceptive use questionnaire. Women were asked which of several contraceptive methods were used during intercourse with their partners: pill (63%), condoms (31%), diaphragm (0%), IUD (0%), sponge/foam (4%) and other (7%).

(Respondents could check more than one.) Because we did not include this questionnaire for the first several ($N$ = 7) couples and several women did not give responses about contraceptive methods, we missed 18 responses to this question. Preliminary analyses revealed that for no conceptive method did users significantly differ from non-users, either with respect to orgasm frequency (all $t$s $\leq$ 1.38, NS) or mate fluctuating asymmetry (all ts $\leq$ 1.48, NS). Any association between female orgasm and male fluctuating asymmetry then, appears not to be due to contraceptive method. Thus, we did not consider this variable in any of our main analyses.

## Fluctuating asymmetry measures

After placing individuals into separate rooms to complete the questionnaires, the two experimenters interrupted each separately to measure fluctuating asymmetry. For these measurements, the subject was escorted to a separate room reserved for measurements alone. One of the two experimenters then measured the subject's left and right sides on seven bilateral characters: foot width, ankle width, hand width, wrist width, elbow width, ear length and ear width (see Thornhill & Gangestad 1994). We choose these traits because they show fluctuating asymmetry that has moderate heritability (Livshits & Kobylianski 1991). We made measurements with steel callipers to the nearest 0.01 mm. Because measurements may involve some measurement error, we re-measured when the left and right sides differed by more than 3 mm (determined from previous studies to be relatively extreme asymmetry; Thornhill & Gangestad 1994, unpublished data). Ten per cent of the measurements met this criterion. On such characters, we averaged the two measurements. We calculated a total fluctuating asymmetry index (FAI) for each subject by taking the absolute difference between the two sides on each character, dividing by the mean size of the character for the subject, and summing these values across the seven characters (Palmer & Strobeck 1986). We measured about one-third of the subjects, and assistants who did not know the research hypothesis took the other measures. In all cases, measures were taken without knowledge of the sexual behaviour of the subject or his/her partner.

Some asymmetry of skeletal characters may be due to breaks or sprains, not developmental instability. In previous research (Thornhill & Gangestad 1994), we did not take these factors into account, but this study attempted to do so. For any feature for which a subject reported a break (ever) or sprain (within the past 3 months) on one side, we assigned either the sex-specific mean asymmetry (in cases in which the asymmetry exceeded the mean) or the subject's measured asymmetry (in cases in which the asymmetry was less than the mean asymmetry). In so doing, we assumed that breaks and sprains more often increased than decreased asymmetry. This procedure affected 4.7% of the fluctuating asymmetry calculations for individual characters (6.5% for men, 2.9% for women). The resulting measure of asymmetry correlated (Pearson's $r$) 0.94 and 0.96 with unaltered measures for men and women, respectively.

Given the small mean asymmetry of individual characters (less than 2 mm), measurement error is a potential concern. To assess inter-rater reliability, two experimenters who were blind to each other's measures took measurements for 47 individuals (23 men and 24 women). The intra-class correlation between the two measurer's FAIs for individuals was 0.63 (0.65 within each sex). Hence, inter-rater reliability for these measures of fluctuating asymmetry appears sufficient for correlational work. For individuals for whom we had two sets of measurements, we averaged across the FAIs. Because a subset of our measures thus had reduced measurement error because of aggregation, our reliability for FAI on the total sample was probably about 0.7.

### Physical attractiveness ratings

After measuring a subject's fluctuating asymmetry, an experimenter took two black-and-white, head-on facial photographs of the subject, who was asked to retain a neutral expression. After processing, we selected one of the two photos of each individual for rating (the one that was most clearly head-on, in which the subject did not close his or her eyes or smile and for which the focus was best). Using these photos, the physical attractiveness of subjects was rated by 10 raters on a scale of 1 (least attractive) to 10 (most attractive). The 10 raters' judgements for each subject were summed to yield a composite attractiveness measure (Cronbach's alphas = 0.74 and 0.89 for men and women, respectively). The raters were all naive to the research hypotheses and did not include individuals who conducted measurements of fluctuating asymmetry. Owing to camera mishaps, processing problems or unsuitable poses, we did not obtain ratings of 13 men and 12 women.

All statistical probabilities are two-tailed unless indicated otherwise. One-tailed values are reported only in cases of patterns predicted a priori.

## RESULTS

Eighty-six of 104 couples reported that their relationship included copulation and completed the orgasm questionnaire. On average (±SD), the 86 women claimed to orgasm during copulation 60 ± 29.5% of the time. Of these, 28 ± 20.8% occurred before male ejaculation, 14 ± 11% with ejaculation, and 19 ± 17.4% after ejaculation. The women claimed to fake orgasm during copulation somewhat more than men perceived them to fake, 13 ± 19.9% versus 10 ± 16.3%, respectively. Owing to missing data, not all couples could be used in all analyses.

To examine the main hypotheses, we performed a series of multiple regression analyses. In a first analysis, we regressed percentage of copulations accompanied by female orgasm on male FAI, male age, male socioeconomic status of family of origin, male perceived future earnings, male social potency, male physical attractiveness, male relationship behaviour measures (Nurturance/Commitment, Non-exclusivity, and Proprietariness) and relationship duration (Table I).

With all other measures controlled, only a single measure significantly predicted reported female orgasm during copulation. As predicted, women with men who possessed lower fluctuating asymmetry were reported by both partners as having more orgasms during copulation (beta (standardized regression coefficient) = –0.31, $t = -2.41$, $df = 59$, one-tailed $P < 0.01$). Follow-up analyses revealed that this effect could not be attributed to the female or male report of copulatory orgasm alone. Two separate regressions replacing the composite measure of copulatory orgasm with either the female or the male report alone yielded significant effects of men's fluctuating asymmetry (betas = –0.26 and –0.24, respectively, *P*s < 0.05) and no other significant effects. Thus the effect of men's fluctuating asymmetry replicated across the two independent reports of copulatory orgasm.

As a group, male investment behaviour measures had particularly low predictive utility. We thus performed a second multiple regression, this time excluding these measures. Men's FAI once again emerged as the only variable significantly predicting female orgasm during copulation (beta = –0.30, $t = -2.54$, $df = 64$, one-tailed $P < 0.01$; all other betas < 0.15, *t*s < 1.19. NS).

**Table I.** Proportion of copulations reported to be accompanied by female orgasm predicted by men's fluctuating asymmetry and other features.

| Predictor variable | Beta | *t* | *P* |
|---|---|---|---|
| **Men's features** | | | |
| FAI | –0.31 | –2.41 | < 0.01 |
| Age | 0.14 | 1.12 | |
| SES | 0.07 | 0.55 | |
| Perceived future earnings | 0.08 | 0.65 | |
| Social potency | 0.09 | 0.69 | |
| Physical attractiveness | 0.15 | 1.16 | |
| **Men's relationship behaviour** | | | |
| Nurturance/commitment | –0.01 | –0.07 | |
| Non-exclusivity | 0.11 | 0.71 | |
| Proprietariness | –0.12 | –0.13 | |
| **Relationship duration** | 0.12 | 0.90 | |

FAI: Fluctuating asymmetry index; SES: socioeconomic status of family of origin. Test on FAI one-tailed, all others two-tailed (although FAI also significant with two-tailed test). $df = 59$. All *P*-values not listed were > 0.10.

To examine HSR and LSR orgasms separately, we repeated the initial multiple regression analysis twice, each time substituting a specific orgasm measure for the total orgasm measure (Table II). With all other variables statistically controlled, men's FAI significantly predicted HSR orgasms (beta = –0.30, $t = -2.37$, $df = 59$, one-tailed $P < 0.02$). Although the sample effect was in the same direction, men's FAI did not significantly predict LSR orgasms (beta = –0.13, $t = -1.01$, NS) nor did any other variable. A comparison of these two effects yielded no significant difference ($t < 1$). Men's physical attractiveness had a non-significant tendency to predict women's HSR orgasms (beta = 0.21, $t = 1.69$, $df = 59$, $P < 0.10$).

Finally we regressed the proportion of all sexual encounters that was reported to have led to female orgasms (including orgasms during intercourse) by men's features (e.g. FAI, age, socioeconomic status of family of origin and relationship behaviour), relationship duration and reported frequency of orgasms during sexual intercourse. By including this last variable, we effectively partialled out the relations between men's features and orgasms during intercourse, and, hence, examined in this analysis whether men's features predict reported female orgasms not during intercourse (e.g. through oral sex or masturbation). No male feature was significantly predictive (all betas < 0.11, *t*s < 1.03). In particular, men's FAI was not predictive of these orgasms (beta = –0.02, $t$ = –0.18, $df$ = 58, NS). We have no evidence, then, that women with men of low fluctuating asymmetry are simply more orgasmic than other women.

**Table II.** Proportion of copulations reported to be accompanied by high sperm retention (HSR) and low sperm retention (LSR) female orgasms predicted by men's fluctuating asymmetry and other features.

| | HSR Orgasms | | | LSR Orgasms | | |
|---|---|---|---|---|---|---|
| Predictor variable | Beta | *t* | *P* | Beta | *t* | *P* |
| **Men's features** | | | | | | |
| FAI | –0.30 | –2.36 | < 0.02 | –0.13 | –1.01 | |
| Age | 0.17 | 1.38 | | 0.02 | 0.18 | |
| SES | 0.08 | 0.60 | | 0.02 | 0.16 | |
| Perceived future earnings | –0.09 | –0.74 | | 0.21 | 1.64 | |
| Social potency | 0.09 | 0.70 | | 0.04 | 0.26 | |
| Physical attractiveness | 0.21 | 1.69 | < 0.10 | –0.01 | –0.06 | |
| **Men's relationship behaviour** | | | | | | |
| Nurturance/commitment | –0.17 | –0.98 | | 0.15 | 0.87 | |
| Non-exclusivity | –0.01 | –0.05 | | 0.17 | 1.04 | |
| Proprietariness | 0.21 | 1.38 | | –0.24 | –1.55 | |
| **Relationship duration** | 0.05 | 0.40 | | 0.11 | 0.86 | |

FAI: Fluctuating asymmetry index; SES: socioeconomic status of family of origin. Test on FAI for HSR one-tailed, all others two-tailed. $df$ = 59. All $P$-values not listed were >0.10. Sperm retention was estimated from the timing of orgasm, after Baker & Bellis (1993).

## Additional Analyses

Possibly, the relation between men's fluctuating asymmetry and their partner's orgasms is attributable not to fluctuating asymmetry or correlated indicators of developmental stability, but rather to correlated aspects of relationships that have their effects through processes having little to do with the theoretical notions we have set forth. Several possibilities come to mind.

(1) Sexual experience. Possibly, low-fluctuating-asymmetry men's partners are more likely to experience orgasm because one or both of the partners have had more sexual experience. In fact, men's fluctuating asymmetry is correlated with their number of

sexual partners in each of three separate samples (Thornhill & Gangestad 1994; Gangestad & Thornhill, in press).

(2) Sexual frequency. Perhaps low-fluctuating asymmetry men have sex with their partners more often than men with high fluctuating asymmetry, leading to greater experience with their partner and thereby greater likelihood of orgasm. Alternatively, perhaps low-fluctuating-asymmetry men have less frequent sex with their partners, such that sex has not become stale in their relationship and hence leads to more frequent orgasms.

(3) Partners' love for one another. Perhaps low-fluctuating-asymmetry men and/or their partners love each other more and hence are able to feel more comfortable about sex with the other, which leads to more frequent female orgasm.

(4) Attitudes towards sex. Perhaps low-fluctuating-asymmetry men or their partner's have more permissive attitudes towards sex even in the absence of love and commitment, and thus are able to relax more during sex, and hence have sex associated with more frequent female orgasm.

(5) Female's investment in the relationship. We had included measures of men's investment in their relationship in analyses reported above. Perhaps, however, men's fluctuating asymmetry is associated with their partners' investment in the relationship, which is associated with their involvement in the relationship, and thereby their likelihood of orgasm.

To attempt to control for these variables, we performed additional analyses. Specifically, we performed another set of regression analyses on reported copulatory female orgasm, with relationship duration, men's FAI, age, and physical attractiveness, both partners' investment measures, both partners' number of lifetime sex; partners, frequency of sexual intercourse in' the last month (mean of partners' reports), both partners' attitudes towards sex without commitment and both partners' love scale (Table III). Results revealed that only one variable predicted reported female orgasm: men's FAI, beta = –0.29, $t = -2.10$, $df = 48$, one-tailed $P < 0.03$.

This analysis included more predictor variables (relative to subjects) than is generally recommended to achieve robust parameter estimates (more than one variable per five subjects), particularly when there exists substantial multicollinearity (Cohen & Cohen 1975). Two points, however, should be made. First, this analysis was not the primary analysis, but rather was a follow-up to one with a much smaller number of variables. It made sense, following the initial finding, to see whether the FAI effect would be removed by including more variables, and it was not. Second, exclusion of any subset of variables added for this analysis did not alter results in any substantive way; men's FAI remains significantly predictive. These facts indicated that the sheer number of variables in this analysis does not account for the FAI effect. Although we cannot rule out all possible confounding variables that may mediate the relationship between men's FAI (or associated markers of developmental stability) and their reports of partners' orgasms, these analyses strongly question a wide variety of major possibilities.

Analyses performed on reported HSR orgasms alone (while controlling for all variables listed above) also revealed a significant effect for men's FAI (beta = –0.36, $t = -2.56$, $df = 48$, one-tailed $P < 0.01$).

**Table III.** Proportion of copulations reported to be accompanied by female orgasm predicted by men's fluctuating asymmetry and additional features

| Predictor variable | Beta | t | P |
|---|---|---|---|
| **Men's FAI** | –0.29 | –2.10 | < 0.03 |
| **Men's age** | 0.12 | 0.80 | |
| **Men's physical attractiveness** | 0.11 | 0.86 | |
| **Relationship duration** | 0.09 | 0.59 | |
| **Sexual experience** | | | |
| Men's number of partners | 0.15 | 0.97 | |
| Women's number of partners | –0.19 | –1.22 | |
| **Sex in the relationship** | | | |
| Sex in the last month | –0.06 | –0.42 | |
| **Partner's love** | | | |
| Men's love | 0.10 | 0.56 | |
| Women's love | –0.11 | –0.74 | |
| **Attitudes towards sex without commitment** | | | |
| Men's attitudes | –0.09 | –0.54 | |
| Women's attitudes | 0.12 | 0.71 | |
| **Partner's relationship behaviour** | | | |
| Men's nurturance/commitment | –0.04 | –0.20 | |
| Men's non-exclusivity | 0.34 | 1.58 | |
| Men's proprietariness | 0.15 | 0.70 | |
| Women's nurturance/commitment | –0.04 | –0.23 | |
| Women's non-exclusivity | –0.15 | –0.95 | |
| Women's proprietariness | –0.28 | –1.26 | |

FAI: Fluctuating asymmetry index. Test on FAI one-tailed, all others two-tailed (although FAI also significant with two-tailed test). $df = 48$. All $P$-values not listed were >0.10.

## Potential Mediators

Although fluctuating asymmetry appears to predict reports of female orgasm independently of a variety of potential confounds, the fluctuating asymmetry measures we used are not observable in normal interaction; hence, certain other features might function as mediators of its relation to female orgasm. Men's FAI appears to predict both men's facial attractiveness (Gangestad *et al.* 1994; R. Thornhill & S. W. Gangestad, unpublished data) and their weight (Manning 1995), such that more symmetrical men are more attractive and heavier. In this sample, men's facial attractiveness did not correlate with their fluctuating asymmetry (with age and age squared partialled out; partial

$r$ = -0.01, $df$ = 86, NS), although an aggregate analysis of three samples (including this one) revealed a small but significant average weighted (by sample size) correlation ($r$ = –0.15, $df$ = 173, $P$ < 0.05: Gangestad & Thornhill, in press). Men's FAI did predict their weight, however (Pearson's $r$ = –0.25, $df$ = 100, $P$ < 0.0I), even when men's height was statistically controlled (partial $r$ = –0.25, $df$ = 99, $P$ = 0.02; height did not predict FAI: $r$ = –0.09, NS).

**Table IV.** Proportion of copulations reported to be accompanied by female orgasm predicted by men's and women's fluctuating asymmetry and other features, including weight and height.

| Predictor variable | Beta | $t$ | $P$ |
|---|---|---|---|
| **Men's Features** | | | |
| FAI | –0.28 | –2.12 | < 0.02 |
| Age | 0.05 | 0.36 | |
| SES | 0.19 | 1.32 | |
| Perceived future earnings | 0.03 | 0.23 | |
| Social Potency | 0.08 | 0.59 | |
| Physical Attractiveness | 0.29 | 2.00 | < 0.06 |
| Weight | 0.42 | 2.61 | < 0.02 |
| Height | –0.33 | –1.99 | < 0.06 |
| **Women's Features** | | | |
| FAI | 0.07 | 0.58 | |
| Physical attractiveness | 0.19 | 1.23 | |
| Weight | 0.30 | 1.59 | |
| Height | –0.44 | –2.55 | < 0.02 |
| **Men's relationship behaviour** | | | |
| Nuturance/commitment | –0.01 | –0.04 | |
| Non-exclusivity | 0.11 | 0.68 | |
| Proprietariness | –0.15 | –0.91 | |
| **Relationship duration** | 0.04 | 0.27 | |

FAI: Fluctuating asymmetry index; SES: SES of family of origin. Test on FAI one-tailed, all others two-tailed (although FAI also significant with two-tailed test). $df$ = 50. All $P$-values not listed were >0.10.

Might men's body mass account for the relationship between their FAI and reported copulatory orgasm? To address this question, we performed another regression analysis, this time including all men's features included in our original analysis (as in Table I), as well as both partners' weight and height, and women's facial attractiveness and FAI (see Table IV). Five variables significantly (or near–significantly) predicted reported frequency of copulatory female orgasm: men's FAI (beta = –0.28, $t$ = –2.12, $df$ = 50, one-tailed $P$ < 0.02), men's weight (beta = 0.42, $t$ = 2.61, $df$ = 50, $P$ < 0.02), men's physical attractiveness (beta = 0.29, $t$ = 2.00, $df$ = 50, $P$ < 0.051), women's height (beta = –0.44, $t$ = –2.55, $df$ = 50, $P$ < 0.02), men's height (beta = –0.33, $t$ = –1.99, $df$ = 50,

$P < 0.052$). Interestingly, then, features found to be associated with men's fluctuating asymmetry (and hence cues of men's developmental stability; specifically, men's weight and physical attractiveness) did independently predict reported female orgasm. None the less, men's fluctuating asymmetry also independently predicted reported female orgasm. The entire relation between men's fluctuating asymmetry and reports of their partner's orgasms is not accounted for by these particular male features. A regression analysis on reported HSR orgasms alone, including all variables listed above, yielded only one significant effect: men's FAI (beta = –0.31, $t = -2.32$, $df = 50$, one-tailed $P < 0.02$).

### Faked Orgasms

As an exploratory analysis, we regressed frequency of reported faked orgasms on all male and female features included in our most expanded analysis above (see Additional Analyses). Only a single feature significantly predicted faked orgasms: female Non-exclusivity (beta = 0.38, $t = -2.38$, $df = 49$, $P < 0.03$). Women who tend to act in less exclusive ways with their mate (e.g. flirted with other men or neglected their partners at social gatherings) tended to fake orgasms more often than other women. In light of the large number of variables included in this analysis and the fact that no a priori prediction was made with regard to this finding, however, this relationship must be interpreted very cautiously.

## DISCUSSION

As predicted, when other potentially relevant male features, female features, and relationship characteristics were statistically controlled, women with men who possessed low fluctuating asymmetry self-reported and were reported by their partners to have had more orgasms during copulation than women with men possessing higher fluctuating asymmetry. Women with men of relatively low fluctuating asymmetry did not simply have more reported orgasms in general than other women; the reports indicate that the former women are only more orgasmic during copulation itself. In our sample, the increased reported copulatory orgasmic response of women with men of relatively low fluctuating asymmetry was greater for orgasms roughly categorized as high sperm retention orgasms by Baker & Bellis (1993), although we did not find a significant difference between the effects of men's FAI on high retention and low retention orgasms. This comparison is not a particularly sensitive one, however, given that our measure of timing was a very crude one. Baker & Bellis (1993) included orgasms occurring less than 1 min before male ejaculation as high sperm retention ones. Given that nearly half of the orgasms before ejaculation recorded by their informants occurred within this time span, many of the orgasms we categorized as low retention orgasms were probably miscategorized. Our findings support Smith's (1984) general notion that female orgasm evolved as a means by which women manipulate sperm competition occurring as a result of facultative polyandry. Furthermore, the pattern of our results is consistent with Baker & Bellis' (1993) specific hypothesis that timing of the orgasm plays a role in the manipulation, although we cannot offer these data as strong support for that claim.

Few other male features significantly predicted the occurrence of reported female orgasm. Our results provide some evidence that men's weight and physical attractiveness predict copulatory female orgasm frequency independent of male body symmetry. Future research may address the robustness and theoretical interpretation of these relations.

Self-report data can be obtained only in animal behaviour research on humans, and provide one way to address hypotheses about humans. Although self-reports alone may contain errors of unknown magnitude, our use of reports by both partners allowed us to examine the validity of individual reports. Women's own reports and the reports of their partners correlated substantially (about 0.6). Unless the errors in reporting covaried across partners (an unlikely scenario), each set of reports (the women's and the men's) must be substantially valid. Moreover, if each set of reports is nearly equally fallible, the average of the two reports should be more valid than either report alone. If we assume that errors in the two reports are uncorrelated, the square root of Cronbach's alpha provides an estimate of the correlation of the averaged reports with the 'true' rate of orgasm (Anastasi 1988): in this case, 0.85, the square root of 0.73. Despite being based on retrospective reports, then, internal analyses of our data suggest that our measures are highly valid. None the less, future studies might attempt to gather more accurate data by asking couples over an extended period to fill out a questionnaire pertaining to the female's sexual response on a day-to-day basis. Such an approach may complement the retrospective approach we used.

Of course, even our internal analyses reveal that our measures are not perfectly valid. For a variety of reasons (e.g. deception, faulty memory) retrospective reports are less than perfect. None the less, it seems unlikely that false reporting of orgasms could account for the relationship between men's body symmetry and female orgasm. To do so, errors in the reports would have to correlate with men's body symmetry, an unlikely situation. Moreover, the reporting errors would have to correlate across women's and men's reports for the effect of men's fluctuating asymmetry replicated across reports. Although we can readily imagine that either partner could falsely report orgasm frequency, it seems unlikely that the reporting errors of one person would substantially correlate with those of the other, and each with men's fluctuating asymmetry. After all, neither men nor their partners knew their body symmetry. Also, although male subjects and their partners can observe the male's physical attractiveness, ratings of male physical attractiveness did not correlate with body symmetry in this study. Once again, however, additional studies using alternative methodologies should complement the procedures we used.

Some concern has been expressed to us about how much can be learned from studies of college undergraduates. Although we appreciate the need to replicate this study on a wider sample of people, there is no theoretical reason to expect that college students do not show species-typical sexual responses. Thus, these results cannot be rejected because they were obtained using college undergraduates.

Our results can be used to examine hypotheses that suggest that female orgasm is an adaptation whose function is other than selective sperm retention. The hypothesis that female orgasm functions to create and maintain the pair bond does not explain variability in the female sexual response. Nor does it account for the pattern we have found between male symmetry and increased orgasm. There is nothing explicit about selective bonding in the traditional pair bond hypothesis for orgasm (Morris 1967; Eibl-Eibesfeldt 1989).

Smith (1984) suggested that female orgasm has two design features: (1) it preferentially retains sperm of males of high phenotypic quality and (2) it promotes differential bonding with such males. As mentioned, our results are consistent with point (1). They may also be consistent with point (2). Oxytocin may mediate both effects. Oxytocin, a hypothalamic hormone stored in the neurohypophysis until its release into circulation, could be directly involved in ejaculate choice by women. Circulating oxytocin controls smooth muscle contractions of the uterus at parturition and smooth muscle contractions that eject milk (Insel 1990). Circulating oxytocin increases during the human female sexual response, peaking at orgasm, and is believed to facilitate orgasmic contractions of the smooth muscles of the uterus and vagina (Carmichael *et al.* 1987; Carter 1992). These contractions may lead to sperm retention. Oxytocin also affects attraction to and acceptance and nurturance of offspring by female mammals (Insel 1990; Panksepp 1992), and recent research implicates oxytocin in heterosexual pair bonding in prairie voles (Carter *et al.* 1992). If oxytocin promotes conditional and selective affiliation with sexual partners, presence or absence of orgasm could be an important proximate cause of female mate choice. We found no evidence that female orgasm is associated with greater investment by women in their romantic relationships or greater love for the male partner. Thus, if oxytocin facilitates differential bonding with men of high genotypic quality, the bonding apparently does not result in greater love or investment.

The hypothesis that female orgasm promotes preferential bonding with males who are capable or willing to invest (Alexander 1979; Alcock 1987; also see Konner 1990) predicts that male love/investment will correlate with women's orgasm frequency. Our results do not support this prediction. Orgasm frequency in women was unrelated to the three measures of relationship investment by men: men's love, men's socioeconomic status of family or origin, and men's perceived future earnings. It is possible that male wealth or status or general male investment potential in our sample of young men varied too little to allow detection of a relationship between female orgasm and male investment. Additional studies are needed to examine this relationship. Recent theory and empirical results suggest, however, that females will have evolved to trade off male investment for male genetic quality (Gangestad 1993); thus there may be little reason to expect female copulatory orgasm to relate to male investment per se.

One version of the hypothesis that female orgasm promotes preferential bonding with investing males could involve male investment in the form of foreplay and extra-copulatory sexual play in general. We did not collect data on foreplay, but there is no reason to believe that symmetrical men engage in more or more stimulating foreplay than asymmetrical men. Our finding that symmetrical men are quicker to copulate in romantic relationships (Gangestad & Thornhill, in press) suggests that courtship investment is reduced in these men, but does not directly address their sexual techniques immediately preceding or otherwise surrounding copulation.

Hrdy (1981) suggested that female orgasm promotes pursuit of multiple partners by females, which either reduces male infanticide of a female's offspring by confusing paternity or results in more material benefits from multiple compared to single mates. This idea appears to be inconsistent with our findings. Hrdy's hypothesis seems to predict that orgasm frequency in women will correlate positively with women's lifetime number of sex partners. Alternatively, it could be argued that women who have many mates are unusually resistant to orgasm and pursue it via sexual variety, which would predict a negative correlation between orgasm frequency and lifetime partner number. We found no relationship, positive or negative, between lifetime sex partner number and orgasm frequency in women. We should add that Hrdy herself has not applied her hypothesis to women, only to other female primates (personal communication). She has proposed that female orgasm is an atavistic trait in women, but is an adaptation in other female primates (Hrdy 1981). Our results suggest, however, that human female orgasm is an adaptation whose evolutionary function is selective sperm retention.

Other ideas concerning the possible causes of human female orgasm have been suggested to us. One view is that female orgasm is dependent upon how relaxed a woman is during sex or on how comfortable she is with her mate in general. We found no evidence that orgasm frequency is related to partners' love for one another, attitudes towards sexual permissiveness of women (or men), female or male investment in the relationship, or copulatory frequency in the last month. A related idea is that the form of contraception will influence how relaxed women are about having sex and thus will affect orgasmic response. We found no evidence for a relationship between contraception type and female orgasm. Further research may examine other hypothetical causes of female orgasm.

Human female orgasm is not functionally designed for conception per se because orgasm is not necessary for conception (e.g. Smith 1984; Baker & Bellis 1993). The notion that human female orgasm positively influences conception ignores the strong selection on males to conceive. The evolutionarily stable strategy of male mating effort is to inseminate fertile females effectively. Selection on males will lead to male sexual adaptations designed for high conception. Female mechanisms that digest, manipulate or nourish sperm are most likely sire-choice mechanisms (Thornhill 1983).

Our findings suggest that human female orgasm is designed to retain sperm of men of high developmental stability, and perhaps the sperm of men who are facially attractive and large in body size. Men's facial attractiveness has been shown to correlate positively with their body symmetry but not in this sample. Men's weight, however, correlated positively with their body symmetry in this sample. When symmetry is controlled, our findings indicate that men's attractiveness and weight each covary positively with female orgasm frequency.

The positive relationship between reported female orgasm and male body symmetry is unlikely to be explained as an incidental effect or by-product of object or mate recognition. This hypothesis has been offered as an alternative explanation to good genes mate choice for female preference of symmetrical males (Enquist & Arak 1994; Johnstone 1994). The body asymmetries that we measured are small (e.g. a small difference in hand width) and are apparently not used in mate choice. Conceivably, body features that are visually assessed in mate choice, for example faces, may correlate with the measured body features; but how a preference for symmetrical body form would cause a greater orgasm response in women mated to symmetrical men remains to be detailed.

Our results place in question the traditional hypothesis that human female orgasm is an incidental effect of male orgasm (Symons 1979; Gould 1987; Fox 1993). Both Symons and Fox emphasized the greater similarity between women's orgasm and the orgasm of pre-pubescent males (boys) than between the orgasms of women and men.

Neither boys' nor women's orgasms involve ejaculation; both boys and women exhibit multiple orgasms and very prolonged periods of high erotic sensation (lengthy plateau phase). According to Symons and Fox, the orgasms of both pre-pubescent males and women are incidental, non-functional consequences of adaptive orgasm in men.

The similarity between women's and boys' orgasms suggest to us that boys' sexual experiences might be a training ground for assessment of true orgasm in sexual partners at maturity. One hypothesized evolutionary function of human feelings is that they allow one to model the minds of other individuals and thereby adaptively increase the predictiveness of others' behaviour (Humphrey 1986; Alexander 1989). This view of feelings leads us to speculate that the orgasmic experiences of boys may be similar to those of women because of design. Boys' orgasms may allow them to personally experience female orgasm, which could improve their ability to distinguish true female orgasm from mere sexual arousal of women and from faked orgasms.

Alexander (1979) proposed that female orgasm increases paternity reliability for the male mate by signalling the female's sexual satisfaction with the mate. Our results support this hypothesis in that they imply that males may have gained reproductively in human evolutionary history by distinguishing female orgasm from mere female arousal because orgasm is associated with sperm retention and thus, presumably, with increased probability of paternity. Also, the results showed moderate positive correlations between men's and women's reports of female orgasm frequency and timing, suggesting that men have knowledge of their partner's orgasms, which is expected if female orgasm conveys information to males. Moreover, although preliminary, our results suggest an advantage to men of distinguishing true from faked orgasm. Women who sexualize males other than the pair-bond mate report more faked orgasms. Thus, faked orgasm may correlate with lower reliability of paternity.

Although copulatory female orgasm may signal enhanced paternity reliability to a mate, our results do not support the idea that copulatory female orgasm is a female signal of phenotypic quality. There was no effect of female body symmetry on reported copulatory orgasmic frequency. The negative relationship between women's height and their copulatory orgasmic frequency is difficult to interpret.

Female orgasm outside humans, especially in non-human primates, has been a topic of controversy (Alexander 1979; Lancaster 1979; Symons 1979; Hrdy 1981; Fox 1993). Our results suggest that female orgasm will occur in mammals in which females mate with multiple males whose ejaculates commonly overlap temporally inside individual females. In female primates, for example, orgasm is predicted in species with multiple sexually active males comprising the social unit (e.g. the chimpanzee, *Pan troglodytes*), but not in harem species in which one male monopolizes sexual access to fertile females (gorilla, *Gorilla gorilla*) or monogamous species with large, dispersed territories (e.g. the siamang, *Symphalangus syndactylus*). The occurrence of female orgasm across mammal species is predicted to be positively related to the development of sperm competition adaptations in males. This prediction is complicated by the fact that orgasm is only one way for females to adjust the outcome of sperm competition during or after mating and thus select the fittest sire. Thornhill (1983) used the term 'cryptic' choice for the many types of female adaptations that may operate during and after mating by which multiply mating female animals might select a fit sire.

## ACKNOWLEDGMENTS

We thank Tara Armijo-Prewitt, Bryant Furlow, Joy Thornhill and Ellen Roots for considerable assistance with data collection. Anne Rice's help with manuscript preparation is greatly appreciated. We thank Robin Baker, Mike Beecher, Sarah Hrdy, Patti Loesche, Meredith West and three anonymous referees for their useful comments on the manuscript. We thank Tara Armijo-Prewitt, Bryant Furlow and Joy Thornhill for help with library work.

## REFERENCES

Alcock, J. 1987. Ardent adaptationism. *Nat. Hist.*, **96**, 4.

Alexander, R. D. 1979. Sexuality and sociality in humans and other primates. In: *Human Sexuality: A Comparative and Developmental Perspective* (Ed. by A. Katchadourian), pp. 81–97. Berkeley, California: University of California Press.

Alexander, R. D. 1989. Evolution of the human psyche. In: *The Human Revolution* (Ed. by P. Mellars & C. Stringer), pp. 456–513. Edinburgh: University of Edinburgh Press.

Anastasi, A. 1988. *Psychological Testing*. 6th edn. New York: Macmillan.

Bailit, H. L.. Workman, P. L., Niswander, J. D. & Maclean, J. C. 1970. Dental asymmetry as an indicator of genetic and environmental conditions in human populations. *Hum. Biol.*, **42**, 626–638.

Baker, R. R. & Bellis, M. A. 1993. Human sperm competition: ejaculate manipulation by females and a function for the female orgasm. *Anim. Behav.*, **46**, 887–909.

Baker, R. R., Bellis, M. A., Hudson, G., Oram, E. & Cook, V. 1989. *Company*, Sept, 60–62.

Bellis, M. A. & Baker, R. R. 1990. Do females promote sperm competition? Data for humans. *Anim. Behav.*, **40**, 997–999.

Benshoof, L. & Thornhill, R. 1979. The evolution of monogamy and concealed ovulation in humans. *J. Soc. Biol. Struct.*, **2**, 95–106.

Buss, D. M. 1988. From vigilance to violence: mate-guarding tactics. *Ethol. Sociobiol.*, **9**, 291–317.

Carmichael, M. S., Humbert, R., Dixen, J., Palmisano, G., Greenleaf, W. & Davidson, J. M. 1987. Plasma oxytocin increases in the human sexual response. *J. Clin. Endocrinol. Metab.*, **64**, 27–31.

Carter. C. S. 1992. Oxytocin and sexual behaviour. *Neurosci. Biobehav. Rev.*, **16**, 131–144.

Carter, C. S., Williams, J. R., Witt, D. M. & Insell, T. R. 1992. Oxytocin and social bonding. *Ann. N. Y. Acad. Sci.*, **652**, 204–211.

Cohen, I. & Cohen, P. 1975. *Applied Multiple Regression: Correlation Analysis for the Behavioral Sciences*. Hillsdale, New Jersey: Lawrence Erlbaum.

Eibl-Eibesfeldt, I. 1989. *Human Ethology*. Hawthorne, New York: Aldine de Gruyter.

Enquist, M. & Arak, A. 1994. Symmetry, beauty and evolution. *Nature*, **372**, 169–172.

Fox, C. A., Wolff, H. S. & Baker, J. A. 1970. Measurement of intra-vaginal and intra-uterine pressures during human coitus by radio-telemetry. *J. Reprod. Fert.*, **22**, 243–251.

Fox, R. 1993. Male masturbation and female orgasm. *Society*, Sept/Oct, 21–25.

Gangestad, S. W. 1993. Sexual selection and physical attractiveness: implications for mating dynamics. *Hum. Nat.*, **4**, 205–236.

Gangestad, S. W. & Thornhill, R. In press. Human sexual selection and developmental stability. In: *Evolutionary Social Psychology* (Ed. by J. A. Simpson & D. T. Kenrick). Hillsdale, New Jersey: Lawrence Erlbaum.

Gangestad, S. W., Thornhill, R. & Yeo, R. A. 1994. Facial attractiveness, developmental stability, and fluctuating asymmetry. *Ethol. Sociobiol.*, **15**, 73–85.

Gould, S. J. 1987. Freudian slip. *Nat. Hist.*, **96**, 14–21.

Grammer, K. & Thornhill, R. 1994. Human (*Homo sapiens*) facial attractiveness and sexual selection: the role of symmetry and averageness. *J. Comp. Psychol.*, **108**, 233–242.

Hamilton, W. D. & Zuk, M. 1982. Heritable true fitness and bright birds: a role for parasites? *Science*, **218**, 384–387.

Hrdy, S. B. 1981. *The Woman That Never Evolved*. Cambridge, Massachusetts: Harvard University Press.

Humphrey, N. 1986. *The Inner Eye*. London: Faber & Faber.

Insel, T. R. 1990. Oxytocin and maternal behaviour. In: *Mammalian Parenting* (Ed. by N. A. Krasnegor & R. S. Bridges), pp. 260–280. New York: Oxford University Press.

Johnstone, R. A. 1994. Female preference for symmetrical males as a by-product of selection for mate recognition. *Nature*, **372**, 172–175.

Konner, M. 1990. *Why the Reckless Survive*. New York: Viking Press.

Lancaster, J. B. 1979. Sex and gender in evolutionary perspective. In: *Human Sexuality: A Comparative and Developmental Perspective* (Ed. by H. Katchadourian), pp. 51–80. Berkeley, California: University of California Press.

Livshits, G. & Kobylianski, E. 1991. Fluctuating asymmetry as a possible measure of developmental homeostasis in humans: a review. *Hum. Biol.*, **63**, 441–466.

Manning, J. T. 1995. Fluctuating asymmetry and body weight in men and women: implications for sexual selection. *Ethol. Sociobiol.*, **16**, 145–153.

Manning, J. T. & Chamberlain, A. T. 1993. Fluctuating asymmetry, sexual selection and canine teeth in primates. *Proc. R. Soc. Lond. B*, **251**, 83–87.

Markow, T. & Ricker, J. P. 1992. Male size, developmental stability and mating success in natural populations of three *Drosophila* populations. *Heredity*, **69**, 122–127.

Mitton, J. B. & Grant, M. C. 1984. Associations among protein heterozygosity, growth rate and developmental homeostasis. *A. Rev. Ecol. Syst.*, **15**, 479–499.

Møller, A. P. 1992*a*. Parasites differentially increase the degree of fluctuating asymmetry in secondary sexual characters. *J. Evol. Biol.*, **5**, 691–700.

Møller, A. P. 1992*b*. Female swallow preference for symmetrical male sexual ornaments. *Nature*, **357**, 238–240.

Møller, A. P. & Pomiankowski, A. 1993. Fluctuating asymmetry and sexual selection. *Genetica*, **89**, 267–279.

Møller, A. P. & Thornhill, R. In press. On the heritability of developmental stability: a review. *J. Evol. Biol.*

Morris, D. 1967. *The Naked Ape*. New York: McGraw-Hill.

Palmer, A. R. & Strobeck, C. 1986. Fluctuating asymmetry: measurement, analysis, patterns. *A. Rev. Ecol. Syst.*, **17**, 392–421.

Panksepp, J. 1992. Oxytocin effects on emotional processes: separation distress, social bonding, and relationships to psychiatric disorders. *Ann. N. Y. Acad. Sci.*, **652**, 243–252.

Parsons, P. A. 1990. Fluctuating asymmetry: an epigenetic measure of stress. *Biol. Rev.*, **65**, 131–145.

Rubin, Z. 1970. Measurement of romantic love. *J. Personality Soc. Psychol.*, **16**, 265–273.

Simpson, J. A. & Gangestad, S. W. 1991. Individual differences in sociosexuality: evidence for convergent and discriminant validity. *J. Personality Soc. Psychol.*, **60**, 870–883.

Singer, I. 1973. Fertility and the female orgasm. In: *The Goals of Human Sexuality* (Ed. by I. Singer), pp. 159–197. London: Wildwood House.

Smith, R. L. 1984. Human sperm competition. In: *Sperm Competition and the Evolution of Animal Mating Systems* (Ed. by R. L. Smith), pp. 601–660. London: Academic Press.

Symons, D. 1979. *The Evolution of Human Sexuality*. Oxford: Oxford University Press.

Tellegen, A. & Waller, N. G. In press. Exploring personality through test construction: development of the multidimensional personality questionnaire. In: *Personality Measures: Development and Evaluation, Vol. 1* (Ed. by S. R. Briggs & J. M. Cheek). Greenwich, Connecticut: JAI Press.

Thornhill, R. 1983. Cryptic female choice in the scorpionfly *Harpobittacus nigriceps* and its implications. *Am. Nat.*, **122**, 765–788.

Thornhill, R. 1990. The study of adaptation. In: *Interpretation and Explanation in the Study of Animal Behaviour, Vol II* (Ed. by M. Bekoff & D. Jamieson), pp. 31–62. Boulder, Colorado: Westview Press.

Thornhill, R. 1992*a*. Female preference for the pheromone of males with low fluctuating asymmetry in the Japanese scorpionfly (*Panorpa japonica*: Mecoptera). *Behav. Ecol.*, **3**, 277–283.

Thornhill, R. 1992*b*. Fluctuating asymmetry and the mating system of the Japanese scorpionfly, *Panorpa japonica*. *Anim. Behav.*, **44**, 867–879.

Thornhill, R. & Gangestad, S. W. 1994. Fluctuating asymmetry and human sexual behaviour. *Psychol. Sci.*, **5**, 297–302.

Thornhill, R. & Sauer, K. P. 1992. Paternal genetic effects on the fighting ability of sons and daughters and mating success of sons in a scorpionfly, *Panorpa vulgaris*. *Anim. Behav.*, **43**, 255–264.

Van Valen, L. 1962. A study of fluctuating asymmetry. *Evolution*, **16**, 125–142.

Watson, P. J. & Thornhill, R. 1994. Fluctuating asymmetry and sexual selection. *Trends Ecol. Evol.*, **9**, 21–25.

Williams, G. C. 1966. *Adaptation and Natural Selection*. Princeton, New Jersey: Princeton University Press.

# LIST OF AUTHORS

Affiliations at time of article publication are indicated at the start of each chapter. Contact details (updated where possible) are provided here.

**R. Robin Baker.** Department of Environmental Biology, University of Manchester, Manchester, M13 9PL, U.K.

**Mark A. Bellis.** Centre for Public Health, Faculty of Health and Applied Social Sciences, Liverpool John Moores University, Liverpool, L3 2AP, U.K.

**Tim Birkhead.** Department of Animal and Plant Sciences, University of Sheffield, Sheffield, S10 2TN, U.K.

**April L. Bleske-Rechek.** University of Wisconsin at Eau Claire, Department of Psychology, Eau Claire, WI 54702-4004, U.S.A.

**Rebecca L. Burch.** Department of Psychology, State University of New York at Oswego, Oswego, NY 13126, U.S.A.

**Randall Comer.** Department of Psychology, University of New Mexico, Albuquerque, NM 87131-1161, U.S.A.

**Harald A. Euler.** Kassel University, FB07 - Institut für Psychologie, Hollaendische Straße, 36-38, 34109 Kassel, Germany

**Matthew J. G. Gage.** School of Biological Sciences, University of East Anglia, Norwich NR4 7TJ, U.K.

**Gordon G. Gallup.** Department of Psychology, State University of New York at Albany, Albany, NY 12222, U.S.A.

**Steve W. Gangestad.** Department of Psychology, University of New Mexico, Albuquerque, NM 87131-1161, U.S.A.

**Aaron T. Goetz.** Department of Psychology, Florida Atlantic University, 2912 College Avenue, Davie, Florida 33314, U.S.A.

**Alexander H. Harcourt.** Department of Anthropology, University of California, Davis One Shields Avenue, Davis, CA 95616-8522.

**Sabine Hoier.** Kassel University, FB07 - Institut für Psychologie, Hollaendische Straße , 36-38, 34109 Kassel, Germany.

**Gregory J. LeBlanc.** Florida Atlantic University, Department of Psychology, 2912 College Avenue, Davie, Florida 33314, U.S.A.

**M. Martin.** Department of Molecular Biology and Biotechnology, University of Sheffield, S10 2TN, U.K.

**Harry D. M. Moore.** Reproductive and Development Medicine, Jessop Wing, University of Sheffield, S10 2SF, U.K.

**Geoffrey A. Parker.** School of Biological Sciences, Biosciences Building, Crown Street, University of Liverpool, Liverpool L69 7ZB, U.K.

**Nicholas Pound.** Centre for Cognition & Neuroimaging, School of Social Sciences & Law, Brunel University, Uxbridge, UB8 3PH, U.K.

**Todd K. Shackelford.** Florida Atlantic University, Department of Psychology, 2912 College Avenue, Davie, Florida 33314, U.S.A.

**Robert L. Smith.** Department of Entomology, The University of Arizona, Tucson, AZ 85721-0036, U.S.A.

**Randy Thornill.** Department of Biology, The University of New Mexico, Albuquerque, NM 87131-0001, U.S.A.

**Nina Wedell.** Centre for Ecology and Conservation, University of Exeter in Cornwall, Tremough Campus, Penryn, Cornwall, TR10 9EZ, U.K.

**Viviana A. Weekes-Shackelford.** Florida Atlantic University, Department of Psychology, 2912 College Avenue, Davie, Florida 33314, U.S.A.

# INDEX